De Bartolo, Curcio, Drioli
Membrane Systems

Also of Interest

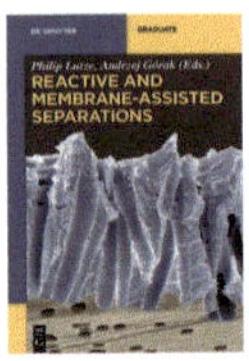

Reactive and Membrane-Assisted Separations
Lutze, Górak (Eds.), 2016
ISBN 978-3-11-030783-2, e-ISBN 978-3-11-030784-9

Membrane Engineering
Giorno, Drioli, 2018
ISBN 978-3-11-028140-8, e-ISBN 978-3-11-028139-2

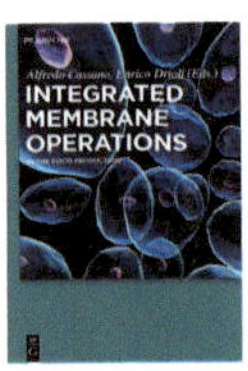

Integrated Membrane Operations.
In the Food Production
Cassano, Drioli (Eds.), 2013
ISBN 978-3-11-028467-6, e-ISBN 978-3-11-028566-6

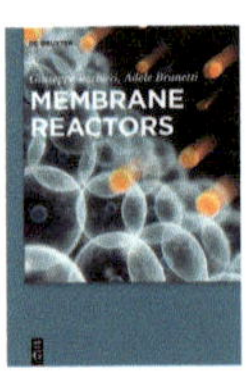

Membrane Reactors
Barbieri, Brunetti, 2018
ISBN 978-3-11-033347-3, e-ISBN 978-3-11-033376-3

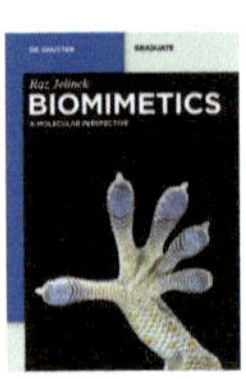

Biomimetics.
A Molecular Perspective
Jelinek, 2013
ISBN 978-3-11-028117-0, e-ISBN 978-3-11-028119-4

Loredana De Bartolo, Efrem Curcio, Enrico Drioli

Membrane Systems

For Bioartificial Organs and Regenerative Medicine

DE GRUYTER

Authors

Dr. Loredana De Bartolo
Institute on Membrane Technology
National Research Council of Italy –
ITM-CNR at University of Calabria
Via P. Bucci 17/C
87030 Rende (CS), Italy
l.debartolo@itm.cnr.it

Dr. Efrem Curcio
Department of Environmental
and Chemical Engineering
University of Calabria
(DIATIC-UNICAL)
Via P. Bucci Cubo 45A
87030 Rende (CS), Italy
e.curcio@unical.it

Prof. Enrico Drioli
Institute on Membrane Technology
National Research Council of Italy –
ITM-CNR at University of Calabria
Via P. Bucci 17/C
87030 Rende (CS), Italy
e.drioli@itm.cnr.it

WCU Energy Engineering Department
Hanyang University
Seoul, South Korea

ISBN 978-3-11-026798-3
e-ISBN (PDF) 978-3-11-026801-0
e-ISBN (EPUB) 978-3-11-039088-9
Set-ISBN 978-3-11-026802-7

Library of Congress Cataloging-in-Publication Data
A CIP catalog record for this book has been applied for at the Library of Congress.

Bibliographic information published by the Deutsche Nationalbibliothek
The Deutsche Nationalbibliothek lists this publication in the Deutsche Nationalbibliografie; detailed bibliographic data are available on the Internet at http://dnb.dnb.de.

Cover image: Human keratinocytes on polycaprolactone membrane
Typesetting: PTP-Berlin, Protago-TEX-Production GmbH, Berlin
Printing and binding: CPI books GmbH, Leck
♾ Printed on acid-free paper
Printed in Germany

www.degruyter.com

Preface

Health care requests for patients with tissue loss or end-stage organ failure are increasing due to the worldwide population aging and the development of pathologies related to organ functions. Transplantation represents the gold standard of lifesaving but is limited by the scarce availability of donor organs, and as a consequence, there has been a growing demand for biological substitutes that are able to restore or to maintain temporarily or improve organ functions. In the last decades advances in biomaterials have generated a range of materials and devices for use either outside the body or through implantation to replace or assist functions which may have been lost through disease or injury, providing a bridge to recovery or transplantation. Membrane processes play a pivotal role in the replacement therapy for acute and chronic organ failure diseases. In fact, all current extracorporeal blood purification and oxygenation methods employ membranes. In these devices membranes act as selective barriers for the removal of endogenous and exogenous toxins from the patient's blood (hemodialysis, hemofiltration, etc.) or for gas exchange with blood (blood oxygenation). Membrane technology offers new, interesting opportunities for the design of bioartificial organs such as the bioartificial liver, pancreas, kidney, or lung. The use of polymeric membranes with different physico-chemical and transport properties is appealing in the creation of a bioartificial organ since these and biomembranes share similarities such as the selective transport of molecules, resistances, and protection. Furthermore, synthetic membranes can easily be mass produced, modulating their morphological and physico-chemical properties for a specific organ or tissue. Membranes must be designed to replicate the properties of the extracellular matrix providing structural support to the cells, which are the functional elements of any tissue or organ. In this context the bioreactor technology represents a promising approach for the regeneration and the production of individualized biological implants. The bioreactor, through the fluid dynamics modulation, may simulate the *in vivo* complex physiological environment ensuring an adequate mass transfer of nutrients and metabolites and the molecular and mechanical regulatory signals. This technology leads to the fabrication of bioartificial organs that are functionally matched to human organs and can be used for clinical organ transplantation.

Membrane Systems: For Bioartificial Organs and Regenerative Medicine reviews the latest developments in membranes and investigates how they can be used to improve the quality and efficiency of bio/artificial organs. The book highlights the design and development of membranes to be used in bio/artificial organs starting from the polymers to the preparation, characterization, and sterilization procedures to be used. Basic issues in membrane separation for biomedical devices such as membrane transport, concentration polarization, and fouling phenomena are specifically discussed. The bioengineering principles, the different strategies pursued for the development of membrane bio/artificial organs including important issues related to blood- and

DOI 10.1515/9783110268010-001

cell-membrane interactions are described with the aim of opening new and exciting frontiers in the next decades. The use of membrane processes and devices can be exciting in helping to find nature's substitutes and to solve the pathogenesis of important human diseases while proposing new and fascinating medical therapies.

Contents

Preface — v

1 Natural and synthetic membranes — 1
1.1 Biomedical polymers used for membranes — **1**
1.2 Synthetic polymers — **4**
1.3 Biodegradable polymers — **9**
1.4 Preparation of membranes — **15**
1.4.1 Thermodynamic principles — **15**
1.4.2 Techniques of membrane preparation — **25**
1.5 Membrane characterization — **35**
1.5.1 Characterization of structural properties — **35**
1.5.2 Characterization of physico-chemical properties — **39**
1.6 Sterilization procedures — **41**
1.6.1 Steam sterilization — **41**
1.6.2 Dry heat — **42**
1.6.3 Ethylene oxide — **42**
1.6.4 Hydrogen peroxide plasma — **43**
1.6.5 Liquid chemical sterilization — **43**
1.6.6 Radiation sterilization — **44**
1.6.7 Ozone — **46**
1.7 References — **46**

2 Basic issues in membrane separation for biomedical devices — 49
2.1 Transport in membranes — **49**
2.2 Membrane transport and nonequilibrium thermodynamics — **51**
2.3 Transport of fluids through porous membranes — **52**
2.3.1 Microfiltration — **52**
2.3.2 Ultrafiltration — **55**
2.3.3 Starling flow — **59**
2.4 Concentration polarization — **61**
2.5 Fouling phenomena — **63**
2.5.1 Cake layer model — **63**
2.5.2 Blocking models — **64**
2.6 The transport of gases through porous membranes — **65**
2.7 Transport of gases through dense membranes — **67**
2.8 Fluid oxygenation — **69**
2.9 Transport through charged membranes — **71**
2.9.1 The Donnan equilibrium — **71**
2.9.2 Nanofiltration — **73**

2.10 Hemodialysis — **75**
2.11 References — **78**

3 Artificial organs — 81
3.1 Membrane artificial organs — **81**
3.2 Hemodialysis — **81**
3.2.1 Kidney functions: What hemodialysis must replace — **81**
3.2.2 Hemodialysis treatment — **83**
3.2.3 History of hemodialysis — **85**
3.2.4 Concepts of hemodialysis — **88**
3.3 Hemofiltration — **93**
3.4 Hemodiafiltration — **96**
3.5 Membranes used in hemodialyzers/hemofilters — **98**
3.6 Plasmapheresis — **99**
3.7 Artificial lung — **102**
3.7.1 Lung functions — **102**
3.7.2 Membrane oxygenator — **103**
3.7.3 History of ECMO — **104**
3.7.4 Extracorporeal membrane oxygenation (ECMO) system — **105**
3.7.5 Intravenous membrane oxygenator (IMO) — **108**
3.7.6 Gas transfer in a membrane oxygenator — **110**
3.7.7 Membrane configuration — **111**
3.8 References — **113**

4 Blood-membrane interactions — 119
4.1 Thrombogenicity of membranes in contact with blood — **119**
4.2 Binding of platelets to the surface — **121**
4.3 Coagulation activation — **123**
4.4 Complement activation — **125**
4.5 Biocompatibility of dialysis membranes — **127**
4.6 The role of protein adsorption — **132**
4.7 Strategies for inhibiting immunoreactions — **133**
4.8 References — **134**

5 Engineering of membrane bio-hybrid organs — 139
5.1 Introduction — **139**
5.2 Fluid-dynamics of membrane bioreactors — **139**
5.3 Reaction kinetics — **150**
5.4 Modeling HF membrane bioreactors — **153**
5.5 Concluding remarks — **159**
5.6 References — **161**

6 Cell-membrane interactions — 165
6.1 Mimicking *in vivo* environment — 165
6.2 Receptors mediating cell interactions — 166
6.3 Cell adhesion — 169
6.4 Protein adsorption to the membrane surface — 170
6.5 Topographical influences on cell adhesion and functions — 172
6.6 Influence of membrane physico-chemical properties on cell adhesion — 175
6.7 Functionalization of membrane surfaces — 177
6.8 Influence of mechanical properties — 179
6.9 Mass-transport influence — 181
6.10 References — 181

7 Membrane bioartificial organs — 187
7.1 Membrane bioartificial organs — 187
7.2 Bioartificial liver — 189
7.2.1 Liver structure and functions — 189
7.2.2 Liver disease — 190
7.2.3 BAL design issues — 192
7.2.4 Membrane bioreactors for BAL systems in clinical applications — 203
7.2.5 Membrane BAL systems in preclinical and *in vitro* evaluation — 204
7.3 Bioartificial pancreas — 208
7.3.1 Anatomy of the pancreas — 208
7.3.2 Pancreatic diseases — 209
7.3.3 Cell source — 211
7.3.4 Membrane bioartificial pancreas — 212
7.4 Bioartificial kidneys — 217
7.4.1 Why bioartificial kidneys? — 217
7.4.2 Cell source — 219
7.4.3 Membranes — 221
7.4.4 Preclinical bioartificial kidney devices — 222
7.5 Bioartificial lungs — 224
7.5.1 Engineering the multiscaled architecture of the bioartificial lung — 226
7.5.2 Cell types — 227
7.5.3 Biomaterials for the bioartificial lung — 228
7.5.4 Bioengineered trachea — 230
7.6 References — 231

8 Regulatory framework and ethical issues — 241
8.1 Regulatory framework of cellular products — 241
8.2 Development and approval process in the US and the EU — 242

8.3 Good manufacturing practice — **245**
8.4 From the laboratory to the market — **249**
8.5 Ethics — **255**
8.6 References — **258**

Index — 261

1 Natural and synthetic membranes

1.1 Biomedical polymers used for membranes

Several compounds can be used for preparing membranes. They can be classified on the basis of: (i) the source of natural or synthetic materials, if they are from nature or synthesized in the laboratory, respectively; (ii) the chemical nature as organic, inorganic, and composite; (iii) degradation properties as biodegradable and nonbiodegradable. The most widely-used compounds are organic polymers or macromolecules. Synthetic polymers have a wide variety of properties and uses and are produced commercially. In particular, biodegradable synthetic polymers have become increasingly popular for use in biomedical applications. Man-made polymers that react to their surroundings are known as smart polymers, or stimulus-responsive polymers, and can also be applied for a variety of purposes in technology and biomedicine.

The first and most important attribute of a polymer is the identity of the monomer residues (repeat units), on which depends the nomenclature. Polymers that contain only a single type of repeat unit are known as homopolymers, while polymers containing a mixture of repeat units are known as copolymers [1]. An example of a homopolymer is poly(styrene), which is composed only of styrene monomer residues, whereas ethylene-vinyl acetate that contains more than one variety of repeat units is a copolymer. Some biological polymers such as polynucleotides (e.g., DNA) are composed of a variety of nucleotide subunits, which are different but structurally related to monomer residues.

In a polymer the number of monomers determines the molecular weight: the larger the number of monomers, the larger the molecular weight. A common means of expressing the length of a chain is the degree of polymerization, which quantifies the number of monomers incorporated into the chain [2, 3]. Synthetic methods include step-growth polymerization, chain-growth polymerization, and plasma polymerization. In chain growth polymerization, monomers are added to the chain one at a time only (e.g., polyethylene), whereas in step-growth polymerization chains of monomers may combine with one another directly (e.g., polyester) [4]. In plasma polymerization monomers are activated by a gas discharge generated by plasma sources in order to initiate polymerization. Synthetic polymerization reactions may be carried out with or without a catalyst. Since synthetic polymerization techniques typically yield a polymer product in a range of molecular weights, the weight is often expressed statistically to describe the distribution of chain lengths present. Common examples are the number average molecular weight and weight average molecular weight. The ratio of these two values is the polydispersity index, commonly used to express the "width" of the molecular weight distribution [5].

DOI 10.1515/9783110268010-002

The physical properties of a polymer are strongly dependent on the size or length of the polymer chains. For example, as chain length is increased, melting and boiling temperatures increase quickly [6]. Impact resistance also tends to increase with chain length, as does the viscosity, or resistance to flow, of the polymer in its melt state. Chain length is related to melt viscosity: a tenfold increase in polymer chain length results in a viscosity increase of over 1000 times. The increase of chain length decreases chain mobility and increases strength, toughness, and the glass transition temperature (T_g) as a result of the increase in chain interactions such as van der Waals attractions and entanglements. These interactions tend to fix the individual chains more strongly in position and resist deformations.

The microstructure of a polymer (sometimes called configuration) relates to the physical arrangement of monomer residues along the backbone of the chain. These are the elements of polymer structure that require the breaking of a covalent bond in order to change. Structure has a strong influence on the other properties of a polymer. For example, two samples of natural rubber may exhibit different durability, even though their molecules comprise the same monomers.

An important microstructural feature of a polymer is its architecture; the polymer can be either linear or branched [3]. A branched polymer molecule is composed of a main chain with one or more substituent side chains or branches. A polymer's architecture affects many of its physical properties including, but not limited to, solution viscosity, melt viscosity, solubility in various solvents, glass transition temperature, and the size of individual polymer coils in solution.

The typical states of polymers are the glassy, rubbery, and semicrystalline state. The threshold temperature below the polymer is glassy (hard and brittle) and above the polymer becomes rubbery (elastic and flexible) is the glass transition temperature or T_g. In the glassy state the mobility of the polymeric chains is very restricted since the segment cannot rotate freely around the main chain bonds. In the rubbery state the segment can rotate freely along the main chain bonds, implying a high degree of chain mobility. The state of the polymer is important for its mechanical, chemical, thermal, and permeation properties. In the case of porous membranes the choice of polymer has an effect on chemical and thermal stability, on wettability as well as on the surface characteristics of adsorption and interactions. In the case of dense nonporous membranes the state of polymer has a strong effect on the permeability of gases and vapors through the membrane. The permeability is generally much lower in the glassy state than in the rubbery state [7].

Semicrystalline materials such as polyamides do not exhibit a clear T_g or "rubbery" region. For these polymers the main transition occurs at T_m when the crystalline regions break down. Some chain rotation in the amorphous regions will occur below T_m, giving some impact resistance at these temperatures. Values of T_g and T_m for a number of polymers are given in Table 1.1. T_g values are affected by the chemical structure of polymers. For example polymers with a main chain characterized by –C–C– bond are more flexible and have low T_g with respect to those having –C–O–

bond. The presence of aromatic groups or unsaturated bonds in the main chain increases the T_g values. In the case of polymers that contains alternating unsaturated and saturated bonds the T_g value does not change significantly, since the rotation around the saturated bonds –C–C– compensates the stiffness of unsaturated bonds –C=C–.

The free volume of the polymer V_f, which is the volume of the polymer mass not actually occupied by the molecules, affects the T_g value. $V_f = V - V_S$, where V is the specific volume, and V_S is the volume of solidly packed molecules. The higher the V_f, the more room the molecules will have in which to move around and the lower is the T_g.

The degree of crystallinity, defined as the volume fraction of crystalline regions in a polymer, also affects the state of a polymer. The polymers that have regular structure units are able to crystallize, while the polymers that are irregular do not crystallize. In the case of unsaturated polymers the crystallisation occurs when all chains have the same conformation. The degree of crystallization affects the optical, thermal, and mechanical properties of a polymer. Semicrystalline polymers are opaque – light is not transmitted through the polymer – and have a definite melting point range at which the crystalline volume is destroyed. On the contrary, amorphous polymers are clear: visible light can pass through them, and they do not have a defined melting range because they are already without any crystallinity. Concerning the mechanical properties, as a polymer's crystallinity increases, the stiffness (elastic modulus) and strength of the polymer increase, while ductility declines as crystallinity increases [4].

The physical structure of the chain is also an important factor that influences the polymer properties. Based on the stereoregularity, which defines the configuration of polymer chains, distinct structures can be obtained: *isotactic*, in which all substituents are on the same side of the polymer chain; *syndiotactic*, in which there are alternating groups; and *atactic*, characterized by a random combination of the groups.

Therefore many polymers are available, but the choice of membrane polymer in biomedical applications is not a trivial task. A polymer must have appropriate characteristics for the intended application. The polymer must be biocompatible, which means that it must avoid host undesirable local or systemic effects for patients. It has to be compatible with the chosen membrane fabrication technology. The polymer has to be a suitable membrane former in terms of its chain rigidity, chain interactions, stereoregularity, and polarity of its functional groups. The polymers can form amorphous and semicrystalline structures affecting the membrane performance characteristics. The polymer has to be obtainable and reasonably priced in order to comply with the low cost criteria of the membrane separation process. Many membrane polymers are grafted, custom-modified, or produced as copolymers to improve their properties.

1.2 Synthetic polymers

The most common polymers used in membrane synthesis for biomedical applications are cellulose acetate (CA), nitrocellulose (CN), and cellulose esters (CE), polysulfone (PSf), polyethersulfone (PES), polyacrilonitrile (PAN), polyethylene (PE), polyetheretherketone (PEEK-WC), and polypropylene (PP), polytetrafluoroethylene (PTFE), polyvinylidene fluoride (PVDF), polyvinylchloride (PVC), ethylenevinyilalcohol (EVAL), polyetherimide (PEI), and polymethylmethacrylate (PMMA) [8].

Cellulose acetate (CA) is the acetate ester of cellulose, a polysaccharide consisting of a linear chain of β (1→4) linked D-glucose units. The hydroxyl groups (–OH) of cellulose can be partially or fully reacted with various reagents in order to obtain derivatives with useful properties, like mainly cellulose esters (e.g., cellulose acetate, cellulose nitrate) and cellulose ethers (–OR). Several type of cellulose acetate are commercially available which differ in their degree of hydroxyl substitution (e.g., cellulose diacetate, cellulose triacetate).

Polysulfone (PSf) and polyethersulfone (PES) are thermoplastic polymers that are known for their toughness and stability at high temperatures. They are characterized by $-SO_2-$ linkages and chain rigidity derived from the relatively inflexible and immobile phenyl sulphone groups, and toughness from the connecting ether oxygen. Both have excellent high-temperature properties and chemical inertness. Polysulfone has a glass transition temperature of 185 °C and is highly resistant in pH ranging from 2 to 13 and to oxidizing agents. Mechanically, polysulfone has high compaction resistance and is also stable in aqueous acids and bases and many nonpolar solvents; however it is soluble in dichloromethane and methylpyrrolidone [9].

Polyacrylonitrile (PAN) is a semicrystalline polymer with the linear formula $(C_3H_3N)_n$ that is prepared by the polimerization of acrylonitrile. This polymer has a T_g of 120 °C and degrades before melting. Almost all polyacrylonitrile are copolymers in which acrylonitrile is the main component.

Polyethylene (PE) consists of long hydrocarbon chains in which the monomer is ethylene. Polyethylene is classified into different categories on the basis of its density and branching. The mechanical properties of the polymer are dependent of the extension and type of branching, as well as of the crystal structure and the molecular weight. The melting point and glass transition varies strongly on the basis of the crystallinity and molecular weight. For the medium- and high-density polyethylene the melting point ranges from 120 to 130 °C, whereas the melting point for low-density polyethylene is around 105–115 °C.

Polyetheretherketone (PEEK) is polymerized by the dialkylation of bisphenolate salts. It has excellent mechanical properties with a Young's module of 3.6 GPa, and outstanding thermal properties and chemical resistance. The polymer is partially crystalline and melts around 350 °C [10], and it has a glass transition temperature of around 143 °C. PEEK is characterized by low solubility in common organic solvents.

Modified polyetheretherketone (PEEK-WC) exhibits chemical stability and excellent thermal and mechanical resistance similarly to traditional PEEKs. Differently from PEEKs, PEEK-WC is soluble in various solvents, owing to a lack of crystallinity for the introduction of isobenzofurane-1,3-dihydro-1-oxo-group into the polymer chain, which is responsible for its amorphous character and solubility in common solvents [11, 12].

Polypropylene (PP) is made from the monomer propylene. It is a hydrophobic polymer. It is mostly isotactic and has an intermediate level of crystallinity. Isotactic PP has a melting point that ranges from 160 to 166 °C, whereas syndiotactic polymer with a crystallinity of 30 % has a melting point of 130 °C. Polypropylene is normally tough and flexible, especially when copolymerized with ethylene. It has high resistance temperature and excellent mechanical properties.

Polytetrafluoroethylene (PTFE) is a synthetic fluoropolymer with high-molecular-weight consisting wholly of carbon and fluorine. PTFE is hydrophobic as it demonstrates mitigated London dispersion forces due to the high electronegativity of fluorine. PTFE is characterised by high melting point (327 °C) and resistance to attack by almost all chemicals. These characteristics are due to the aggregate effect of carbon-fluorine bonds.

Polyvinylidene fluoride (PVDF) is produced by the polymerization of vinylidene difluoride ($CH_2{=}CF_2$). PVDF has a glass transition temperature (T_g) of about −35 °C and is typically 50–60 % crystalline. Compared to other fluoropolymers, it has a relatively low melting point of around 171 °C, which makes melting processes easier.

Polyvinylchloride (PVC) is produced by polymerization of vinyl chloride monomers, which are mainly arranged head-to-tail. In the linear polymer chain there are chlorides on alternating carbon centers, which are responsible for the very different properties of the polymer with respect to the polyethylene. PVC has high hardness and mechanical properties, expressed by Young's module of 3.4 GPa. Heat stability of PVC is very poor; in fact its melting temperature is 160 °C.

Ethylenevinylalcohol (EVAL) is a copolymer of ethylene and vinyl alcohol. It has T_g of about 69 °C and a melting point of 181 °C. The structure of EVOH copolymers affects the material's ability to limit gas or hydrocarbon diffusion, thus resulting in an excellent barrier against gases and hydrocarbons [13]. It has a hydrophilic character due to the presence of the hydroxyl side groups in the polymer chain. The polymer has a good mechanical strength.

Polymethylmethacrylate (PMMA) is synthesised by polymerization of methyl methacrylate. It can be atactic, syndiotactic, and isiotactic, although the molecular chains are mostly syndiotactic meaning the molecular chain's substituents alternate uniformly. This uncrystallized polymer shows remarkable transparency (92 % light transmission). The glass transition temperature (T_g) of atactic PMMA is 105 °C. The polymer, at ambient temperature, is hard, rigid, and brittle with little elongation. PMMA swells and dissolves in many organic solvents; it also has poor resistance to many other chemicals on account of its easily hydrolyzed ester groups.

Tab. 1.1: Properties of non-biodegradable polymers used for membrane preparation.

Polymer	T_g [°C]	T_m [°C]	Young modulus [GPa]	Structure
CA	80	210	1.4	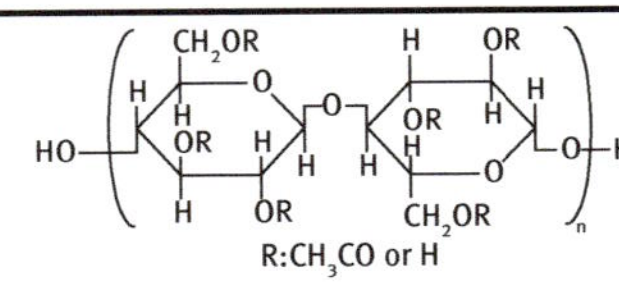
CN	53	240	1.4	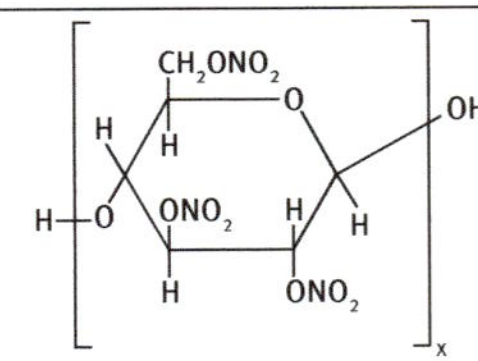
PSf	185	200	2.61	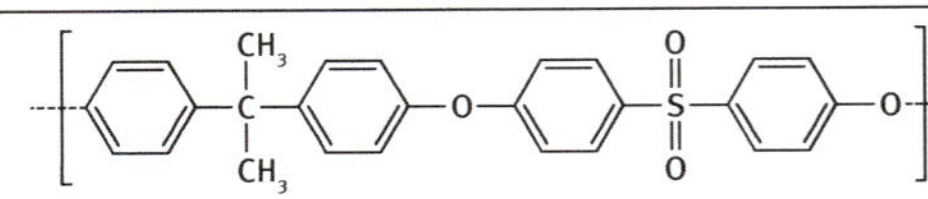
PES	224	360	2.5	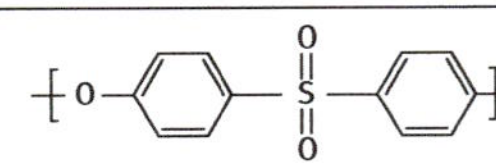
PTFE	20	325	0.5	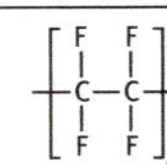
PVDF	−35	171	1.5	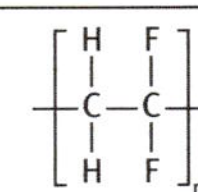

Tab. 1.1: (continued)

Polymer	T_g [°C]	T_m [°C]	Young modulus [GPa]	Structure
PEI	218	204–232	2.3–12.5	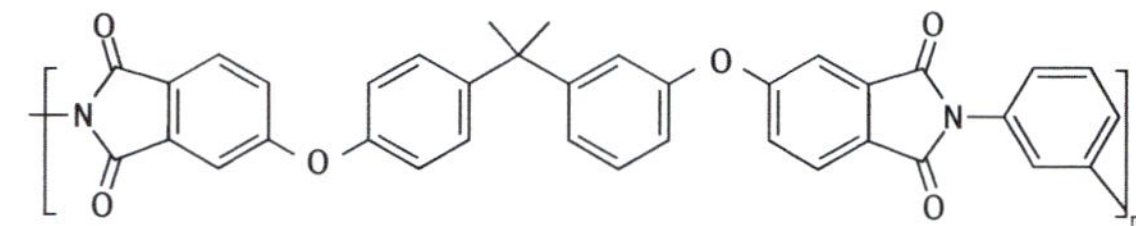
PP	−10	175	1.3	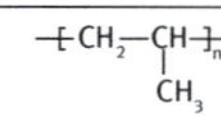
PAN	120	322	8–12	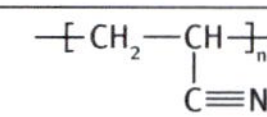
PMMA	110	175	3.2	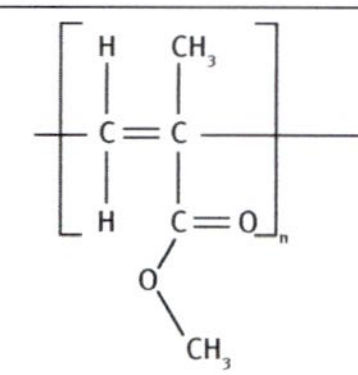
PE	−68	135	0.7	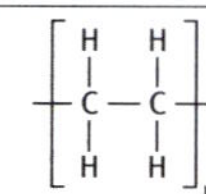
PEEK	143	350	3.6	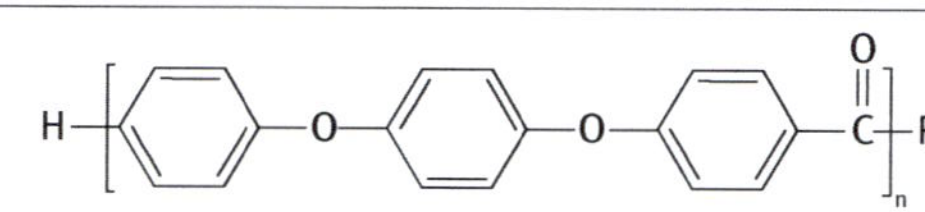

Tab. 1.1: (continued)

Polymer	T_g [°C]	T_m [°C]	Young modulus [GPa]	Structure
PEEK-WC	225	400		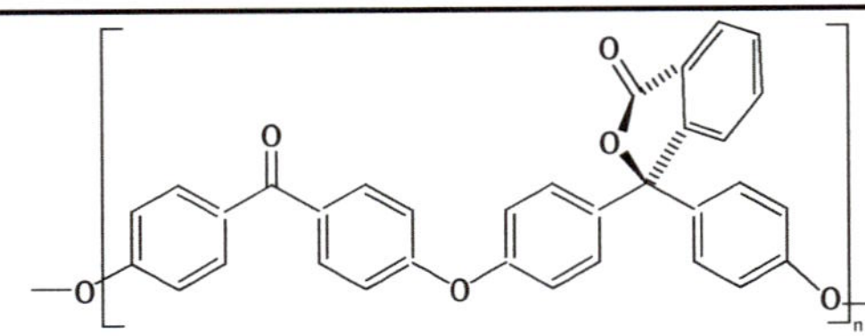
PVC	65	160	3.4	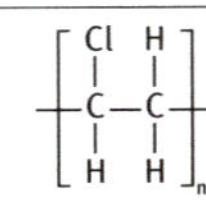
PVA	70	225	0.6	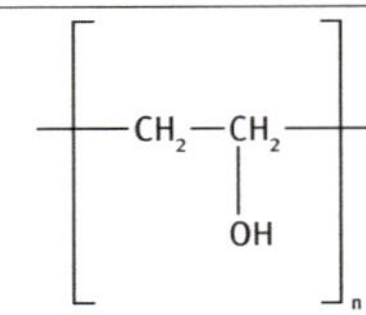
EVAL	48.0–72.0	156–195	3.1	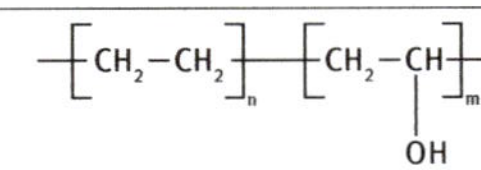

Polyimide (PI) is a polymer consisting of imide monomers. According to the composition of their main chain, polyimides can be aliphatic (linear polyimides), semiaromatic, or aromatic. The latter are the most used polyimides because of their thermostability. They have also good chemical resistance and excellent mechanical properties.

Linear and matrix smart polymers exist with a variety of properties depending on reactive functional groups and side chains. These groups might be responsive to pH, temperature, ionic strength, electric or magnetic fields, and light [14, 15]. Some polymers are reversibly cross-linked by noncovalent bonds that can break and reform depending on the external conditions [16]. Nanotechnology has been fundamental in the development of certain polymeric nanoparticles such as dendrimers and fullerenes that have been applied to drug delivery. Traditional drug encapsulation has been done using lactic acid polymers. More recent developments have seen the formation of lattice-like matrices that hold the drug of interest integrated or entrapped between the polymer strands [17].

Smart polymer matrices release drugs by a chemical or physiological structure-altering reaction, often a hydrolysis reaction resulting in cleavage of bonds and release of the drug as the matrix breaks down into biodegradable components. The use of natural polymers has given way to artificially synthesized polymers such as polyanhydrides, polyesters, polyacrylic acids, poly(methyl methacrylates), and polyurethanes. Hydrophilic, amorphous, low-molecular-weight polymers containing heteroatoms (i.e., atoms other than carbon) have been found to degrade the fastest. Scientists control the rate of drug delivery by varying these properties, thus adjusting the rate of degradation.

1.3 Biodegradable polymers

Biodegradable polymers are very interesting for making membranes used in tissue engineering and regenerative processes, because these membranes are able to be broken down and excreted or resorbed without removal. Natural polymers such as components of extracellular matrix (e.g., collagen) have been widely used for many years; the synthesis of these polymers started about 50 years ago. In the design of biodegradable polymers, important issues must be addressed: (i) they must not evoke an inflammatory response; (ii) they possess a degradation time coinciding with their function; (iii) they have appropriate mechanical properties; (iv) they produce nontoxic degradation products that can be readily resorbed or excreted; and (v) they include appropriate processability for biomedical membranes. These properties are dependent on material chemistry, molecular weight, hydrophobicity, surface charge, water adsorption, degradation, and erosion mechanisms [18]. Polymers undergo hydrolytical degradation due to the presence of hydrolytically unstable bonds, hydrophilic enough for the water access. Hydrolytically degradable polymers possess bonds that are susceptible to hydrolysis including esters, anhydrides, acetals, carbonates, amides, urethanes, and phosphates [19]. Enzymatically degradable polymers possess bonds that require enzymatic reactions to undergo significant degradation under physiological conditions. Most of these polymers contain ether or amide bonds, which have very low hydrolytic degradation rates. The most common hydrolytically biodegradable polymers are poly(α-esters), which contain aliphatic ester bond in their backbone, which is responsible of their hydrolytic degradation.

Polyglycolide or poly(glycolic acid) (PGA) One of the polyesters widely used is the polyglycolide or poly(glycolic acid) (PGA) prepared starting from glycolic acid. This polymer has a glass transition temperature between 35–40 °C and a melting point in the range of 225–230 °C [20]. It has rapid degradation and insolubility in many common solvents. The rapid degradation leads to the loss of mechanical strength and significant local production of glycolic acid, which can create an undesired inflammatory response. This drawback limits its application.

Polylacide (PLA) is more hydrophobic than PGA and stable against hydrolysis because of the presence of methyl groups. The polymer exists in four forms: poly(L-lactic acid) (PLLA), poly(D-lactic acid) (PDLA), poly(D,L-lactic acid) (PDLLA) – a racemic mixture of PLLA and PDLA, and mesopoly(lactic acid). Only PLLA and PDLLA have shown to be promising in tissue engineering applications and have been extensively studied. PLLA is produced by polymerization of L,L-lactide. It has a T_g of 60–65 °C, a melting temperature around 175 °C and a mechanical strength of 4.8 GPa [21]. High molecular weight PLLA has been shown to have a slow degradation time. For this reason PLLA is modified or blended or copolymerized with other degradable polymers. An interesting modification technique of PLLA consists of the use of radiation which induces branching and crosslinking and thus causing a decrease in crystallinity. In PDLLA the random positions of its two isomeric monomers within the polymer chain yields a slightly lower T_g of 55–60 °C and lower mechanical strength of 1.9 GPa [20]. PDLLA degradation is faster than that of PLLA but it still takes over a year to properly disappear.

Poly(lactic-co-glycolic acid) (PLGA) is synthesized by means of the copolymerization of glycolic acid and lactic acid, which are linked together by ester linkages, thus yielding a linear, aliphatic polyester as a product. Depending on the ratio of lactide to glycolide used for the polymerization, different forms of PLGA can be obtained, and consequently it is possible to modulate the degradation times from 1–2 months for 50 : 50 PLGA, to 5–6 months for 75 : 25 PLGA, or 4–5 months for 85 : 15 PLGA [22]. All PLGAs are amorphous rather than crystalline and show a glass transition temperature in the range of 40–60 °C. Unlike the homopolymers of lactic acid (polylactide) and glycolic acid (polyglycolide), which show poor solubilities, PLGA can be dissolved by a wide range of common solvents.

Polycaprolactone (PCL) is synthesised by ring opening polymerization of ε-caprolactone. The polymer is semicrystalline with great organic solvent solubility; it has a melting temperature of 55–60 °C and a T_g of −54 °C [23]. PCL has a very low *in vivo* degradation rate and moderate wettability. It has also a low tensile strength (0.023 GPa), but very high elongation at breakage (4700 %), making it a very good elastic biomaterial.

Polycarbonate (PC) are characterized by the presence of carbonate groups (–O–(C=O)–O–). Although this bond is hydrolytically stable, the observed *in vivo* degradation is due to enzymatic degradation. The polymer has a glass transition temperature of about 147 °C and a melting point of 155 °C. It has low tensile strength (55–75 MPa).[1] The most extensively studied polycarbonate is poly(trimethylene carbonate) (PTMC), which is an elastomeric aliphatic polymer with great flexibility and slow degradation profile, but poor mechanical strength. Other polycarbonates are tyrosine-derived polycarbonates, which are variations of poly(amino acids) in which amino acid-like backbones are connected by carbonate bonds such as poly(desaminotyrosyl-tyrosine alkyl ester carbonates) (PDTEs). PDTEs have a variable pendant alkyl chain allowing for modulation of their thermal and mechanical properties.

Polyurethanes (PU) are produced by reaction of isocyanate with polyol in the presence of a catalyst. As a result the polymer is constituted by a combination of monomers with isocyanate (R–N=C=O) and hydroxyl (–OH) functional groups. The properties of a polyurethane are greatly influenced by the types of isocyanates and polyols used to make it. The polymer has a mechanical strength of about 40–60 MPa [24, 25]. Polyurethanes are composed of hard and soft segments and can undergo microphase separation allowing for these polymers to handle physical stress very well. T_g and T_m vary with the variation of the percentage of soft and hard segments. Hard and soft segments are high T_g and low T_g portions, respectively.

Polyphosphazenes (PZ) are hybrid inorganic-organic polymers in which the backbone is completely inorganic consisting of phosphorous and nitrogen bonded linearly through alternating single and double bonds. They contain two organic or organometallic side groups attached to each phosphorus atom. Side groups affect the degradation rate and physical properties of the polymer. Thermal and mechanical properties can vary with T_g from −10 to 35 °C and tensile strength 2.4–7.6 MPa [26]. Side groups also affect properties such as hydrophobicity, hydrophilicity, biodegradation, or ion transport properties [27].

Polyamide (PA) is an organic polymer constituted of monomers of amides joined by peptide bonds. They can be produced naturally (e.g. polypeptides) and artificially (e.g. nylons, aramids, and sodium poly(aspartate)). According to the composition of their main chain, polyamides are classified as aliphatic, semiaromatic, or aromatic. On the basis of the number of repeating unit types, polyamides can be homopolymers

1 City Plastics Polycarbonate, www.cityplastics.com.au/materials-polycarbonate/

or copolymers. Polyamides can be semicrystalline or amorphous. Polyamides contain the same amide bond as in polypeptides, but form polypeptides differently and have strong chain interactions and a low rate of biodegradation. Enzymes and microorganisms can degrade low molecular weight oligomers [28]. Biodegradation may be improved by the introduction of various side groups, such as benzyl, hydroxyl, and methyl groups. For example copolymers with amide and ester groups are found to be more easily degradable [29].

Chitosan (CHT) is a linear polysaccharide composed of randomly distributed β-(1-4)-linked D-glucosamine (deacetylated unit) and N-acetyl-D-glucosamine (acetylated unit). It is the N-deacetylated derivative of chitin and is naturally degraded by body enzymes. The chitosan degradation rate is dependent on the degree of acetylation and crystallinity. Chitosan with lower acetylation percentages has been shown to last *in vivo* up to several months [30]. The polymer has reactive amino groups and hydroxyl groups and chelates many transitional metals. It has poor solubility in some solvents. As chitosan by itself is mechanically weak, it is often crosslinked or combined with other degradable polymers such as PLA, PLGA, PEG, collagen, and alginate.

Hyaluronic acid (HA) is a linear anionic polysaccharide consisting of alternating units of N-acetyl-D-glucosamine and glucuronic acid. HA homopolymer is mechanically weak, but its mechanical properties can be improved by crosslinking with ethyl esters or benzyl esters. Depending on the extent of esterification, the degradation rate can be varied from 1–2 weeks to 4–5 months [31]. Hydrolytic degradation causes scission of the ester bond, converting it back to HA, which is degraded *in vivo* by hyaluronase enzymes.

Poly(p-dioxanone) (PDO) is prepared by ring opening polymerization of *p*-dioxanone. PDO is semicrystalline, with a low glass transition temperature in the range from −10 to 0 °C and a T_m of 115 °C. An increase in the molecular weight can increase the tensile strength and modulus and improve the thermal stability of PDO. Its biodegradation is due to the ester bonds in the polymer chains. PDO has a slow degradation rate (6–12 months for complete mass loss) and a low modulus (1.5 GPa) but good flexibility and strength maintenance [20].

Collagen is a long, fibrous structural protein composed of a triple helix consisting of two identical chains (α1) and an additional chain that differs slightly in its chemical composition (α2). The amino acid sequence of collagen is glycine-proline-X and glycine-X-hydroxyproline, where X is any amino acid other than glycine, proline, or hydroxyproline. It is used to prepare membranes especially for use in dental surgical procedures as a resorbable material for placement in the area of dental implants, bone defects, or ridge reconstruction [21].

Tab. 1.2: Properties of biodegradable polymers used for membrane preparation.

Polymer	T_g [°C]	T_m [°C]	Tensile strength [GPa]	Degradation	Structure
PGA	35–40	225–300	12.5	Hydrolitic degradation: rapid	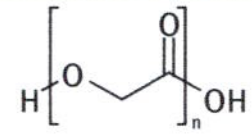
PLLA	60–65	175	4.8	Hydrolitic degradation: fast	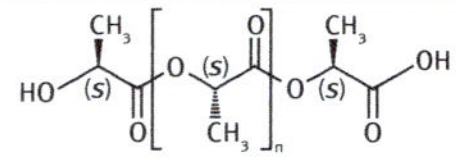
PDLLA	55–60	175	1.9	Hydrolitic degradation: fast	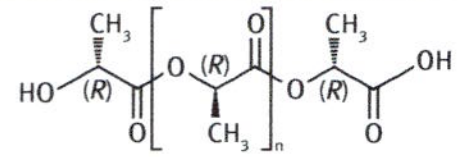
PLGA	45–55	–		Hydrolitic degradation: fast	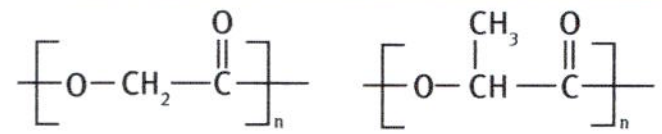
PCL	−54	60	0.023	Hydrolitic degradation: low	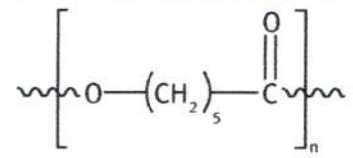
PC	147	155	0.055–0.075	Hydrolitic degradation: limited	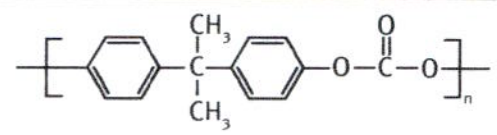
PU	−19/−60	72–130	0.085–0.125	Hydrolitic degradation: limited	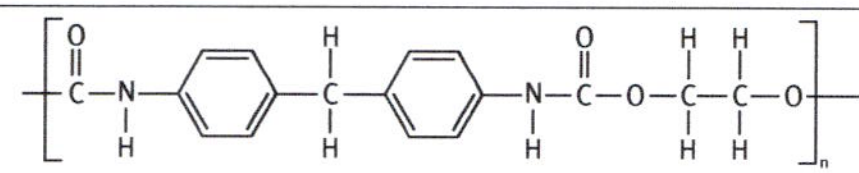
POP	−100/+149	217–244	0.0024–0.0076	Hydrolitic degradation: low or fast	

Tab. 1.2: (continued)

Polymer	T_g [°C]	T_m [°C]	Tensile strength [GPa]	Degradation	Structure
PA	50	215	2.4	Enzymatic degradation: limited	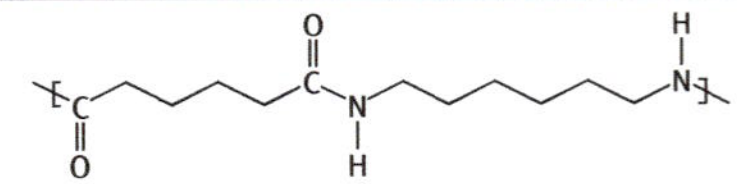
PDO	−10/0	115	1.5	Hydrolitic degradation: low	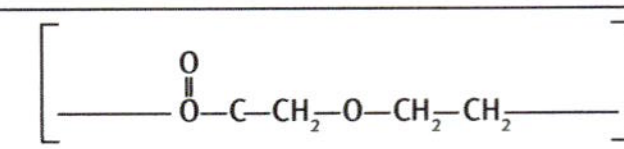
HA	20–25		0.000023	Enzymatic degradation	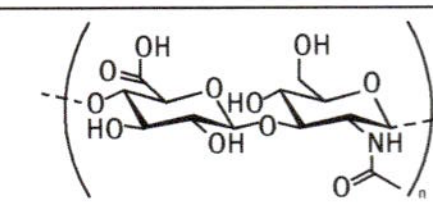
CHT	118	88–101	0.043	Enzymatic degradation	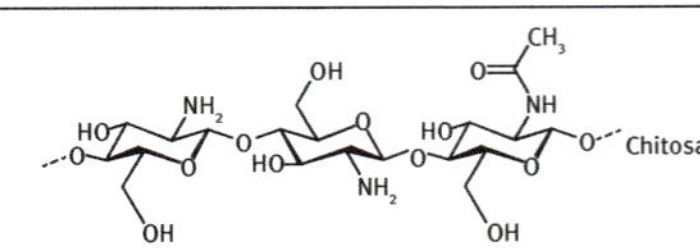
Collagen	55	109		Enzymatic degradation	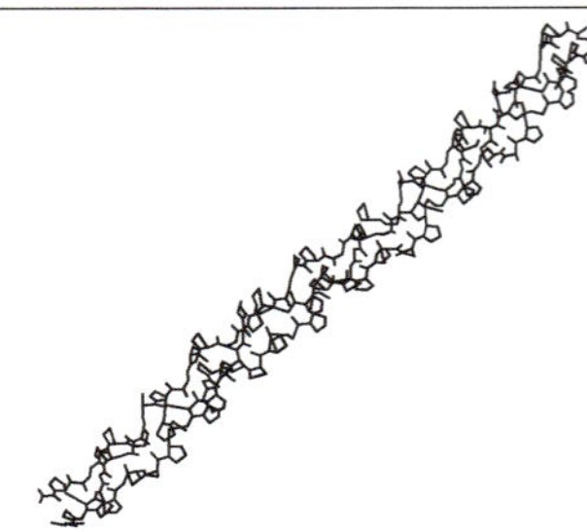

1.4 Preparation of membranes

1.4.1 Thermodynamic principles

Phase inversion processes allow the production of membranes which dissolve one or more polymers in a unique solvent (polymer/solvent system). In order to ensure a perfect combination of membrane structure and chemical-physical properties, the polymeric solution must be thermodynamically stable, which means that at the standard condition (constant temperature and pressure), in the system all the chemical components should be in a perfect equilibrium state in a sole phase and should not subject to the de-mixing process. In fact two or more reagents in solution can originate a liquid phase perfectly homogeneous or can re-arrange in phases immiscible with each other in which all the components are shared differently. Knowing the characteristics of the elements and the proper condition that serves the complete control of the reaction provides the understanding of the thermodynamics information about how to blend together different elements. Although for a pure substance the thermodynamic properties of the system are directly functions of two parameters only, temperature and pressure, for a mixture all variables in the composition, number of moles of the components, intermolecular forces (dispersion forces, polar forces, hydrogen bonding, etc.), activity, and the chemical potential of the polymer are really important. Binary and/or a ternary polymeric systems spontaneously turn into a state of minimum energy called Gibbs free energy of mixing ΔG_m which describes the equilibrium properties of the solution [32]. Knowing the parameters and the independent variables which ΔG_m is dependent on, the thermodynamic properties of the system can be derived. At constant temperature and pressure ΔG_m is defined by the expression

$$\Delta G_\mathrm{m} = G - \sum_i n_i \bar{G}_i^0, \tag{1.1}$$

where n_i is the number of moles of the component i and $\bar{G}_i^0$ refers to the Gibbs molar function of the pure component; but it can be expressed as the result of the entropic (ΔS_m) and the enthalpic (ΔH_m) contribution in the system:

$$\Delta G_\mathrm{m} = \Delta H_\mathrm{m} - T\Delta S_\mathrm{m}, \tag{1.2}$$

where ΔH_m is the enthalpy of mixing and ΔS_m is the entropy of mixing. Two components will spontaneously blend together if the Gibbs free energy of mixing is negative ($\Delta G_\mathrm{m} < 0$), as a consequence of the complete dissolution of the solute in the solvent. For a polymer/solvent solution the effect of the entropy of mixing on the system is much smaller than the free enthalpy of mixing determined by the ΔH_m only. The enthalpy of a system is defined as the sum of the internal/external motion liberty degree and the intermolecular forces contribution, the latter being considered the main important parameter for the determination of the enthalpy of mixing of an ideal system. In terms of energy, those forces are due to the movement of all the electrons in the molecules that originate the fluctuations of the dipoles in all the chemicals and consequentially the coordination of the charges in the system that causes an energy decrease. When the dispersion and repulsion forces between the positive and the negative charge of the dipoles are balanced, the energy of the system reaches a state of minimal value defining the so-called energy of interaction ε, namely the level of energy requested to separate the molecules of a component in a given volume of liquid, which is expressed in terms of cohesive energy density e as a function of the enthalpy of vaporization (ΔH_{vap}) of the component involved:

$$e = \frac{\Delta H_{\text{vap}} - RT}{V_i}, \tag{1.3}$$

where V_i is the molar volume of the liquid. Considering a binary system, the enthalpy of mixing is done by the Hildebrand expression

$$\Delta H_{\text{mix}} = V_{\text{m}}[(e_1)^{0.5} - (e_2)^{0.5}]^2 \phi_1 \phi_2. \tag{1.4}$$

Since the square root of the cohesive energy density is the solubility parameter δ of a molecule, the expression (1.4) can be simplified as

$$\Delta H_{\text{mix}} = V_{\text{m}}(\delta_1 - \delta_2)^2 \phi_1 \phi_2, \tag{1.5}$$

where V_{m} is the molar volume of the solution, ϕ are the volume fraction and δ are the solubility parameters of the solvent and the polymer referred with number 1 and 2, respectively. When $\delta_1 \approx \delta_2$, the value of ΔH_{m} approaches zero, then the components are miscible (because ΔS_{m} is always positive). In reality except when particular interactions are involved, the polymers are usually soluble just in a few organic solvents because of their heavy chains and volume; in fact, with respect to the low molecular weight molecules they cannot freely move in the system, and just a few rearrangements in space are possible when they are mixed with other components. This concept is well explained by the Flory–Huggins theory of the lattice model (Figure 1.1).

In a binary system where A (solvent) and B (solute) are the components with small molecular weight, each molecule of A can be imagined to be surrounded by z particles of A or B, where z is the coordination number of the reticule, and the particle motion depends on the translation, rotation, and vibration liberty in the system AB. Three different kind of interactions are possible: A–A , B–B, and/or A–B. For each one of them a specific energy of interaction ε can be calculated, and the enthalpy of the system is the results of their coordination:

$$\Delta H_{\text{m}} = H - H_{\text{A}} - H_{\text{B}}, \tag{1.6}$$

where H is the contact energy for the solution and the components, respectively. Assuming a homogeneous molecules surface and knowing the total number of moles N in the system, the probability for a molecule of A to contact with B depends on the molar fraction of B; but when it is a polymer to be dissolved in a solvent, the probability

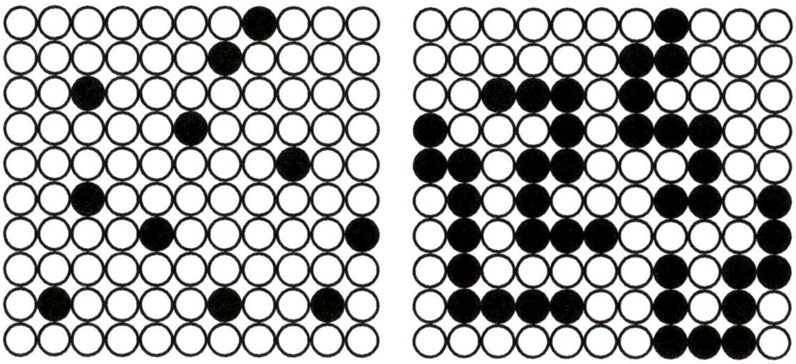

Fig. 1.1: Flory–Huggins lattice model: molecular distribution in a binary system of (a) monomers and (b) polymers in a low-molar-mass solvent.

that a portion of space is occupied by a segment of the macromolecule is higher the more concentrated and heavier the polymer is (Figure 1.1 (b)), then the equation of the enthalpy of mixing has to be expressed as a function of the volume fraction, i.e.

$$\Delta H_\mathrm{m} = n_1 \phi_2 N_\mathrm{av} z \,\Delta\varepsilon_{12} = RTn_1\phi_2 X_{12}, \tag{1.7}$$

where the numbers 1 ad 2 refer to the solvent and the polymer, n is the number of moles, N_av is the Avogadro's number, and X_{12} is the Flory–Huggins interaction parameter of the system. For ideal solutions, $\Delta H_\mathrm{m} = 0$ and the ΔG_m is determined by the ΔS_m only. As said earlier, in a binary system AB, each molecule of A is surrounded by z particles, where z is the coordination number of the reticule, and the ability of A to move in the system is linked to the parameters that define an entropic degree of disorder: translation, rotation, and vibration liberty. Assuming again an ideal system, the entropy of mixing is due to the molecular disposition of the component in space, thus the only entropic contribution is the entropy of configuration (S)

$$S = K \ln W, \tag{1.8}$$

where K is the Boltzmann constant and W, depending on how the molecules of A (N_A) can arrange with respect to B, is the number of all the microscopic states possible. Expressing the number of molecules N as a function of both the molar fraction of the particles and the number of moles n of the components, the expression of the entropy of configuration (equation (1.8)) becomes

$$S = -R(n_\mathrm{A} \ln x_\mathrm{A} + n_\mathrm{B} \ln x_\mathrm{B}). \tag{1.9}$$

When the solute is a macromolecule, the degree of motion of the particles is strongly lower due to the dimension and the concentration of the molecule in the solution, then assuming that the polymer is linear and composed by σ segments, the entropy is expressed again in terms of volume fraction ϕ and refers directly to the mixing process [33]. For a total number of sites equal to

$$n_\mathrm{t} = n_1 + \sigma n_2, \tag{1.10}$$

where σ is the number of segments in the chain of polymer, the volume fraction for a system polymer/solvent is

$$\phi_1 = \frac{n_1}{n_1 + \sigma n_2} \tag{1.11}$$

and

$$\phi_2 = \frac{\sigma n_2}{n_1 + \sigma n_2}; \tag{1.12}$$

where 1 and 2 are the solvent and the polymer. Then the equation of ΔS_{m} is

$$\Delta S_{\mathrm{m}} = -R(n_1 \ln \phi_1 + n_2 \ln \phi_2). \tag{1.13}$$

When two polymers are mixed together the equation is rearranged considering

$$n_1 = \left(\frac{\phi_1}{\sigma_1}\right) n_{\mathrm{t}} \tag{1.14}$$

and

$$n_2 = \left(\frac{\phi_2}{\sigma_2}\right) n_{\mathrm{t}}. \tag{1.15}$$

For a polymer/solvent system the Gibbs free energy of mixing resultant is

$$\Delta G_{\mathrm{m}} = RT(n_1 \ln \phi_1 + n_2 \ln \phi_2 + n_1 \phi_2 X_{12}). \tag{1.16}$$

For a pure component i, the molar partial Gibbs function is the expression of the chemical potential μ_i^0 of the component

$$\mu_i^0 = \frac{\partial G_i}{\partial n_i}; \tag{1.17}$$

this means that the *activity* coincides with the molar fraction of the chemicals in the system

$$\mu_i - \mu_i^0 = RT \ln x_i. \tag{1.18}$$

Expressing the Gibbs free energy of mixing as function of equation (1.18), we obtain

$$\Delta G_{\mathrm{m}} = RT(n_1 \ln x_1 + n_2 \ln x_2), \tag{1.19}$$

where the molar fraction x will be substituted by the volume fraction ϕ when a binary polymer/solvent system is considered [34].

In Figure 1.2 we see the Gibbs free energy of mixing of binary solutions as a function of the volume fraction (ϕ) of the solute: the low molecular components are shown in green, the polymer/solvent system in red, and the polymer/polymer systems in blue. It is clear that the simpler the system is, in terms of nature and size of the components, the lower are the values of the ΔG_{m}. The trend of the three curves is symmetric, and this means that in ideal conditions the miscibility of the elements is certain for all the compositions possible.

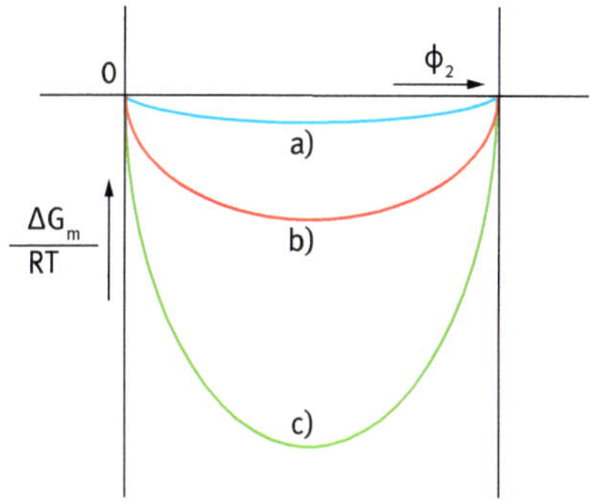

Fig. 1.2: Diagram of the Gibbs free energy of mixing ΔG_m for (a) a polymer/polymer system, (b) a polymer/solvent system, and (c) low molecular components.

As explained earlier, when polymers are in solution, the effect of the entropy of mixing on the system is very small, and then the Gibbs free energy of mixing is determined by the sole enthalpy (ΔH_m); even the slightest change in the temperature of the system can determine the occurrence of the phases separation, and the solution will eventually demix. Analyzing the same polymer/solvent system (equation (1.19)) at two different temperature T_1 and T_2 (Figure 1.3), where $T_1 > T_2$, the thermodynamic behavior of the system followed a different trend with respect to what is described for ideal solutions. When the components are blended at high temperature (T_1 near the melting point of the polymer), negative values of ΔG_m are obtained. Even if the curve

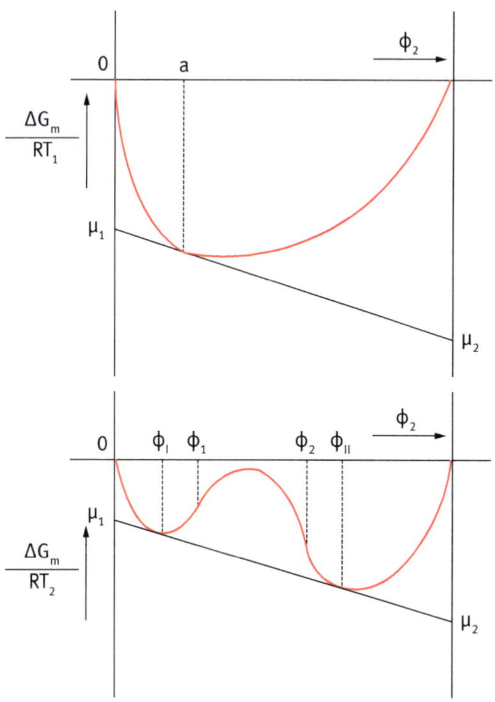

Fig. 1.3: Diagram of ΔG_m for a polymer/solvent system at two different temperatures T_1 and T_2 as a function of the volume fraction of the components.

is not symmetric, at all the points of the graph it is possible to draw a tangent, and all the points have the same derivative:

$$\Delta\mu_i = \frac{\partial \Delta G_m}{\partial n_i}.$$

Since the intercept at the axes meet the value of μ_i for the components the solute results miscible in the solvent over the whole composition range. When a lower temperature is used (T_2) in the graph, two points of minima and two points of inflections are recognizable where the curve shifts its convexity/concavity. The points of the minima lie on the same tangent, unique for the graph, and the intercept crosses the axes at specific values of chemical potentials for the components. At every temperature a new tangent can be drawn, and eventually the points intercepted on the components axes will coincide defining the so-called critical point, i.e. the temperature up to which the system is stable at all the compositions.

The points of minima and inflections described earlier are important to understand the stability of the system when a membrane has to be obtained from a polymeric solution. They are derived when the first and the second derivate of the ΔG_m are both equal to zero (Figure 1.4).

Specific regions and ranges of concentrations are bounded, and different trends of the curves are defined. For compositions where the solvent concentration is higher than the polymer, i.e. $0 > \phi < \phi_I$, the values of ΔG_m are descendent, and the solution in this volume fractions range is thermodynamically stable, due to the perfect miscibility of the components.

With increasing polymer concentration, the miscibility starts to decrease, as explained by the Flory–Huggins theory, and the Gibbs free energy of mixing rises, leaving a point of minimum (ϕ_I) and reaching the first point of inflection at a volume fraction equal to ϕ_1. From this point the highest values of ΔG_m are recorded, and the solution for volume fractions included between ϕ_1 and ϕ_2 is thermodynamically instable and incurs in demixing. When the concentration of the polymer exceeds ϕ_2 a phase rich in solute is formed and the system results in stability again. The plots of the first (in blue) and the second derivate (in green) underlined what was stated above. In particular for volume fraction ϕ included between ϕ_1 and ϕ_2 the second derivate (lowest part of Figure 1.4) is negative, implying that the solution is instable, and demixing incurs spontaneously. Plotting the points of minima and the points of inflection for a polymeric solution, the binodal and the spinodal curves can be depicted as a function of the volume fraction, and a temperature/composition diagram can be obtained (Figure 1.5). Referring to the Flory–Huggins theory the less wide system is more complex and the binodal and the spinodal curves are symmetric, and the critical point shifts towards the left of the graph near the solvent axis.

The point where the binodal and spinodal coincide is the critical point, as the temperature up to which the system is stable at all the compositions. In fact for all the temperature higher than T_c, every single ratio between the polymer and the solvent in

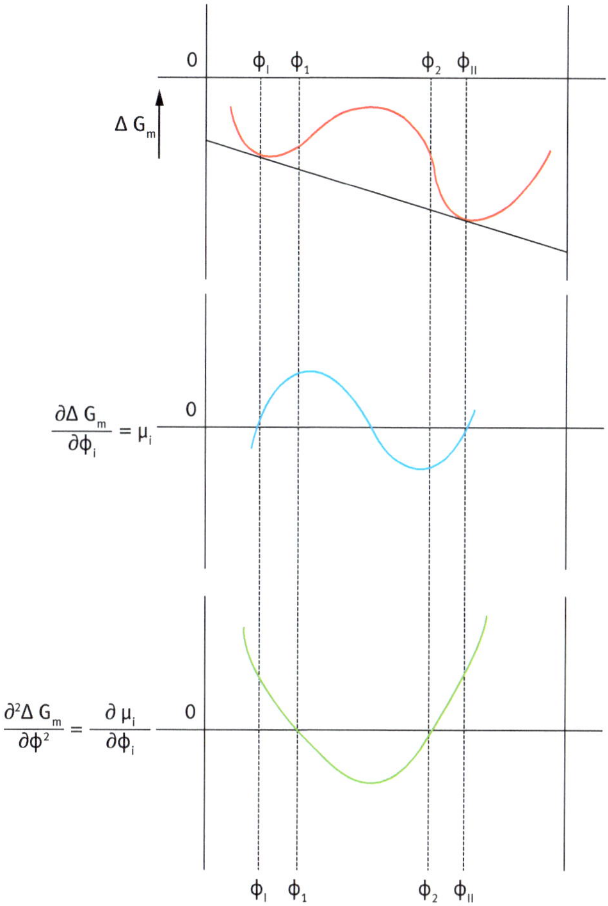

Fig. 1.4: Diagrams of ΔG_m, first and the second derivatives for a polymer/solvent system as a function of the volume fraction of the components.

the system allow us to obtain a homogeneous solution, and as described in Figure 1.3, the Gibbs free energy of mixing is always negative. When a lower temperature is used (T_2), with the increase of the polymer concentration the binodal curve is reached (ϕ_I), and the stability of the system starts to be reduced. At that point, for all the composition included in the range ϕ_I–ϕ_1, the phenomenon called nucleation occurs, and the solution is characterized by the presence of little drops of polymer dispersed in the solvent. Those drops are in equilibrium, and then the system is stable due to the absence of driving forces that can destabilize it. The same is valid for the ratio included in the range ϕ_{II}–ϕ_2, where in the polymeric phase little drops of solvent are dispersed. Those two regions are called metastable, and they are particularly important for the membrane preparation with a phase inversion process: when a polymer/solvent solu-

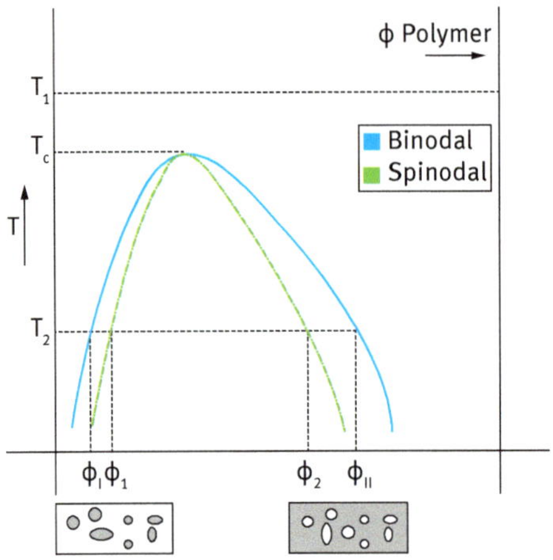

Fig. 1.5: Diagrams of the binodal and the spinodal curves for a polymeric solution as a function of the volume fraction.

tion with composition included in the subscribed ranges is used, if the droplets in the solution coalesce before the polymer solidifies following the precipitation process, an open porous system will result. But when the composition is in the region bounded by $\phi_1 - \phi_2$ under the spinodal curve, demixing is incurred due to the growth of the dispersed drops: two different phases, one rich in solvent and one in polymer, are formed, and the energy of the system decreases in order to stabilize the system. Therefore, the demixing process is due to the destabilization of the system further to variation in the nature and complexity of the components, the temperature, and the final composition. Adding a third element in the system is another important cause of demixing: the stability of a ternary system then depends not only on the polymer/solvent ratio, but also on this third element, the nonsolvent, used in few phase inversion methods [35]. From the Flory–Huggins theory discussed earlier (equation (1.16)), a Gibbs free energy of mixing expression for three elements can be derived:

$$\Delta G_{\text{m}} = RT(n_1 \ln \phi_1 + n_2 \ln \phi_2 + n_3 \ln \phi_3 + n_1 \phi_2 X_{12} + n_1 \phi_3 X_{13} + n_2 \phi_3 X_{23}). \quad (1.20)$$

Then new diagrams and plots may be depicted, but in that case the system depends on three parameters and the temperature/composition graph assumes a three-dimensional conformation. As a function of the temperature, three axes referring to the volume fraction of polymer, solvent and nonsolvent build the diagram for a ternary system, as represented in Figure 1.6.

It is clear that even in that case where the two curves binodal (blue) and spinodal (green) coincide, the critical point is determined, and at all the temperatures higher than T_c all compositions between the elements are possible, whereas when lower T are used the demixing area is reached and the phase separation occurs. Crossing the diagram transversely, a bidimensional representation of the binodal and spinodal curves is obtained, and the thermodynamic trend of the system can be analyzed at a determined temperature (isothermal system). The corners of the triangle depicted refer to

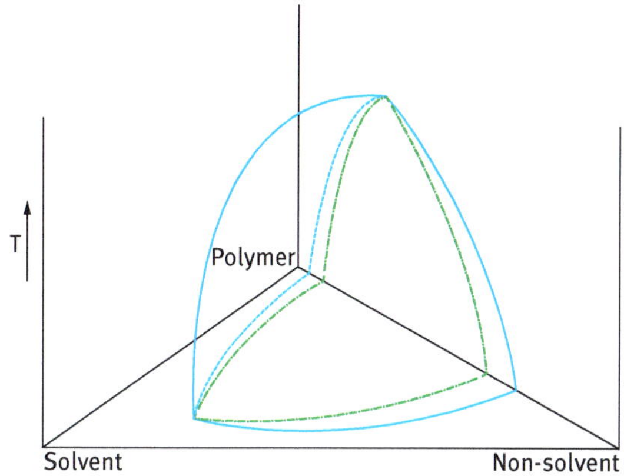

Fig. 1.6: 3D-diagram of the binodal and the spinodal curves for a ternary system polymer/solvent/nonsolvent as function of the temperature.

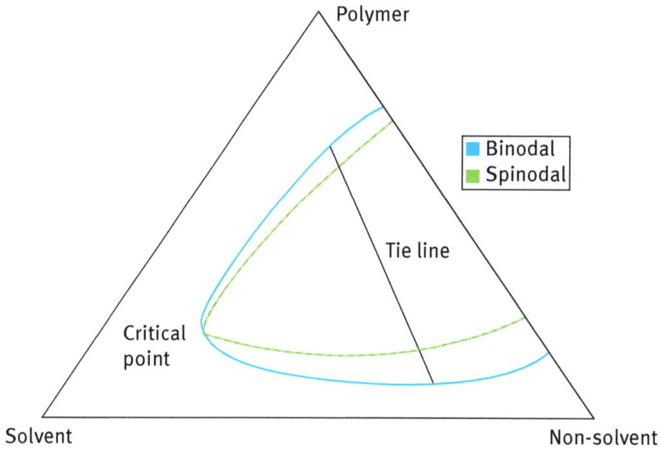

Fig. 1.7: Isothermal graph of a ternary system.

the three components at their maximum concentration (100 %), and consequently each point on the sides of the diagram represents a specific composition of the solution and a determined ratio (ϕ) between the two components placed on the ends of the same side, whereas all the points in the triangle area ascribe to a specific composition of the three elements together (Figure 1.7).

As described for the binary system, the metastable regions bounded by the binodal and the spinodal curves contain the volume fractions values that refer to the solution in which the nucleation occurs, and two separated liquid phases are created. Droplets of solvent and nonsolvent in the polymer-rich phase and *vice-versa* are formed and start to grow and coalesce, leaving a porous structure after the solidification of the continuous phase: then a porous membrane is obtained with a ternary metastable system as with a binary one. When the volume fractions are included under the spinodal curves, demixing will occur. When a point is placed under the binodal, on the curve itself two points can be located referring to the phase composition in equilibrium, which generate a tie line if linked together. For each ternary phase diagram a set of tie lines is depicted, and as soon as the top of the binodal is reached, they narrow to a point called the plait point, where the composition of the two phases coincide (Figure 1.8).

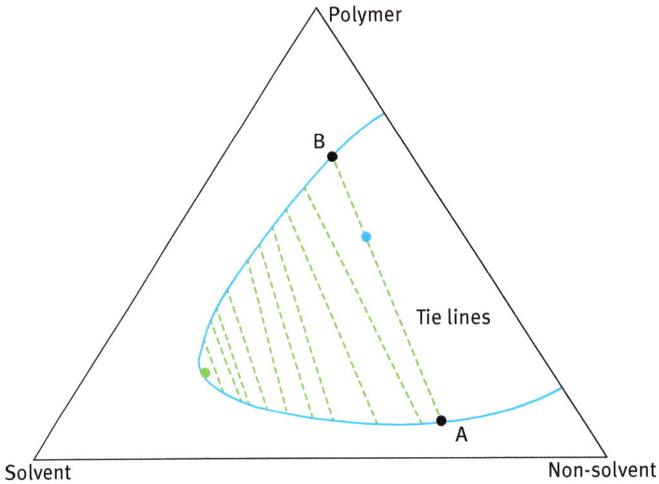

Fig. 1.8: Tie lines for an isothermal graph of a ternary system: two points (A and B) are identified on the binodal curve for a specific composition of the solution considerate (point in blue).

The characteristics and structure of a membrane are functions of the composition of the polymeric solution as much as the nature and the affinity between the elements of the system. Using the phase diagram the process for a membrane production from a ternary system foresees four steps (Figure 1.9):

1. preparation of a homogeneous polymeric solution (red point in the graph);
2. adding of the nonsolvent: the rate of the exchange between the solvent and the nonsolvent depends on their affinity. The higher the affinity is, the faster is the exchange;
3. crossing of the binodal region and beginning of the demixing process. This can be instantaneous when the solvent/nonsolvent affinity is high and the diffusion process followed by the exchange is quick, or it can be retarded otherwise;
4. final composition (A) of the ternary system with two phases in equilibrium: one rich in solvent (A′) and one rich in polymer (A″). Their composition in terms of volume fraction lies on the tie line.

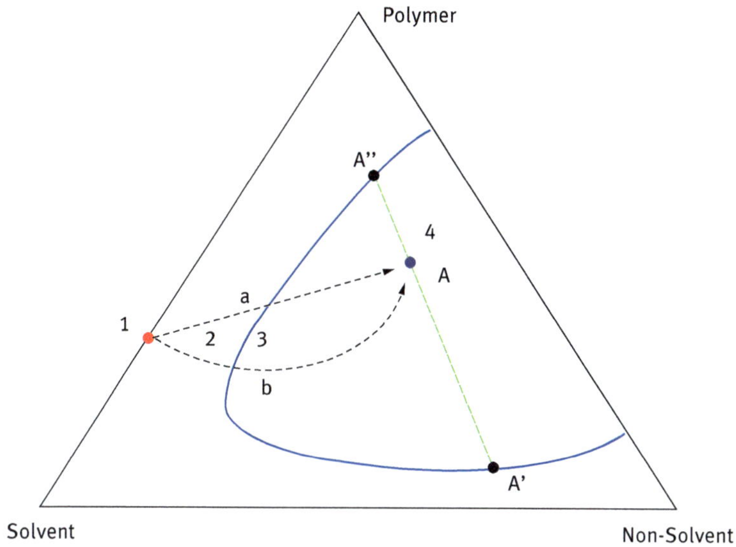

Fig. 1.9: Graphic representation of the four steps (1–4) involved in non-solvent induced membrane preparation. The rate of exchange between solvent and nonsolvent is marked with lower case letters a and b on the arrows.

The choice of the proper solvent and nonsolvent depends on the nature of the polymer requested for the membrane preparation, and they not only have to be similar but must be completely miscible with each other. In that way the exchange process will occur, and a membrane with a specific structure can be obtained. Generally the quicker the exchange rate is, the higher is the final porosity of the membrane, but it is possible to control the process by adding small quantity of one of the liquids: more solvent in the system moves the equilibrium of the reaction, decreasing the rate of exchange and raising the density of the final membrane, whereas small amounts of nonsolvent achieve the opposite effect, and a porous membranes is obtained.

1.4.2 Techniques of membrane preparation

Natural or synthetic, organic or inorganic, electrically charged or neutral materials may be used for the preparation of membranes. Polymers, ceramics, glass, metals, and liquids with their chemical-physical characteristics and properties determine the chemical stability and the mechanical properties of the final membrane, and depending on the application for which a membrane is designed, a suitable preparation technique must be selected in order to achieve a proper structure and conformation (flat, tubular, hollow fiber, etc.), function and transport properties. Several techniques can be used for processing a starting material and obtaining a membrane. Changing or modifying the working parameters like temperature, humidity, solvent evaporation rate, pH, and the supports used for the moulding and choosing the materials concerning their distinctiveness (molecular weight, solubility, density, etc.) allow us to produce a variety of different membranes, classified as porous, dense, symmetric or asymmetric, integral or composite [35]. Polymeric membranes are nowadays the most widely used for industrial chemical and biomedical applications, and they can be obtained with a few different techniques depending on the starting material involved (Figure 1.10).

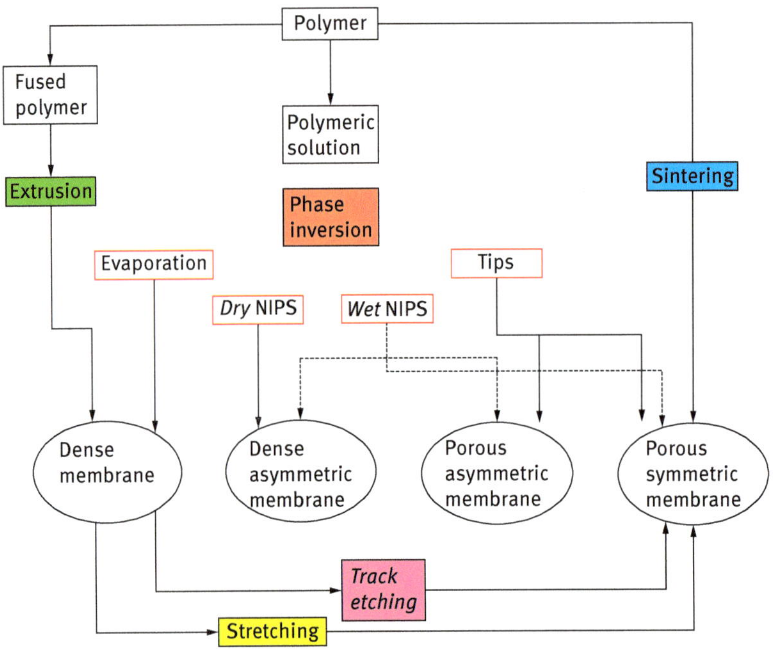

Fig. 1.10: Scheme of membrane preparation techniques.

Nowadays the most popular and widely used technique for the preparation of polymeric membranes is the **phase inversion technique**. Starting from almost any polymer (natural or synthetic) and dissolving it in a proper solvent, films or plates with specific thickness and structure can be obtained through a precipitation process. A few different procedures are available, but independently from them, each phase inversion technique causes the separation of the polymeric solution in two distinct phases: the polymeric solid film that originates the final membrane, and a liquid phase due to the solvent that can be removed. Changing the polymer, its concentration, the solvent, and the precipitation conditions, different kinds of membranes are obtained with symmetric or asymmetric structure and a pore size in the range of 0.1–20 μm. Generally the precipitation of the polymer from the starting solution is due to the film treatment with a nonsolvent as soon as the membrane is moulded. The nonsolvent substitutes the initial solvent in the structure, allowing the formation of the pores, which size and shape are dependent on the time and the rate of the nonsolvent treatment; the technique is then called **nonsolvent diffusion induced phase separation (DIPS or NIPS)** (Figure 1.11 (a)). When the separation of the two phases and the precipitation of the film are due to a change in the temperature, the method is defined as **temperature induced phase separation (TIPS)** (Figure 1.11 (b)).

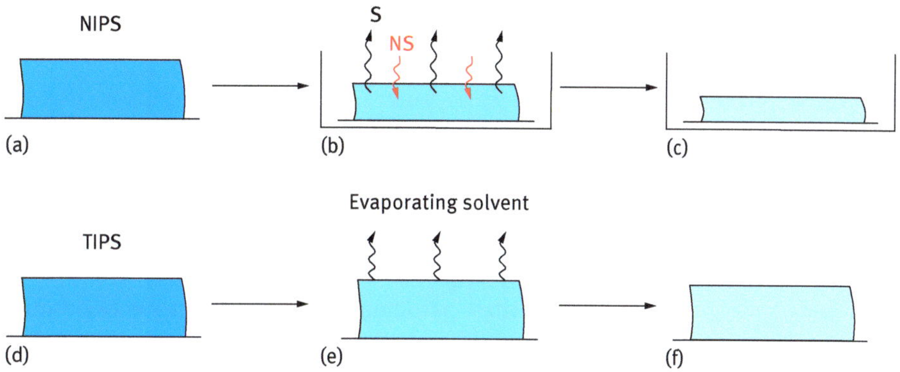

Fig. 1.11: Schematic representation of the nonsolvent diffusion induced phase separation (NIPS) and temperature induced phase separation (TIPS). S: solvent; NS: nonsolvent.

The starting solution is prepared by dissolving a polymer at high temperature in a solvent that evaporates as soon as the solution is spread as a film. Homogeneous pore sizes and distributions are obtained [32, 36]. Although DIPS and TIPS are the most widely used phase inversion techniques, membranes can also be created by controlled evaporation. In this case the precipitation is simply due to the solvent evaporation that controls the structural characteristics of the film obtained from a two-component (polymer/solvent) or a three-component solution (polymer/precipitant/solvent). Therefore the methods are called **precipitation by solvent evaporation process**. These methods allow us to produce membranes with a low degree of porosity and a dense structure in which pores can be defined with a size generally in the nanometers range (Figure 1.2 (a)), but depending on the polymer characteristics and properties, the formation of a *gelation* phase may occur. The polymer freezes into a three-dimensional network as result of chemical or physical crosslinking, and further removal of solvent may result in a porous structure.

One or more of the techniques described earlier are used for the preparation of asymmetric and composite membranes. Asymmetric membranes are formed of a polymeric porous layer on which a homogeneous thin film is placed, usually dense and selectively permeable, which allows the separation process. An asymmetric membrane is also obtained, precipitating a polymeric solution in the form of a continuous solid

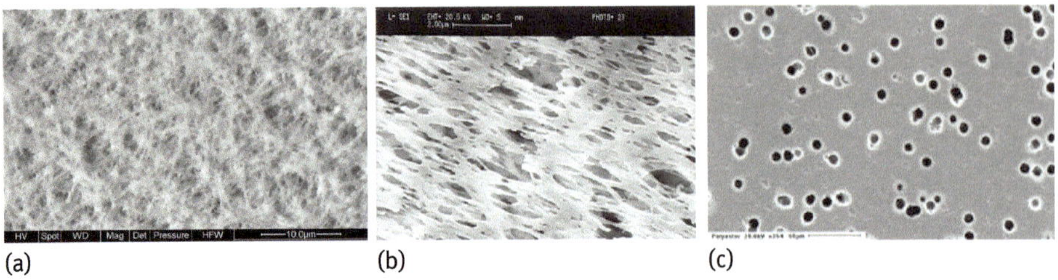

Fig. 1.12: Scanning electron micrographs of (a) modified polyetheretherketone prepared by non-solvent diffusion induced phase separation (NIPS); (b) polypropylene membranes prepared by stretching; (c) polyester membrane prepared by track-etching.

phase through the cooling process, solvent evaporation, or addition of a nonsolvent. All of these processes lead the liquid phase formation that originates the membrane pores, and both the polymeric support and the thin film are made from the same polymer. When the film is deposited on a different porous support, the results is a composite membrane. Their advantage is due to the chance of using different materials together combining their properties, particularly when a polymer is not usable for a specific membrane process if used alone. Four different methods are available for the composite membrane preparation: (1) casting of the barrier (film) and lamination on the porous support film; (2) coating of the support by a polymer followed by heat treatment or radiation; (3) gas phase deposition of the thin layer on the support; (4) interfacial polymerization of reactive monomers on the support [37]. Membranes for gas separation, pervaporation, and osmosis processes are obtained through them.

The liquid membranes consist of material with low vapor pressure, immiscible with aqueous solutions that ensures long-term stability. They can be obtained filling the pores of a specific support with a liquid material or stabilizing the liquid in form of a sol–gel system obtained through an emulsion process with a surfactant (Figure 1.13) [38].

Heavy metals, polluting organics, are generally separated from a matrix or a stream by these membranes. The support used must be hydrophobic and highly porous with a pore size dimension which is homogeneous and particularly wide;

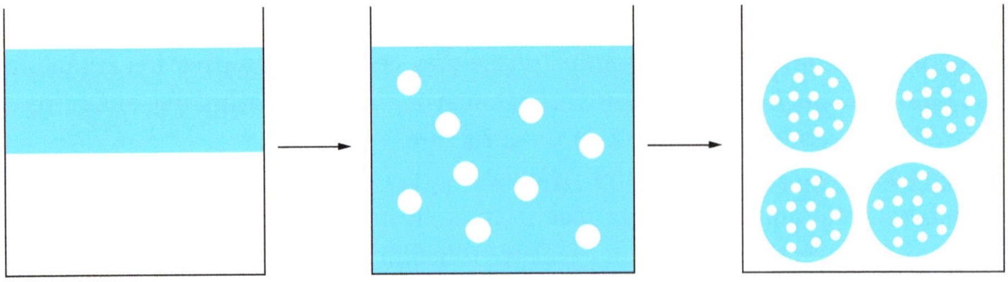

Fig. 1.13: Drawing of the preparation method of an emulsion-type liquid membrane.

but it is possible to obtain the liquid membranes even without it by means a double emulsion system: an aqueous solution is dispersed in an oily one, forming drops that are eventually dispersed in a further aqueous solution. The oil drops constitute the separating membrane. Ion-exchange membranes consist of gels, highly swollen, carrying an electrical charge. As suggested by their names, these membranes are able to separate ions in a mixture or a matrix, substituting them with the charged particles present on it. A membrane is defined as cation exchange if there is a negative charge fixed on it, and consequentially an anion exchange in the other case. To be perfectly functional these membranes have to be highly permselective, with low electrical resistance and high chemical and mechanical stability; then the nature of the charge fixed, their distribution, and their position in the membrane (as a cross-linked system) must be carefully designed. A ion-exchange membrane is obtained by mixing a resin electrically charged with a binder polymer, which are extruded as a film at a temperature higher than the melting point of the polymer; but is even possible to polymerize charged monomers and insert them in a proper polymeric solution or a preformed film obtaining membranes without resins. The ion exchange level is due to the final concentration of the ions fixed thanks to the resin. N^+H_2R, NR_3^+, S^+R_2 and SO_3^-, COO^-, and PO_3^- are just some of the functional groups used as charging sources for the membranes. The flat configuration is not the only one available for a polymeric membrane. Tubular membranes, capillaries, and hollow fibers have wide applications in industrial, chemical, and biomedical processes. They are generally classified by dimension: the tubular membranes have the largest diameter ($d > 5$ mm); the capillary membranes are included in a diameter range of 0.5–5 mm, whereas the hollow fibers have a diameter < 0.5 mm. Since the structures of this kind of membranes are so different from the equivalent flat ones, even the preparation technique is different. Depending on the final results desired, specific methods and proper conditions are used. The extrusion methods presented earlier for the flat membrane preparation is one of the techniques available to obtain that kind of membrane. The method is called in this case **melt spinning,** because hollow fibers are obtained by it and the protocol used is exactly the same (Figure 1.14).

The polymer is melted at a temperature just lower than the melting point until it is fused in an extruder and is pushed out at high pressure through an annular opening. Since the final product can collapse on itself, an inert gas or air is introduced often through the inner ring of the extruder in order to support the structure. Another method for producing hollow fibers is the **dry-wet spinning** method, and this can be considered as a colliding form of both the NIPS and the extrusion process (Figure 1.15) [32].

It foresees that in place of air or an inert gas in the annular opening of the spinneret a fluid consisting of a nonsolvent is pumped. A viscous polymeric solution and the bore fluid are pushed at determined pressure in the tube of the spinneret and left for a short period of time in the air, in order to induce polymer precipitation and fiber formation. After this short period (the dry step) the fibers are immersed in a nonsol-

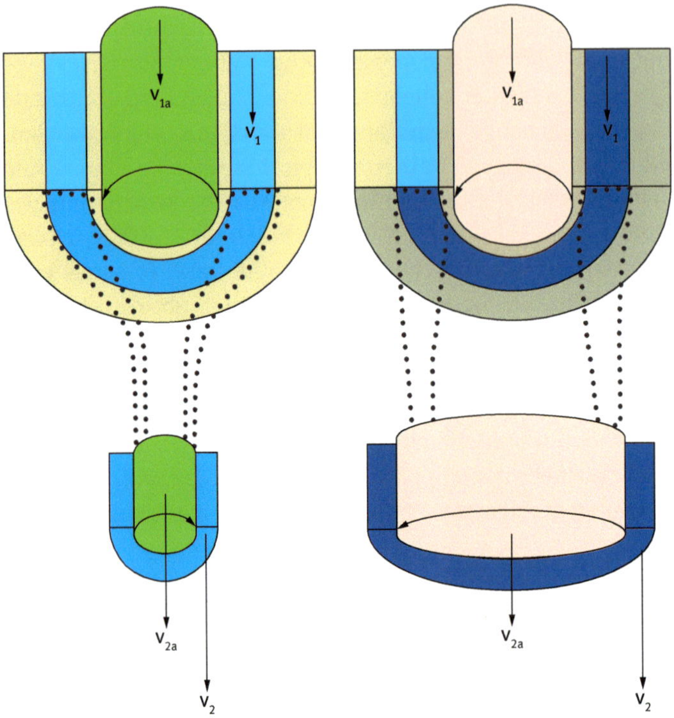

Fig. 1.14: Melt spinning processes for the preparation of capillary and hollow fiber membranes. The polymer is presented in blue, whereas inert gas or air introduced in order to support the structure are depicted in green and pink. High pressure and speed can induce the increase of the final fiber diameter, eventually fixed through a coagulation bath.

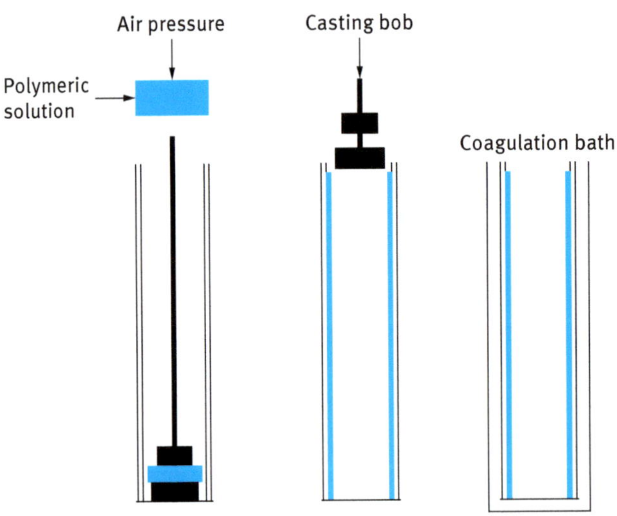

Fig. 1.15: Tubular membrane preparation.

vent coagulation bath and then collected. When the bath is not used the technique is called dry spinning. Although the wet–dry method allows us to produce fiber with specific dimension and diameter fixed after the coagulation bath, in the extrusion or the melt spinning techniques the dimension can be manipulated by changing the rate of the process and the collection, as mentioned earlier, and by the power used to push the air in the annular opening. One of the main characteristics of a hollow fiber or a capillary is that they are self-supporting; then their dimension is important and small and stable fibers can be prepared. The tubular membranes, on the contrary, are not self-supporting, and their preparation foresees the casting of the polymeric solution on a tubular support. The device (Figure 1.15) used consists of a reservoir in which the polymeric solution is contained, and a hollow tube, the central part of the system, containing at the end a "casting bob" with porous walls. Thanks to the proper pressure applied on the reservoir, the solution is pumped in the tube and forced through the holes of the bob, which is moved vertically mechanically or by gravitation until it is completely out of the system. This movement produces the casting of the polymeric solution on the walls of the tube and after a coagulation bath the tubular fiber is obtained (Figure 1.16).

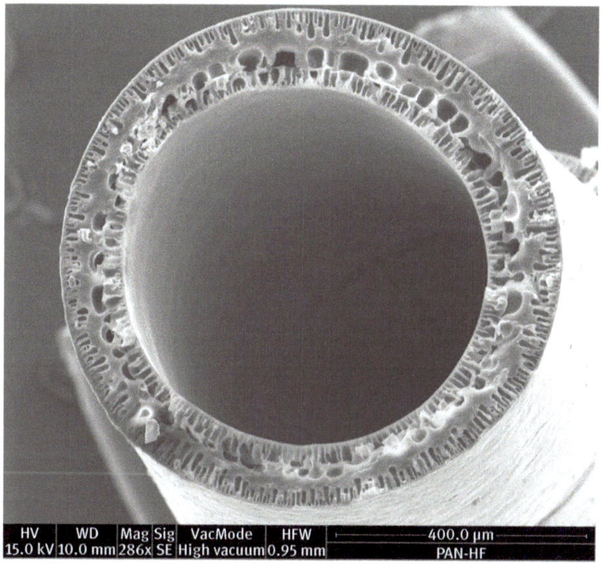

Fig. 1.16: SEM image of polyacrylonitrile membrane prepared by the dry-wet spinning process.

Starting from the polymer the **extrusion** method is the one which allows us to use the polymer powder without dissolving it in a solvent [33]. It is considered to be the simpler and the fastest way to prepare a dense membrane (Figure 1.17).

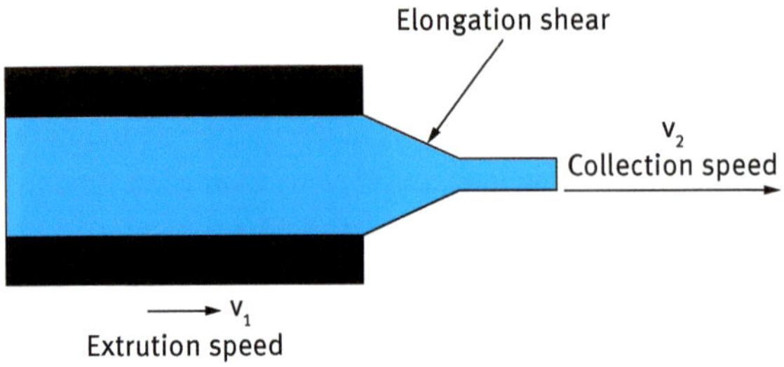

Fig. 1.17: Schematic representation of the extrusion method.

The polymer is melted at a temperature just lower than the melting point until it fused in an extruder and pushed out at high pressure through a thin opening in order to obtain a flat membrane. The rate of the process is very high ($V \gg 1000\,\text{m/min}$) and the collection speed of the membranes affects the final thickness: the faster the collection is, the thinner is the membrane produced; moreover, selective membranes are produced due to the elongation shear that induces the orientation of the polymeric chain, obliging them to assume a liner disposition instead of a chaotic one.

The **sintering** method is one of the most widely used technique to obtain porous membranes from organic or inorganic materials selected according to the application of the final membrane. The methods involve pressing a powder consisting of particles of a given size into a film or a plate and then sintered just below the melting point of the starting material (Figure 1.18). The pore size of the membrane depends on the size of the particles that are in the range of 0.2–20 μm and on the size particle distribution. The process produces membranes with low porosity ranging from 10 to 40 % and with a pore size of 0.1–10 μm. Only with metal is it possible to obtain a porosity of 80 %. Polymers such as polyethylene, polytetrafluoethylene, and polypropylene can be used for preparing a membrane by this technique [32].

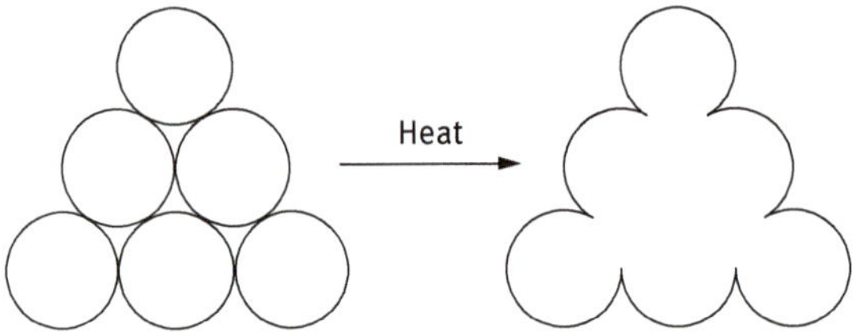

Fig. 1.18: Schematic representation of the sintering method.

When a polymeric film of partial crystallinity is obtained extruding a hydrophobic polymer powder at a temperature close to the melting point, a rapid drawdown of the porous polymeric film takes place, and in order to increase the porosity of the system changes in its structure are made through **stretching** (Figure 1.19). In that case the procedure is always coupled with a rapid drawdown of the porous polymeric film, and then it is stretched perpendicularly to the direction of the drawing in order to obtain a uniform pore shape and distribution, with a final membrane highly permeable to

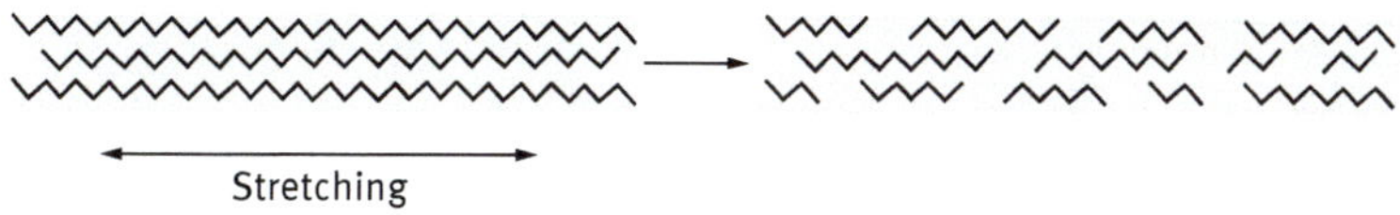

Fig. 1.19: Schematic representation of the stretching method.

gases and vapors but, below a specific hydrostatic pressure, impermeable to aqueous solutions. For this reason the membranes are usable for oxygenation and removal of carbon dioxide and sterile filtration. This technique allows preparing membranes with pore sizes ranging from 0.1 to 3 µm (Figure 1.12).

Nearly perfect porous membranes are obtained by the **track-etching** technique (Figure 1.20), even if they show a low porosity degree (10 %) and pore size of 0.2–10 µm (Figure 1.12 (c)). The polymeric film is exposed to collimated beam of high energy charged particles irradiated perpendicularly to the film [32, 36]. Passing through the film, the particles weaken the chemical bonds between the atoms damaging the polymer backbone. The film is then placed in an etching bath (acid or alkaline) and all the hit-sensitized areas leave tracks that originate uniform cylindrical pores. The pore density and diameters are determined by the resident time of the film in the irradiator and the time in the etching bath, respectively.

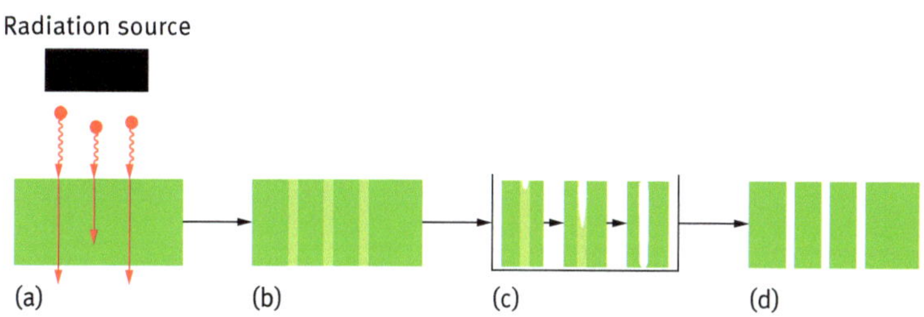

Fig. 1.20: Schematic representation of the track-etching method.

Microlithography and **template leaching** are two other methods for obtaining porous membranes usable for microfiltration processes, in particular from glass, ceramics, and metals. The procedure foresees two different starting materials mixed together, of which one type is completely dissolved, leaving a template or a network in the structure of the remaining undissolved material. The minimum pore size obtainable with this technique is 0.05 µm and silica membranes are the main examples.

The **electrospinning** method is another preparation method of fibrous mats from a liquid or a polymeric solution. The advantages of the electrospinning technique are the production of very thin fibers, on the order of few nanometers or micrometers, with a large specific surface areas and superior mechanical properties. The process does not require the use of coagulation agents or high temperatures to produce solid threads from solutions. This makes the process particularly suited to the production of fibers using large and complex molecules. A high electric field is applied to the droplet of a fluid which may be a melt or solution coming out from the tip of a die, which acts as one of the electrodes (Figure 1.21). This leads to droplet deformation and finally to the ejection of a charged jet from the tip of the accelerating cone towards the counter electrode leading to the formation of continuous fibers [39, 40].

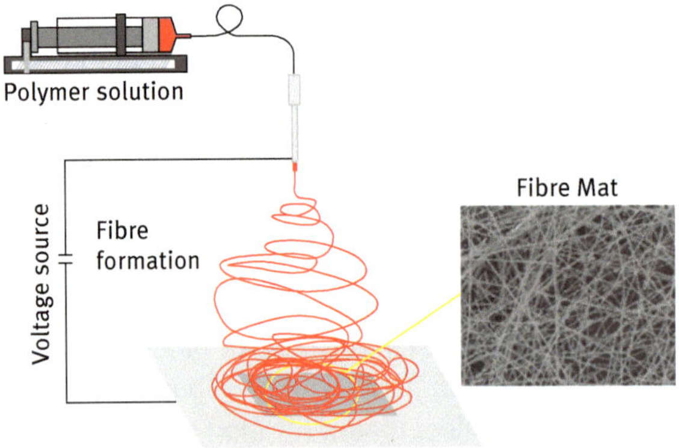

Fig. 1.21: Scheme of electrospinning process (adapted from [39]).

The standard laboratory setup for electrospinning consists of a spinneret (typically a hypodermic syringe needle) connected to a high voltage (5–50 kV) direct current power supply, a syringe pump, and a grounded collector. A polymer solution, sol-gel, particulate suspension, or melt is loaded into the syringe, and this liquid is extruded from the needle tip at a constant rate by a syringe pump. Alternatively, the droplet at the tip of the spinneret can be replenished by feeding from a header tank providing a constant feed pressure. Specifically electrospinning occurs when the electric forces at the surface of a polymer solution or melt overcome the surface tension and cause an electrically charged jet to be ejected. Modification of the spinneret and/or the type of solution can support the creation of fibers with unique structures and properties: a coaxial setup for example, uses a multiple solution feed system which allows for the injection of one solution into another at the tip of the spinneret; emulsions can be used to create core shell or composite fibers without modification of the spinneret by simply adding surfactants. Electrospinning of polymer melts eliminates the need for volatile solvents in solution electrospinning, and even polymers generally immiscible with each can be processed with this technique by using a very similar setup to that employed in conventional electrospinning processes.

1.5 Membrane characterization

When a membrane is developed or when it has to be chosen for a specific application, knowing the characteristics in terms of chemical-physical and mechanical properties is essential. Membranes can show different performances if prepared following diverse processing methods even if made with the same polymer. Then different techniques are required for a characterization, depending on the kind of membrane considered. Permeability and selectivity are the parameters to be investigated for each kind of membrane, but specifically porous membranes are characterized in terms of flux, pore size and distribution, and molecular weight cut-off. A dense and homogeneous membrane is investigated in terms of diffusion rate, rejection coefficient, and separation coefficient, whereas the characterization of a ion-exchange membrane is focused on the hydraulic permeability determination, the fixed charged density, the ion exchange capacity, and the electrical resistance, then all properties directly related to its preparation and structure.

1.5.1 Characterization of structural properties

Regardless of the membrane being considered, the starting point of a characterization is the structure analysis by means of microscopic techniques such as scanning electron microscopy, field emission electron microscopy, transmission electron microscopy, and atomic force microscopy. Each sample (surface or a cross section) before the analysis must be treated specifically; then they have to be stored in the perfect condition for not damaging them. Each technique gives a specific analysis, and even the determination of the pore size is possible: scanning electron microscopy can visualize pores of 5 nm, whereas the field and the transmission electron have a resolution of 0.6–0.7 nm and 0.4–0.5 nm, respectively. The pure water flux, or "pure water permeability", is another important investigation for a membrane, particularly when used for micro- or ultrafiltration processes. The flux J through pores is dependent on the driving force involved, the hydrostatic pressure, and following Darcy's law is inversely proportional to the solvent viscosity. Prior to measuring the flux the membrane must be washed in order to eliminate preservatives and residuals, and then pressurized by filtration of pure water at higher pressure than the operating conditions to stabilize the system before analysis. The equation used is

$$J = \frac{L_{\mathrm{p}}\,\Delta P}{\eta}, \tag{1.21}$$

where L_{p} is the hydrodynamic solvent permeability, P is the pressure, and η is the viscosity of the solution passing through the pores of the membrane being investigated. The separation properties of various components is another parameter always analyzed in addition to the permeability. It is based on a sieving mechanism by which particles smaller than the pore sizes are allowed to pass through the membrane. This is valid for filtration tests and for determining molecular weight cut-off. The relation between the particle size and the pore size for the retention properties of a membrane is described by the Ferry equation for the rejection [37]

$$R = \left[1 - \left(1 - \frac{r_p}{r_s}\right)^2\right]^2, \tag{1.22}$$

where r_p is the radius of the pores and r_s is the radius of particles if considered as spheres. The sharpness of the cutoff of a membrane is determined by measuring the retention of the membrane for components with different molecular weights and shapes (dextrans and proteins) and plotting the retention versus one of them. In that case the retention of the particles is expressed as a percentage:

$$R = \left(1 - \frac{c_p}{c_f}\right) \times 100, \tag{1.23}$$

where c is the concentration, and suffix p and f indicate permeate and feed solution. The obtained profile (Figure 1.22) is dependent on the pore distribution in the membrane, and since the molecular weight cutoff (MWCO) measurement is sensitive to the experimental condition it is very important to run the analysis at the standard parameters (transmembrane pressure of 100 kPa, solution concentration of 0.1 %, and test temperature 25 °C). In addition to the structural and filtration tests the determination of the pore size and pore size distribution in the membrane are also important.

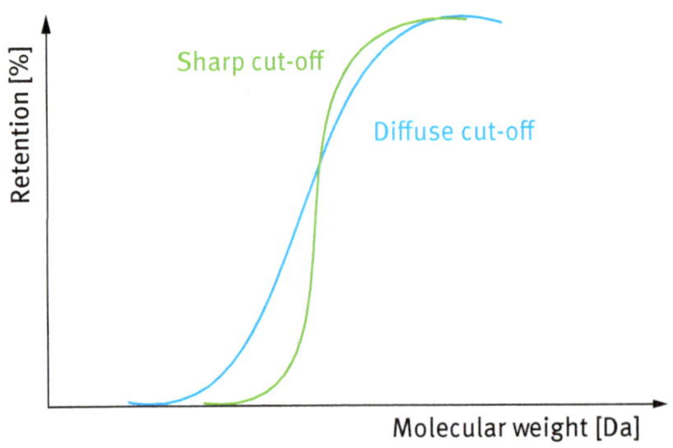

Fig. 1.22: Comparison of rejection characteristics between membranes with a sharp cutoff and a diffuse cutoff.

Several techniques are available: mercury porosimetry, gas-liquid displacement, liquid-liquid displacement, etc.; but the bubble-point test and the perm-porometry are the most widely used. The bubble-point test is a structure related characterization methodology which makes available the classification of the maximum pore size in a membrane. It is based on the capillary effect due to surface tension forces. The membrane is placed on a filter and immersed in a liquid (e.g. water) which fills all its pores. From the bottom side of the same filter, air or nitrogen gas is introduced at increasing pressure. At a specific pressure value, the air replaces the liquid in the largest pores, permeates the membrane, and a bubble rising from the surface can be detected. The relationship between the pressure and the pore size (radius) is done by the Laplace equation

$$r_{\mathrm{p}} = \frac{2\gamma}{\Delta P} \cos \theta, \tag{1.24}$$

where r_{p} is the radius of a capillary shape pore, γ is the surface tension at the liquid/air interface, and θ is the contact angle of the bubble observed. Other techniques to determine the pore size distribution are also used based on the same principle of the bubble-point test, and they cover different pore size ranges when different pressure ranges are applied. Perm-porometry is one of these techniques. Perm-porometry is based on the principles of capillary condensation as adsorption-desorption hysteresis. The adsorption and the desorption isotherm of an inert gas is determined as a function of the relative pressure. The adsorption isotherm starts at a low relative pressure. At a certain minimum pressure the smallest pores will be filled with the gas used, and as the pressure increases, the largest pores are also filled. Near the saturation pressure all the pores are blocked. In perm-porometry the blockage of pores by means of a condensable gas is linked with the simultaneous measurement of gas flux through the membrane, and the method proceeds recording not only the dried flow-through described above, but also the wet flow, in which a starting membrane filled with a liquid, is subjected to a reduction of the pressure. In this case, the condensed vapor is removed from the largest pores, and the diffusive gas flow through these open pores is measured. On reducing the relative pressure still further, smaller pores become available for gas diffusion, and when it is reduced to zero, all the pores are open, and gas permeate through all them [32, 37] (Figure 1.23).

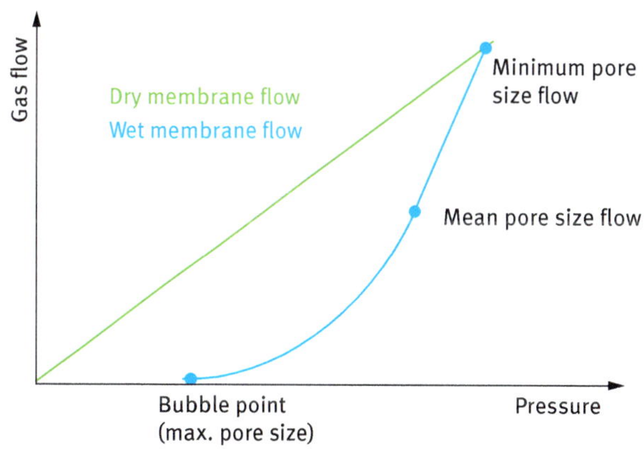

Fig. 1.23: Gas flow as a function of pressure with a dry and a liquid filled membrane.

Because a certain pore radius (Kelvin radius r_k) is related to a specific pressure, a measurement of the gas flow provides information about the number of these specific pores. The chemical-physical characteristics of the polymer used to prepare a membrane affect not only the structure and its perm-selectivity, but also the mechanical stability and swelling behavior. They are both dependent on the crystallinity and the cross-linking of the polymer matrix and are related to each other. The mechanical characterization concerns the elastic or plastic properties and deformation of the membrane, and it is obtained plotting the level of stress applied to the sample versus the strain. Generally the tensile strength test is run and defined in terms of a Young modulus and elongation parameter. Young's modulus is a measure of the stiffness of a material. It is defined as the ratio of the uniaxial stress over the uniaxial strain in the range of stress in which Hooke's law holds, predicting how much a material sample extends under tension or shortens under compression, and its values are normally indirectly proportional to elongation parameter values (Figure 1.24).

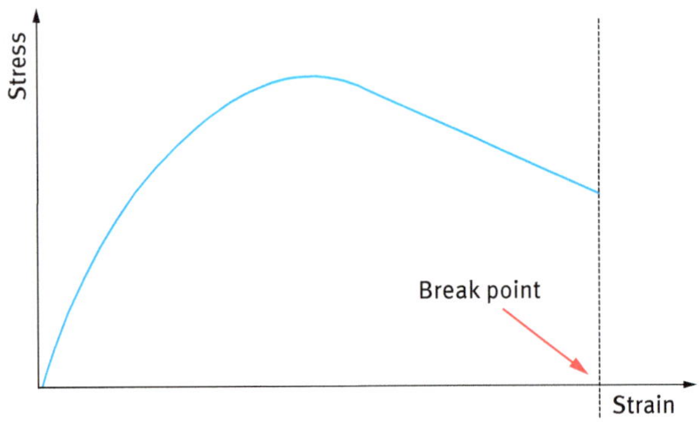

Fig. 1.24: Typical stress/strain diagram indicating the break point.

Since the water content in a membrane greatly affects the mechanical properties, for this analysis dried samples must always be used. The water sorption of a membrane is dependent on a few parameters – the nature of the polymer, cross-linking density, ions and their concentration, etc. – and is often determined by osmotic processes. The total water uptake, in weight percent, is defined in terms of weight differences between a membrane in its wet and dry state according to the equation

$$\text{wt\% swelling} = \frac{W_{\text{wet}} - W_{\text{dry}}}{W_{\text{wet}}} \times 100, \tag{1.25}$$

where W refers to the weight in the dry and wet state of the sample. Last but not the least important, the dissolution behavior, better defined as degradability, is another very important parameter which allows us to understand how long in time the stability of a membrane is under chemical stress conditions: the weight loss after treatment with salted solutions, acids or alkali, or biochemicals and enzymes is defined by the equation

$$\text{wt\% degradation} = \frac{W_0 - W_t}{W_t} \times 100, \tag{1.26}$$

where W_0 refers to the sample weight in the dry state at the beginning of the test, and W_t is the weight recorded at the time point during the analysis.

1.5.2 Characterization of physico-chemical properties

The physicochemical surface properties (i.e. wettability) of the membranes are responsible for the interfacial interactions between the membrane surface and molecules of the solution contacting membrane. These interactions are mainly controlled by the interfacial tension between liquid and solid. The wettability of the membranes plays an important role in many processes, especially in biocompatibility and fouling. Consequently the characterization of the surface properties and especially the surface free energy components of the solids are recognized as the key to understanding the mechanism of surface-based phenomena. The surface free energy (γ) is the measurement of the cohesive (excess) energy present at a gas/liquid interface. The molecules of a liquid attract each other. The interactions of a molecule in the bulk of a liquid are balanced by an equally attracting force in all directions [41]. Molecules on the surface of a liquid experience an imbalance of forces. The net effect of this situation is the presence of free energy at the surface. The excess energy is called surface free energy and can be quantified as a measurement of energy/area. The surface free energy has an apolar (γ^{LW}) and a polar component (γ^{AB}):

$$\gamma = \gamma^{LW} + \gamma^{AB}. \tag{1.27}$$

At present the most accurate method for determing the interactions between a liquid and surface is the contact angle. Young's equation can be used to determine the contact angles and is based on the surface energies of the three interfaces. At equilibrium three interfacial tensions corresponding to solid/gas (γ^{SG}), solid/liquid (γ^{SL}), and liquid/gas (γ^{LG}) interfaces are counterbalanced (Figure 1.25).

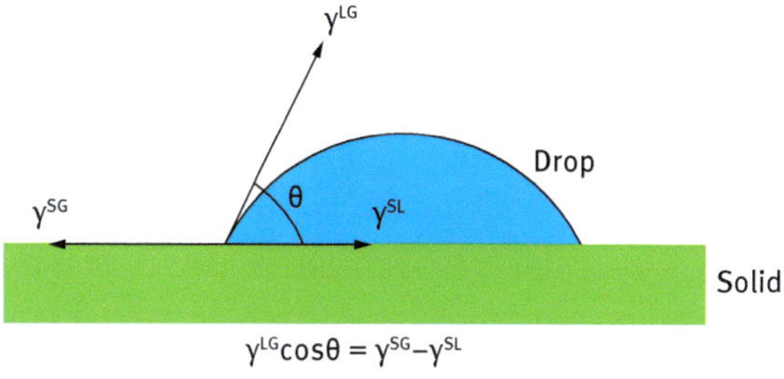

Fig. 1.25: Schematic representation of the contact angle formed between a liquid droplet and a solid surface.

Hence the contact angle (θ) can be evaluated from Young's equation [42], which is given as

$$\gamma^{LG}\cos\theta = \gamma^{SG} - \gamma^{SL}. \tag{1.28}$$

Young's equation assumes that the surface is completely smooth, homogeneous, and rigid. The surface must be also physically and chemically inert with respect to the liquid used. This corresponds to an ideal surface, but the membranes are real surfaces with roughness and porosity, and therefore the actual contact angle deviates from the estimated contact angle obtained from Young's equation.

The contact angle depends on the molecular forces that exist within the liquid (cohesive) and between a liquid and solid (adhesive). Thus, if the molecular adhesion between a solid and a liquid is greater than the cohesion between the molecules of the liquid, the liquid drop spreads over the solid surface, and the contact angle will be close to 0°. On the contrary, if the cohesive force is greater than the energy of adhesion between a liquid and a solid, the liquid drop tends to give a finite (nonzero) contact angle.

Figure 1.26 illustrates membrane surfaces with different wettability. Generally, a surface is considered to be hydrophilic if the water contact angle is less than 90°, while it is considered to be hydrophobic if the water contact angle is greater than 90°. Several methods can be used to measure the contact angle static sessile drop: the pendant drop method, and the dynamic sessile drop method. The sessile drop contact angle is measured with a goniometer using high resolution cameras and software to capture the profile of a pure liquid on a solid substrate. Angles measured in such a way are often quite close to the advancing contact angles. The pendant drop method is much more complicated than for sessile drops, due to the inherent unstable nature of inverted drops. This method allows for the deposition of multiple microdrops on the underside of a textured substrate, which can be imaged using a high resolution CCD camera. An automated system allows for tilting the substrate and analyzing the images for calculating the advancing and receding contact angles.

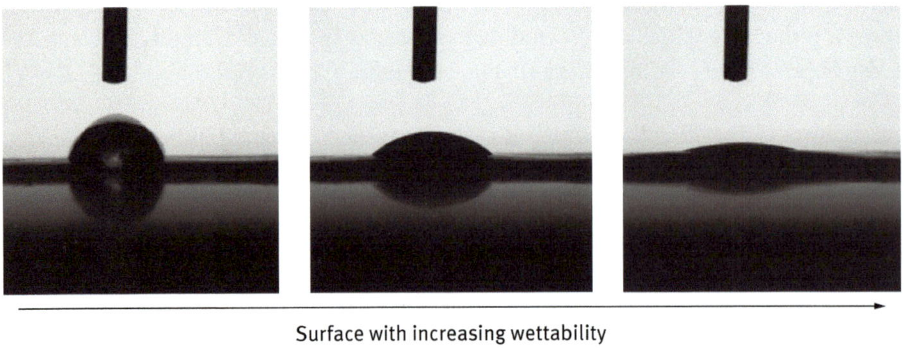

Fig. 1.26: Water drop on surfaces with increasing wettability.

The dynamic sessile drop is similar to the static sessile drop, but requires the drop to be modified by adding volume and removing dynamically. The maximum angle formed by adding liquid is the advancing angle (θ_a). Then the volume is removed to produce the smallest possible angle, the receding angle (θ_r). In general, θ_a is used as a measure of apolar domains of surface (low surface energy sites), while θ_r of its polar domains (high surface energy sites), as θ_a is always larger than θ_r. The difference between the advancing and receding angle is the contact angle hysteresis $\Delta\theta = \theta_a - \theta_r$. The hysteresis is usually attributed to chemical and morphological heterogeneity of surface, roughness, swelling, rearrangement, interdiffusion, and/or surface deformation [41]. For an ideal surface the values of θ_a and θ_r should be very close.

1.6 Sterilization procedures

All membranes and devices which are developed for biomedical applications and clinical use have to be sterilized before use.

Sterilization involves physical or chemical procedures to destroy all microbial life, including highly resistant bacterial spores [43]. The Association for the Advancement of Medical Instrumentation (AAMI) defines sterilization as:

> A process designed to remove or destroy all viable forms of microbial life, including bacterial spores, to achieve an acceptable sterility assurance level.

Sterilization methods can be classified as (a) high temperature/pressure sterilization (autoclave), (b) chemical sterilization, and (c) radiation sterilization. These methods are based on the use of sterilizing agents such as saturated steam, ethylene oxide gas, irradiation γ e β, hydrogen peroxide gas plasma, and liquid chemicals [44, 45]. A new sterilizing agent, ozone, has recently become available for use in the US.

1.6.1 Steam sterilization

Saturated steam under pressure is the oldest, safest, and most cost-effective and reliable method of sterilization available. The steam sterilizer consists of a pressurized chamber, in which the temperature reaches values around 120 °C. The sterilizing chamber and all its content must be free of any air entrapment to ensure direct contact of the steam to all surfaces to be sterilized. Steam sterilizers are designed to eliminate all air from the chamber during the conditioning phase in the sterilization cycle. Steam is vaporized water and serves as the conduit to rapidly permeate packaging, delivering high temperature moist heat to all contents and destroying microorganisms. Steam kills microorganisms through heat denaturation of cell walls and proteins. The essential conditions for steam sterilization are temperature, saturated steam, time, and pressure.

The sterilization period is dependent on the temperature and size of the load and can range from 10 to 60 min. Normally temperatures around 120 °C and times of about 20–30 min are used for sterilization of materials. As temperature is increased, the time may be decreased.

Usually thermal sterilization is considered to be the best sterilization method because it avoids all risks of chemical residues and changes of the material due to chemical treatment or irradiation. In the case of polymeric biodegradable membranes, which are sensible to the high temperature, this method of sterilization cannot be applied.

1.6.2 Dry heat

In dry heat sterilization higher temperatures (e.g., 180 °C) and/or times are necessary, although this method is not commonly employed in medical devices. Dry heat sterilization is accomplished by conduction, where heat is transferred from molecule to molecule or from the exterior surface of an item to its internal parts. The destruction of organisms occurs by oxidation, which is a slow burning process of coagulating the cell protein. This sterilization process is long, because unlike steam sterilization there is no moisture present, which speeds up heat penetration, and for this reason materials require more time to reach the appropriate temperatures. Variations in times and temperatures are based on volume, density, packaging, and sterilization apparatus. Usually the sterilization process may vary from 30 min at 180 °C to 6 h at 120 °C.

1.6.3 Ethylene oxide

Ethylene oxide (EtO) is a low-temperature sterilization method which can be used for membranes and devices that cannot tolerate the high temperatures and moisture associated with steam sterilization. EtO is a gas that destroys microorganisms by alkylation of sulphur-containing proteins. The EtO penetrates the cell membranes and reacts with the nuclear material, rendering it unable to metabolize and reproduce.

The sterilization procedure using EtO requires a relatively long sterilization cycle and there is a need for aeration for a specified amount of time. Membranes and devices sterilized by this process must be packaged with wraps and be aerated. The aeration time may be long and is needed to make sterilized items safe for handling and patient use.

EtO is extremely flammable in its pure form and requires special precautions for storage and use. Furthermore, EtO is also a known carcinogen.

Important parameters for sterilization with EtO are gas concentration, humidity, temperature, and time. Exposure time varies with the size of the sterilizer and the load of the materials to be sterilized. The exposure time also varies with the temperature, since the higher the temperature the faster is the diffusion of EtO. Operating temperatures range from 21 to 65.5 °C. By increasing the temperature it is also possible to reduce the gas concentration. For example the sterilization of dialyzers is achieved with 10 % EtO and 90 % CO_2 for 4 h at 40 °C at 40 % humidity.

Disadvantages of EtO gas are that it can leave toxic residues on sterilized items, and it possesses several physical and health hazards to personnel and patients, which merits special attention.

1.6.4 Hydrogen peroxide plasma

Plasma is a low temperature sterilization alternative suitable for heat- and moisture-sensitive or moisture-stable medical devices. Plasma is ionized gas made up of ions and electrons and is distinguishable from solid, liquid, or gas. In this process, hydrogen peroxide is activated to create a reactive plasma or vapor. Unlike EO sterilization, plasma sterilization does not have the disadvantages with regard to toxic residues and patient safety. For this reason medical devices are ready for use directly following sterilization without any need for aeration. Plasma is also less expensive, and the total cycle times are significantly less than EtO. This method disperses a hydrogen peroxide solution in a vacuum chamber, creating a plasma cloud; reactive species are generated from the hydrogen peroxide, which is reactive with microorganisms. Plasma sterilizes by oxidizing key cellular components of microorganisms, which inactivates and destroys them. There are some restrictions on lumen devices validated for sterilization by this method based on the gauge and length of the device. Gas plasma is not compatible with highly porous absorbers such as cellulose and paper products, and cannot be utilized to process liquids. Temperature range varies but is maintained between 40–55 °C. Total cycle time ranges between 28–75 min depending on the sterilizer model and size.

1.6.5 Liquid chemical sterilization

Liquid chemical sterilization is utilized for the sterilization of heat-sensitive materials and devices which are immersed in the liquid. This method employs the use of a germicidal solution and requires the complete immersion of items in the solution for a prescribed period of time to kill microorganisms.

Peracetic acid is a chemical sterilant used in conjunction with a self-contained automated processor designed for this method of sterilization. The process is achieved at temperatures of 50–56 °C. The sterilization time is about 30 min. Peracetic acid, by itself, is an oxidant and disinfecting agent for liquid immersion. It maintains its effectiveness when high levels of organic debris are present. It is an acetic acid plus an extra oxygen atom that reacts with most cellular components to destroy cells.

Glutaraldehyde is also a liquid chemical sterilant when used according to the manufacturer's directions for sterilization; however, it is usually used as a high-level disinfectant. There are several disadvantages of using glutaraldehyde as a sterilant. These include toxic fumes, long exposure time, and potential for contamination of sterile devices during rinsing and transfer to the area of use, and there is no method to biologically monitor the sterilization process. Glutaraldehyde requires 10 h of exposure.

1.6.6 Radiation sterilization

Membranes and devices can be sterilized through exposure to beta and gamma rays. High energy rays ionize atoms of the material, and free radicals are formed. The free radicals made inside a cell's nucleus can cause damage to the DNA structure (the radiation may also directly cause ionizing breaks in the DNA's chemical chain). Microorganisms are killed by disruption of the DNA molecule, which prevents cell division and propagation of biologic life. The product to be sterilized is exposed to radiation for 10–20 h, depending on the strength of the source. The highest temperatures reached in gamma sterilization are usually 30–40 °C. Gamma radiation is popular for sterilizing before shipment, and it can be done through the packaging. Both beta and gamma rays are suitable for sterilization, but they have difference in depths of material penetration and dose rates. Beta rays have a high dose rate and limited penetration depth, while gamma rays have high penetration capability and relatively low dose rates. Beta radiation is a stream of high energy electrons that are generated by electrical energy and accelerated by an accelerator. These electrons penetrate into matter before being stopped by collisions with other atoms, and their usefulness in sterilizing an object is therefore limited by the density, thickness of the object, and by the energy of the electrons. These free electrons produce their effect by ionizing the atoms they hit, producing secondary electrons that kill microorganisms. Gamma rays are electromagnetic waves that travel with the speed of light; they must pass through a thickness measuring several feet before making sufficient collisions to lose all of their energy. For this reason they have the ability to penetrate a much greater distance than beta rays before losing their energy from collision. Cobalt 60 is the radioactive isotope source most commonly used for irradiation sterilization. Usually the material to be sterilized is placed close to the radiation source for a time lasting a few seconds for beta radiation and minutes to hours for gamma radiation.

Radiation sterilization is thus the only process that enables products, including their packaging, to be sterilized without any appreciable increase in temperature – and without the use of chemicals, which always brings in the problem of residues. Even the contaminated inner surfaces of closed packages, or components with complex geometrical shapes, are safely and reliably sterilized by high-energy radiation.

Tab. 1.3: Sterilization methods of membranes.

	Thermal sterilization		Chemical				Radiation	
	Steam sterilization	dry heat	EtO	Plasma	Liquid		Gamma	Beta
Sterilizer	Steam	heat	EtO gas	Hydrogen peroxide vapor	Peracetic acid	Glutar-aldehyde	Radiation Energy	High energy electrons
Tempera-ture [°C]	120	120–180	21.1–65.5	45–50	50–56		30–40	30
Time	20–60 min	30 min/180 °C to 6 h/120 °C	4–18 h	55–70 min	30 min	10 h	10–20 h	Few seconds
Effect on membranes and devices	High thermal stress for heat susceptible and highly porous membranes		Low mechanical stress	not compatible with highly porous mem-branes, cellulose	Chemical degradation for some polymers		Inter-mediate material stress	—
Advantages	Safe; effective; economical: reliable; fast	Safe; effective; economical; reliable	Effective; low tem-perature; reliable; suited for heat sensitive materials	Safe; effective; low temperature; no toxic residues; no aeration; suitable for heat sensitive materials	Safe; non-toxic; short cycle time		Simple; reliable; cost com-petitive; fast	Very fast
Disadvan-tages	Wets and heats materials; pos-sible corrosion; not suited for heat sensitive membranes	Not suited for heat sensi-tive materials	Long cycles; leaves toxic residuals; car-cinogenic; costly; long aeration; physical, health hazards	Damages nylon based materials; very expensive	Handles only small items; has to soak in solution		Induce changes in some poly-mers (e.g. PVC, PTFE, cellulose)	Limited depth of pene-tration, uneven absorption

These benefits make radiation sterilization a simple, effective, and environmentally friendly alternative to other methods.

Radiation can change the properties of materials, including polymeric membranes. Literature data shows that gamma radiation induces cross-linking in the case of PA, PE, and silicone, and chain scission in the case of cellulose, PMMA, and PVC.

1.6.7 Ozone

Ozone sterilization is a new low-temperature sterilization method, recently introduced in US. Ozone sterilizes by oxidation, a process that destroys organic and inorganic matter. In this process, oxygen is converted in ozone by means of a generator. Ozone is formed by applying electrical energy to oxygen molecules, which splits some portion of those oxygen molecules in half into singlets of O. Therefore ozone molecules contain three atoms of oxygen and are unstable. Ozone penetration is controlled by vacuum pressure or by adding humidity. The cycle time may be up to 60 min, depending on the size of the chamber or the number of items to be sterilized. Ozone is inexpensive, nontoxic, and environmentally friendly. Due to ozone gas being corrosive and it being able to damage moisture sensitive equipment, there has not been much use of it in the medical industry. The cycle time is approximately 4.5 h, at a temperature of 85–94 °F.

1.7 References

[1] Ravve A. Principles of Polymer Chemistry, 2nd edition. New York: Plenum Publishing; 2000.
[2] McCrum NG, Buckley CP, Bucknall CB. Principles of polymer engineering. Oxford, New York: Oxford University Press; 1997.
[3] Rubinstein M, Colby RH. Polymer physics. Oxford, New York: Oxford University Press; 2003.
[4] Sperling, LH. Introduction to physical polymer science. Hoboken, NJ: Wiley; 2006.
[5] Painter PC, Coleman MM. Fundamentals of polymer science: An introductory text. Lancaster, Pa.: Technomic Pub. Co.; 1997.
[6] De Gennes PG. Scaling concepts in polymer physics. Ithaca NY: Cornell University Press; 1979.
[7] Hiemenz PC. Polymer chemistry – the basic concepts. New York: Marcel Dekker Inc.; 1984.
[8] Brandrup J, Immergut EH, Grulke EA, Abe A, Bloch DR. Polymer Handbook, 4th edition. New York: John Wiley & Sons; 2005.
[9] Parker D, Bussink J, van de Grampel HT, Wheatley GW. Dorf E-U, Ostlinning E, Reinking K. Polymers, High-Temperature. In: Ullmann's Encyclopedia of Industrial Chemistry. Weinheim: Wiley-VCH; 2002
[10] McKeen W. Effect of Temperature and other Factors on Plastics and Elastomers. Norwich, NY: William Andrew Inc.; 2008.
[11] Liu KJ, Zhang HC, Chen TL. Chin. Pat. CN 85,101,721; 1987.
[12] Zhang HC, Chen TL, Yuan YG. Chin. Pat. CN 85,108,751; 1987.
[13] Lagaron JM, Powell AK, Bonner G. Permeation of water, methanol, fuel and alcohol-containing fuels in high-barrier ethylene-vinyl alcohol copolymer. Polym Testing. 2001; 20: 569–577.
[14] Lendlein A, Kelch S. Shape-memory polymers. Angew Chem Int Ed. 2002; 41: 2034–2057.

[15] Lendlein A, Jiang H, Jünger O, Langer R. Light-induced shape-memory polymers. Nature. 2005: 434: 879–882
[16] Purohit K, Khitoliya P, Purohit R. Recent Advances in Nanotechnology. Int J Sci Eng Res. 2012; 3: 1–11.
[17] Galaev I, Mattiasson B. Smart Polymers: Applications in Biotechnology and Biomedicine. Boca Raton, London, New York: CRC Press; 2010.
[18] Ulery BD, Lakshmi SN, Laurencin CT. Biomedical applications of biodegradable polymers. J Polymer Sci B: Polymer Physics. 2011; 49: 832–864.
[19] Burkersroda FV, Schedl L, Gopferich A. Why degradable polymers undergo surface erosion or bulk erosion. Biomaterials. 2002; 23: 4221–4231.
[20] Maurus PB, Kaeding CC. Bioabsorbable implant materials review. Oper Techn Sports Med. 2004; 12: 158–160.
[21] Jiang B, Wu Z, Zhao H, Tang F, Lu J, Wei Q, Zhang X. Electron beam irradiation modification of collagen membrane. Biomaterials. 2006; 27: 15–23
[22] Middleton JC, Tipton AJ. Synthetic biodegradable polymers as medical devices. Med Plast Biomater. 1998. Available at: www.mddionline.com/article/synthetic-biodegradable-polymers-medical-devices.
[23] Patlolla A, Collins G, Arinzeh TL. Solvent-dependent properties of electrospun fibrous composites for bone tissue regeneration. Acta Biomater. 2010; 6: 90–101.
[24] Chapman TMJ. Models for polyurethane hydrolysis under moderately acidic conditions: A comparative study of hydrolysis rates of urethanes, ureas, and amides. Polym Sci Part A: Polym Chem. 1989; 27: 1993–2005.
[25] Tatai L, Moore TG, Adhikari R, Malherbe F, Jayasekara R, Griffiths I, Gunatillake A. Thermoplastic biodegradable polyurethanes: the effect of chain extender structure on properties and in-vitro degradation. Biomaterials. 2007; 28: 5407–5417.
[26] Singh A, Krogmas NR, Sethuraman S, Nair LS, Sturgeon JL, Brown PW, Laurencin CT, Allcock HR. Effect of side group chemistry on the properties of biodegradable L-alanine cosubstituted polyphosphazenes. Biomacromolecules. 2006; 7: 914–918.
[27] Richards M, Dahiyat BI, Arm DM, Brown PR, Leong KW. Evaluation of polyphosphates and polyphosphonates as degradable biomaterials. J Biomed Mater Res, 1991; 25: 1151–1167.
[28] Chandra R, Rustgi R. Biodegradable polymers. Progr Polym Sci. 1998; 23: 1273–1335.
[29] Vroman I, Tighzert L. Biodegradable Polymers. Materials. 2009; 2; 307–344.
[30] Li D-H, Liu L-M, Tian K-L, Liu J-C, Fan X-Q. Synthesis, biodegradability and cytotoxicity of water-soluble isobutylchitosan. Carbohydr Polym. 2007; 67: 40–45.
[31] Lepidi S, Abatangelo G, Vindigni V, Deriu GP, Zavan B, Tonello C, Cortivo R. In vivo regeneration of small-diameter (2 mm) arteries using a polymer scaffold. FASEB J. 2006; 20: 103–105.
[32] Mulder M. Basic principles of membrane technology. Dordrecht, Boston, London: Kluwer Academic Publishers; 1991.
[33] Clark AH. Direct analysis of experimental tie line data (two polymer–one solvent systems) using Flory–Huggins theory. Carbohydrate Polymers. 2000; 42(4): 337–351.
[34] Casarino P, Vicini S, Pedemonte E. Thermodynamics of polymer mixtures: study on the mixing process of the poly(styrene)/poly(vinylmethylether) system. Thermochimica Acta. 2001; 372(10): 59–66.
[35] Jyh-Ping H, Sung-Hwa L. Thermodynamic properties of polymer solutions: effect of free volume fraction of solvent molecules. Polymer. 2003; 44(26): 8201–8207.
[36] Baker RW. Membrane technology and applications. 3rd edition. Chichester, West Sussex: John Wiley & Sons; 2012.
[37] Strathman H, Giorno L, Drioli E. An introduction to membrane science and technology. CNR-Servizio Pubblicazioni; 2006.

[38] Porter MC. Handbook of industrial membrane technology. Park Ridge, NJ: Noyes Publications; 1990.
[39] Merritt SR, Exner AA, Lee Z, von Recum HA. Electrospinning and Imaging; Advanced Engineering Materials. 2012; 14: B266–B278.
[40] Greiner A, Wendorff JH. Electrospinning: a fascinating method for the preparation of ultrathin fibers. Angew Chem Int Ed Engl. 2007; 46: 5670–5703
[41] Good RJ. Contact angles and surface free energy of solids, In: Surface and colloids science. New York: Plenum Press; 1979.
[42] Young T. An essay on the coesion of fluids. Phylosophical transaction of the royal society of London; 1805; 95: 65–87.
[43] Anderson JM, Bevacqua B, Cranin AN, Graham LM, Hoffman AS, Klein M, Kowalski JB, Morrissey RF, Obstbaum SA, Ratner BD, Schoen FJ, Sirakian A, Whittlesey D. Implants and Devices. In: Ratner BD, Hoffman AS, Schoen FJ, Lemons JE (eds.). Biomaterials Science. London: Academic Press; 1996. pp. 415–420.
[44] Lanfield H. Sterilization of medical devices based on polymer selection and stabilisation techniques. In: Szyzher M (ed.). Biocompatible polymers, metals and composites. Lancaster USA: PA Technom, Publ Inc; 1983. pp. 975–999.
[45] Simmons A. Sterilisation of Medical Devices. Business briefing: Medical device manufacturing & technology, 1–4; 2004.

2 Basic issues in membrane separation for biomedical devices

2.1 Transport in membranes

A membrane is a phase (solid or liquid) interposed between two systems that allows the selective transport of matter and energy. The transport is generally promoted by a difference in chemical potential (or, more generally, electrochemical potential) between two systems; in a passive transport, the direction of the transmembrane flux is from the high-potential to the low-potential system (Figure 2.1).

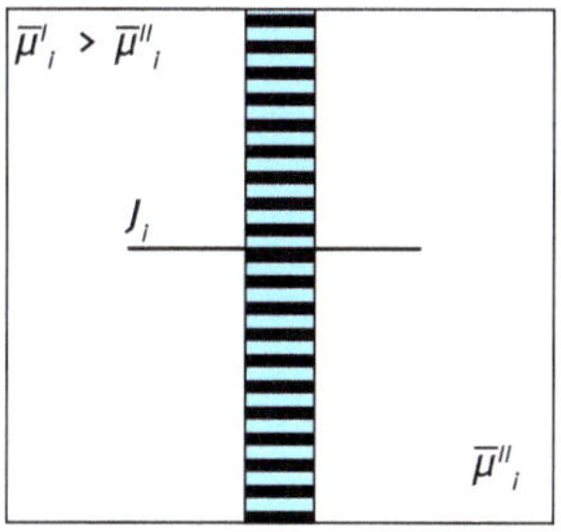

Fig. 2.1: Passive transport of i-th component under an electrochemical potential gradient.

A general expression for the electrochemical potential is given by

$$\bar{\mu}_i = \mu_i + z_i F \phi_i, \tag{2.1}$$

where subscript i refers to the i-th species, $\bar{\mu}_i$ is the electrochemical potential, μ_i is the chemical potential, z is the charge, F the Faraday constant, and ϕ the electrostatic potential. The transmembrane flux of the i-th species (J_i) is usually related to the electrochemical potential difference $\Delta\bar{\mu}_i$ as

$$J_i = A\frac{\Delta\bar{\mu}_i}{\delta}. \tag{2.2}$$

In equation (2.2), A is a proportionality constant and δ is the membrane thickness. However, the use of such a general relationship is not useful in practical applications. A potential gradient between two systems arises from differences in pressure, concentration, temperature, electrical potential; as a result, membranologists commonly identify a specific variation responsible for the transmembrane transport as "driving force". For applications in artificial organs, the role of electric potential is – so far – limited; therefore, the following discussion is restricted to chemical potential.

By definition, the chemical potential μ_i of the i-th species in a multicomponent system is defined as the Gibbs free energy increase per mole of the i-th substance added to a system at constant temperature (T), pressure (P), and numbers of moles

DOI 10.1515/9783110268010-003

of other substances ($n_{j\neq i}$) present in the system

$$\mu_i = \left(\frac{\partial G}{\partial n_i}\right)_{T,P,n_{j\neq i}} . \tag{2.3}$$

For liquid or gaseous mixtures of ideal solutions

$$\mu_i = \mu_i^0 + RT \ln x_i , \tag{2.4}$$

where μ_i^0 is the chemical potential of the i-th pure substance at 1 atmosphere, and x_i the mole fraction of the i-th substance in solution. Specifically:

- for ideal gaseous mixtures (the behavior of gases used in cell cultures applications well approaches ideality), x_i is the mole fraction indicated in the Dalton law of partial pressures:

  $$p_i = Px_i , \tag{2.5}$$

  where p_i is the partial pressure of the i-th gaseous component in a mixture having total pressure P;
- for ideal liquid solutions, x_i is the mole fraction indicated in the Raoult law of vapour pressures:

  $$p_i = p_i^0 x_i , \tag{2.6}$$

  where p_i is the vapor pressure of the i-th component in solution, and p_i^0 the vapor pressure of the pure i-th substance.

Liquid solutions usually diverge from ideality; in order to take into account for such deviations, an activity coefficient (y_i) is introduced in equation (2.4):

$$\mu_i = \mu_i^0 + RT \ln x_i y_i . \tag{2.7}$$

The term $x_i y_i$ is known as the "activity" of the i-th component, and for diluted solutions ($x_i \to 0$): $y_i \to 1$.

For a system in chemical equilibrium, the chemical potentials of the substance in all phases must be equal. In membrane technology, a prominent importance is given to osmotic pressure (Π), defined as the pressure required to maintain equilibrium between a solute(s)-solvent mixture (system I) and the pure solvent (system II) across a semi-permeable membrane through which the solvent, but not the solute, can diffuse. The osmotic pressure is determined from the chemical potential at equilibrium. For the solvent, on both sides of the membrane:

$$\mu^{\mathrm{I}}(T, P + \Pi, x) = \mu^{\mathrm{II}}(T, P) . \tag{2.8}$$

From equation (2.7), assuming a diluted solution,

$$\begin{aligned} \mu^{\mathrm{I}}(T, P + \Pi, x) &= \mu^{0}(T, P + \Pi) + RT \ln x_i , \\ \mu^{\mathrm{II}}(T, P) &= \mu^{0}(T, P) . \end{aligned} \tag{2.9}$$

Therefore,

$$\mu^0(T, P + \Pi) + RT \ln x_i = \mu^0(T, P). \tag{2.10}$$

On the basis of the Gibbs–Duhem equation adapted to an isothermal system ($\mathrm{d}T = 0$):

$$\mathrm{d}\mu = \bar{V}\,\mathrm{d}p, \tag{2.11}$$

where $\bar{V}$ is the volume per mole of i-th substance, and

$$\mu^0(T, P + \Pi) - \mu^0(T, P) = \int_P^{P+\Pi} \bar{V}^0\,\mathrm{d}p, \tag{2.12}$$

or, considering equation (2.10),

$$\int_P^{P+\Pi} \bar{V}^0\,\mathrm{d}p + RT \ln x_i = 0. \tag{2.13}$$

For liquid solutions, the solvent can be assumed to be incompressible, so that the molar volume $\bar{V}^0$ remains constant. Under this assumption, a simple expression for Π arises by integrating equation (2.13):

$$\Pi = -\frac{RT \ln x_i}{\bar{V}^0}. \tag{2.14}$$

Considering diluted solution in which a single solute 2 is dissolved in a solvent 1,

$$\ln x_1 = \ln(1 - x_2) \approx x_2 = -\frac{n_2}{n}, \tag{2.15}$$

where n_2 is the number of moles of solvent and n the total number of moles. Substituting equation (2.15) into equation (2.14),

$$\Pi \approx \frac{n_2 RT}{n\bar{V}^0} = cRT, \tag{2.16}$$

where c is the solute concentration.

2.2 Membrane transport and nonequilibrium thermodynamics

Nonequilibrium thermodynamics provides the most general description of solute transport through a membrane; the mathematical approach does not require any information, neither on the nature of the membrane nor on the specific separation mechanism. Assuming that the flux J_i of the i-th species is linearly related to the driving force X_j, the following equation applies:

$$J_i = \sum L_{ij} X_j, \tag{2.17}$$

where L_{ij} are proportionality coefficients. For the simple case of two-components transport,

$$\begin{aligned} J_1 &= L_{11}X_1 + L_{12}X_2, \\ J_2 &= L_{21}X_1 + L_{22}X_2. \end{aligned} \tag{2.18}$$

Coupling coefficients L_{12} and L_{21} are equal according to Onsager reciprocal relations [1]; therefore, three phenomenological coefficients need to be considered: L_{11}, L_{22}, and L_{12}.

Nonequilibrium thermodynamics is applied to membrane processes treating diluted solute-solvent solutions in the form of Kedem–Katchalsky–Spiegler equations [2]. The set of equations for the volumetric flux J_v and the solute flux J_s through a membrane is

$$\begin{aligned} J_\text{v} &= \sum_i J_i \bar{V}_i = -l_\text{p}\left(\frac{\text{d}P}{\text{d}x} - \sigma RT\frac{\text{d}c}{\text{d}x}\right), \\ J_s &= -k_\text{d}\frac{\text{d}c}{\text{d}x} + c(1-\sigma)J_\text{v}, \end{aligned} \tag{2.19}$$

where J_i is the transmembrane flux of the i-th species, $\bar{V}_i$ is the corresponding partial molar volume, l_p is the local hydraulic permeability of the membrane, P is the hydrostatic pressure gradient, x is the direction perpendicular to the membrane surface, σ the Staverman reflection coefficient, R the gas constant, T the absolute temperature, c is the solute concentration, and k_d the local diffusive permeability of the membrane. For a homogeneous membrane, parameters l_p, k_p, and σ are independent from x; under this assumption, the integration of equations (2.19), taking into account equation (2.16), gives

$$J_\text{v} = L_\text{p}(\Delta P - \sigma\,\Delta\Pi), \tag{2.20}$$

$$J_s = K_\text{d}\Delta c + \bar{c}_\text{s}(1-\sigma)J_\text{v}, \tag{2.21}$$

where ΔP, $\Delta\Pi$, and Δc are the differences in hydrostatic pressure, osmotic pressure, and concentration across the membrane, L_p is the hydraulic permeability coefficient of the membrane (being L_p the ratio between l_p and the membrane thickness δ), K_d is the diffusive permeability coefficient of the membrane (being K_p the ratio between k_p and the membrane thickness δ), and $\bar{c}_\text{s}$ is the mean solute concentration in the membrane phase.

2.3 Transport of fluids through porous membranes

2.3.1 Microfiltration

In microfiltration (MF), membranes separating a feed solution from the permeate show a symmetric structure with average pore size between 0.1 and 10 μm. Selective separation takes place due to size exclusion mechanism: particles with sizes larger than the pore diameter are rejected (Figure 2.2). The driving force for the mass transfer is a pressure gradient imposed across the membrane, usually in the range of 0.05–0.2 MPa, and the prevalent transport mechanism is convection (viscous flow through the pores).

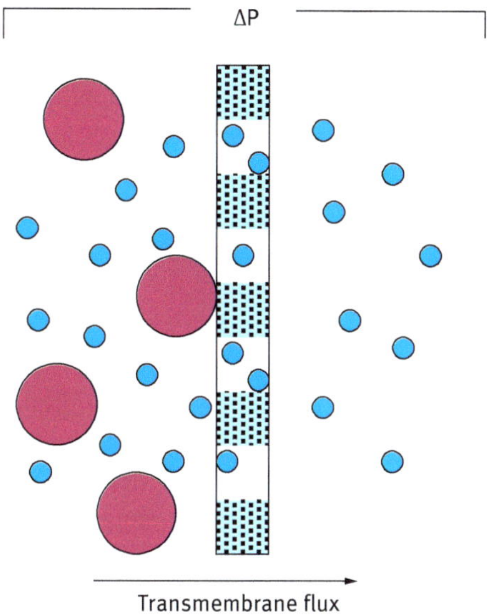

Fig. 2.2: Schematic representation of a microfiltration process.

Since only quite large particles with diameters above hundreds of nanometers are rejected, the Stavermann coefficient approaches zero, and the diffusion mechanism and osmotic pressure differences between feed and permeate become negligible. Under these reasonable approximations, equation (2.20) reduces to

$$J_V \cong L_P \Delta P. \tag{2.22}$$

In order to determine L_P, the polymeric matrix is considered impermeable, and the volumetric flux J_V is obtained by the Hagen–Poiseuille's law.

The Hagen–Poiseuille's law

The Hagen–Poiseuille's law is the basic equation for the convective flow through a porous media, and it is valid for a Newtonian fluid in laminar flow. Let us refer to Figure 2.3 to illustrate a cylindrical pore of radius R oriented along the z-axis.

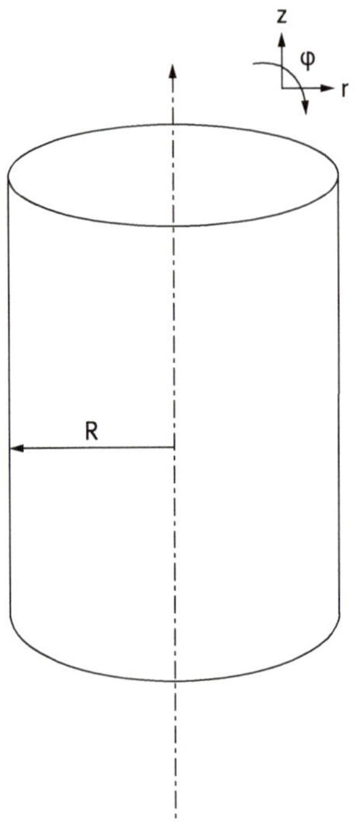

Fig. 2.3: Convective mass transport through a cylindrical pore.

Assuming axial symmetry and steady-state condition, the third equation of Navier–Stokes gives

$$\frac{1}{r}\frac{\partial}{\partial r}\left(r\frac{\partial u_z}{\partial r}\right)=\frac{1}{\mu}\frac{\partial P}{\partial z}, \tag{2.23}$$

where u_z is the z-component of the fluid velocity, P is the hydrostatic pressure, and μ the dynamic viscosity of the fluid.

Deriving equation (2.23),

$$\frac{\partial^2 u}{\partial r^2}+\frac{1}{r}\frac{\partial u}{\partial r}=\frac{1}{\mu}\frac{\mathrm{d}P}{\mathrm{d}z}, \tag{2.24}$$

with constant pressure drop along the pore. A general solution for the differential equation (2.24) is reported below:

$$u = A + Br^2 + C\ln r, \tag{2.25}$$

with A, B, and C constant.

Therefore,

$$\begin{aligned}\frac{\partial u}{\partial r}&=2Br+\frac{C}{r},\\ \frac{\partial^2 u}{\partial r^2}&=2B-\frac{C}{r^2}.\end{aligned} \tag{2.26}$$

By substituting (2.26) in (2.24), we get

$$B = \frac{1}{4\mu}\frac{\mathrm{d}P}{\mathrm{d}z}. \tag{2.27}$$

In order to avoid the presence of singularity on the z axis ($r = 0$) it is requested that $\frac{\partial u}{\partial r} = 0$ and $C = 0$. From the "no-slip" condition on the pore wall ($r = R$, $u = 0$),

$$A = -\frac{1}{4\mu}\frac{\mathrm{d}P}{\mathrm{d}z}R^2. \tag{2.28}$$

Therefore

$$u = -\frac{1}{4\mu}\frac{\mathrm{d}P}{\mathrm{d}z}R^2\left(1 - \frac{r^2}{R^2}\right) \tag{2.29}$$

and

$$-\frac{\mathrm{d}P}{\mathrm{d}z} = \text{constant} = \frac{\Delta P}{\delta}, \tag{2.30}$$

where δ is the membrane thickness and ΔP the hydrostatic pressure gradient.

Considering equations (2.29) and (2.30), the parabolic profile of the velocity along the z axis is

$$u_z(r) = v_{\max}\left(1 - \frac{r^2}{R^2}\right) \quad \text{with } v_{\max} = \frac{\Delta P}{\delta}\frac{R^2}{4\mu}. \tag{2.31}$$

The volumetric flowrate through a single pore Q_1 is calculated as

$$Q_1 = \int_0^R u_z(r)\,2\pi r\,\mathrm{d}r = \frac{\Delta P\,\pi R^4}{8\delta\mu}, \tag{2.32}$$

while the total flowrate Q_{tot} through the membrane is $Q_{\text{tot}} = Q_1 N_{\text{pores}}$. Assuming all pores to be uniform in size with radius R, and indicating with A the membrane surface area, and with ε the porosity,

$$N_{\text{pores}} = \frac{A\varepsilon}{\pi R^2}. \tag{2.33}$$

When removing the assumption of cylindrical pores by introducing the tortuosity parameter τ, the volumetric flux J_{v} is

$$J_{\text{v}} = \frac{Q_{\text{tot}}}{A} = \frac{\varepsilon R^2}{8\mu\tau}\frac{\Delta P}{\delta}. \tag{2.34}$$

2.3.2 Ultrafiltration

Ultrafiltration (UF) is driven by a hydrostatic pressure gradient applied through an asymmetric membrane, having pores with smaller diameters (usually between 2 and 10 nm) facing the feed side. Ultrafiltration membranes are typically characterized in terms of MWCO, defined as the lowest molecular weight solute in which 90 % of the solute is retained by the membrane. Typical MWCO of UF membranes ranges between 5 and 500 kDa.

The volumetric flux is properly described by equation (2.20); for a diluted feed solution, the filtration rate J_V can be adequately approximated by the flux of solvent.

In UF, the mass transport can be evaluated according to the nonequilibrium thermodynamics as described in Section 2.2. When the solute is transported through a porous membrane, the coefficients L_P, K_P, and σ equations (2.20) and (2.21) are evaluated on the basis of correlations provided by the so-called "pore model". The value of r_S is determined by the Stokes–Einstein equation

$$r_S = \frac{k_B T}{6\pi \eta D^0}, \tag{2.35}$$

where k_B is the Boltzmann constant, η is the viscosity, and D^0 the diffusion coefficient in infinitely diluted solution. The radius r_S varies with solute molecular weight and geometry of the molecule; Table 2.1 reports three experimental correlations between r_S and molecular weight (M_W) as from tests with dextrans, linear DNA and globular proteins.

Tab. 2.1: Experimental correlations between Stokes radius (r_S) of different molecules and molecular weight (M_W).

Target molecules	Correlation	Reference
Globular proteins	$r_S = 0.0402(M_W)^{0.395}$	[3]
Linear DNA	$r_S = 0.0187(M_W)^{0.582}$	[4]
Linear dextrans	$r_S = 0.0251(M_W)^{0.489}$	[5]

The diffusive permeability of the solute, K_d, can be calculated by the modified version of the Renkin equation as

$$K_d = \frac{D_S^m f(q) S_D(q)}{\delta}, \tag{2.36}$$

where δ is the membrane thickness, D_S^m is the effective diffusion coefficient of the solute in the membrane, f is the friction coefficient between the pore walls and the solute, and S_D is the steric hindrance factor at the pore inlet under diffusive motion. Both f and S_D coefficients depends of the ratio (q) between the solute radius r_S – evaluated from equation (2.35) – and the pore radius of the membrane r_p:

$$q = \frac{r_S}{r_p}. \tag{2.37}$$

Specifically [6],

$$f(q) = 1 - 2.104q + 2q^3 - 0.95q^5 \tag{2.38}$$

and

$$S_D(q) = (1 - q)^2. \tag{2.39}$$

The term $f(q)\,S_D(q)$ is also indicated as diffusive effectiveness factor.

The Stavermann reflection coefficient σ is a function of the friction coefficient between water and the solute $g(q)$, and of the steric hindrance factor at the pore inlet under convective motion $S_F(q)$:

$$\sigma = 1 - g(q)\,S_F(q), \tag{2.40}$$

where

$$g(q) = \frac{1 - 0.67q^2 - 0.20217q^5}{1 - 0.75857q^5} \tag{2.41}$$

and

$$S_F = 2(1 - q)^2 - (1 - q)^4. \tag{2.42}$$

The term $g(q)\,S_F(q)$ is also indicated as the convective effectiveness factor. Factors S_D, S_F, and σ versus q are diagrammed in Figure 2.4.

Free diffusion coefficients in water of some species of interest in cell culture experiments are reported in Table 2.2.

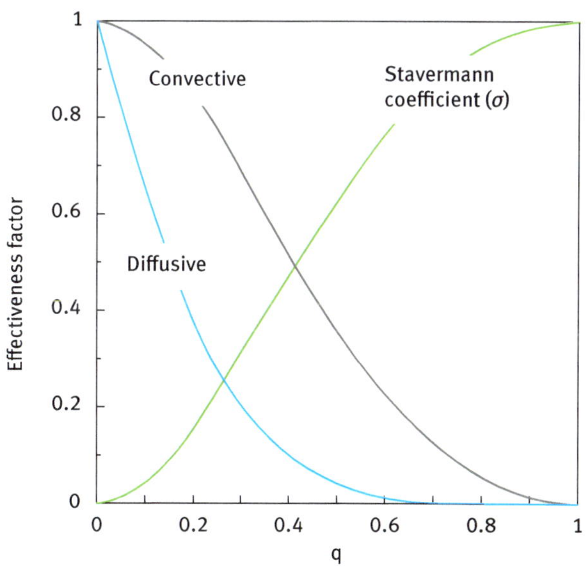

Fig. 2.4: Diffusive and convective effectiveness factors, and Stavermann coefficient σ, evaluated according to the pore model as a function of the ratio of solute radius to membrane pore radius.

Tab. 2.2: Free diffusion coefficients in water [7].

Species	Molecular weight [Da]	Stokes radius [nm]	Diffusion coefficient [cm^2/s]
Oxygen	32	0.28	2.4×10^{-5}
Glucose	186	0.35	9.24×10^{-6}
Bovine serum albumin	67 000	3.61	9.64×10^{-7}
Apotransferrin	81 000	3.78	9.52×10^{-7}
IgG	155 000	5.13	6.29×10^{-7}

When solution components diffuse through a polymeric membrane or a cellular tissue, free diffusion coefficients need to be corrected, and an effective diffusion coefficient D_{eff} is introduced. Realistic evaluation of D_{eff} from theoretical considerations is a challenging issue; commonly, the following correlation applies:

$$D_{\text{eff}} = D_{\text{water}} \frac{\varepsilon}{\tau}, \tag{2.43}$$

where D_{water} is the free diffusion coefficient of the species in water, and ε and τ are the porosity and the tortuosity of the medium where diffusion takes place. For molecules diffusion through membranes, ε and τ are parameters usually provided by the manufacturer.

In case of diffusion through interstitials of cellular tissues, the effective diffusion can be estimated – as a first approximation – by considering cells as uniformly-sized hard spheres in a close-packing configuration; for this kind of system, the porosity ε, defined as the ratio of the interstitial volume to the total volume, is given by

$$\varepsilon = 1 - \frac{\pi}{6}\rho\theta^3, \tag{2.44}$$

where ρ is the cell number density and θ the cell diameter.

For hard spheres, intraparticle tortuosity τ is predicted by the Maxwell equation [8]

$$\tau = 1 + 0.5\phi. \tag{2.45}$$

The volume fraction ϕ, i.e. the fraction of space covered by spheres, can be assumed equal to $\pi/\sqrt{18}$ (≈ 0.74) if considering the maximal volume fraction attainable for arrays of spheres in a cubic lattice [9].

Mass transfer is, in general, related to the physico-chemical properties of membranes and molecules. In membrane bioreactors for cell culture purposes, medium-soluble components span over a wide range of molecular sizes (from tens of Dalton for oxygen and small electrolytes, to thousands of Dalton in case of large proteins and macromolecules), while membranes can be manufactured and provided in a large variety of morphological structures. SEM micrographs of three different hollow fiber membranes, a microporous polyethersulfone (PES), and ultrafiltration polysulfone (PSf) and modified poly-etheretherketone (PEEK-WC), are shown in Figure 2.5.

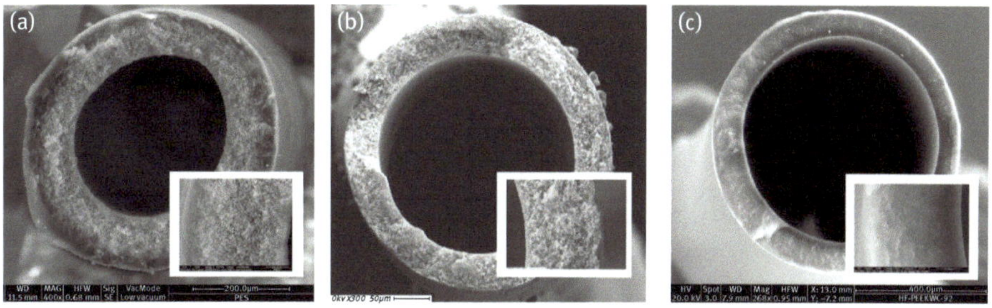

Fig. 2.5: SEM micrographs of hollow fiber membranes made in (a) poly-ethersulfone (PES), (b) polysulfone (PSF), and (c) modified poly-etheretherketone (PEEK-WC).

From equation (2.20), the hydraulic permeability coefficient of a membrane L_p is evaluated from filtration flux of pure water ($\Delta\pi = 0$) at different transmembrane pressures:

$$L_p = \left(\frac{J_{solvent}}{\Delta P}\right)_{\Delta\pi=0} . \tag{2.46}$$

The linear trend of hydraulic permeability for PES, PSF, and PEEK-WC hollow fiber membranes shown in Figure 2.6 is illustrated in Figure 2.7. The microporous structure of PES is associated to the highest hydraulic permeance (15.21 l/m^2 h mbar), while ultrafiltration PSF and PEEK-WC membranes show a comparable value of L_p (0.97 and 0.76 l/m^2 h mbar, respectively).

2.3.3 Starling flow

Starling's equation (2.20), illustrating the importance of the hydrostatic and osmotic pressures on the volumetric flux across hollow fibers and capillaries membranes, has a significant impact on clinical filtration.

Due to the pressure drop along a cylindrical porous fiber, in the presence of a significantly high reflection coefficient (σ) and osmotic pressure gradient ($\Delta\pi$) far away from the inlet section of a module, the transmembrane flux reverses its direction if $\sigma\,\Delta\pi > \Delta P$. This event, called the Starling flow, must be avoided in order to preserve the proper functionality of the biomedical device (Figure 2.7).

Karode [10] evaluated the pressure drop ΔP along porous cylindrical tubes with radius R and length L by modifying the Hagen–Poiseuille solution for impermeable tube wall. Assuming a constant wall velocity u_w,

$$\Delta P = \frac{8\mu Q_i z}{\pi R^4}\left(1 - \frac{2\pi R u_w z}{2Q_i}\right), \tag{2.47}$$

where μ is the viscosity, z the coordinate denoting the direction of the fluid flow, and Q_i the inlet volumetric flowrate.

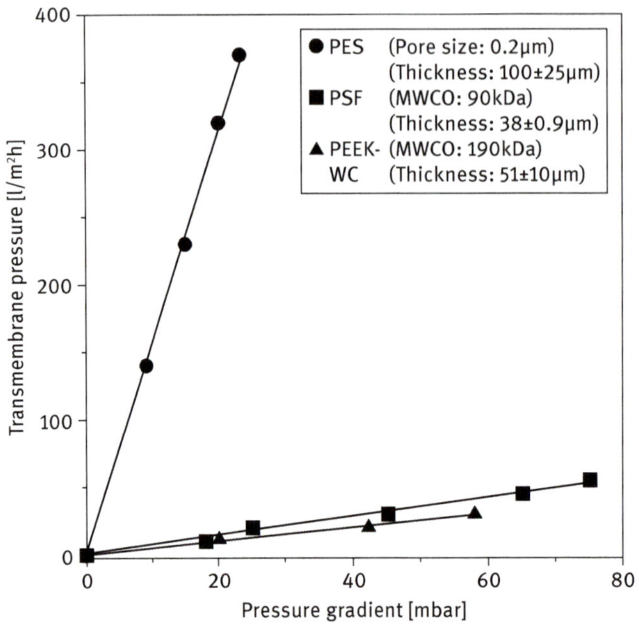

Fig. 2.6: Hydraulic permeability measurements of different hollow-fiber membranes.

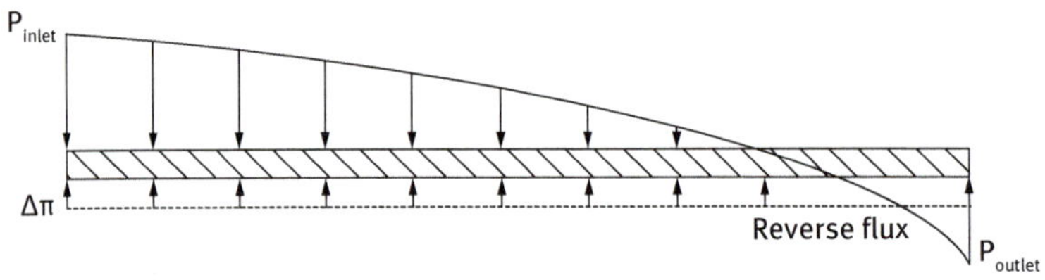

Fig. 2.7: Schematic representation of the Starling flow, occurring when $\sigma\Delta\pi > \Delta P$.

At the exit of the tube ($z = L$),

$$\Delta P = \frac{8\mu Q_\mathrm{i} L}{\pi R^4}\left(1 - \frac{Q_\mathrm{p}}{2Q_\mathrm{i}}\right), \tag{2.48}$$

where Q_p is the permeate flowrate ($Q_\mathrm{p} = 2\pi R L u_\mathrm{w}$).

For constant wall permeability (wall velocity being proportional to the trans-membrane pressure difference),

$$\Delta P = \frac{1}{2}\frac{8\mu Q_\mathrm{i}}{2\pi R^4 \lambda}(\mathrm{e}^{\lambda z} - \mathrm{e}^{-\lambda z}) + (P_\mathrm{i} - P_\mathrm{p})\left(1 - \frac{\mathrm{e}^{\lambda z} - \mathrm{e}^{-\lambda z}}{2}\right), \tag{2.49}$$

with

$$\lambda = \sqrt{\frac{16\mu\Psi}{R^3}}.$$

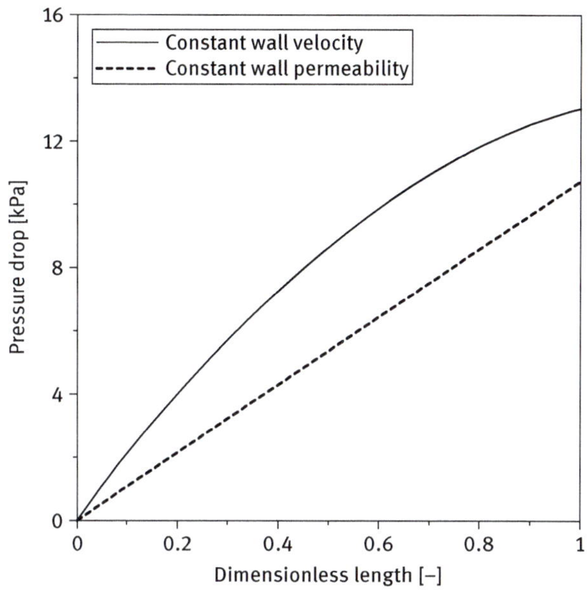

Fig. 2.8: Pressure drop in a tubular channel predicted for constant wall velocity (equation (2.47)) and constant wall permeability (equation (2.49)). Simulation conditions: $R = 250\,\mu m$, $Q_i = 10$ ml/min, $Q_p = 8$ ml/min.

In equation (2.49), P_i and P_p are the inlet and permeate side pressure (usually atmospheric), and Ψ the membrane permeability (m/s Pa).

2.4 Concentration polarization

In a membrane process, the progressive withdrawn of solvent and/or specific components in the permeate stream causes a gradual raise of the concentration of the rejected components within the boundary layer of the feed stream. This phenomenon, known as concentration polarization, can significantly affect the performance of the membrane process in terms of flux decline. Referring to Figure 2.9, as a consequence of the concentration gradient generated between the bulk of the solution and the membrane interface, nonpermeating solute molecules will diffuse away from the membrane surface.

Under steady-state conditions, the diffusive flux of solute is balanced by the convective flux through the membrane. Moreover, assuming negligible the concentration polarization phenomenon at the permeate side,

$$D_i \frac{dc_i}{dz} = -J_v(c_i - c_i^p). \tag{2.50}$$

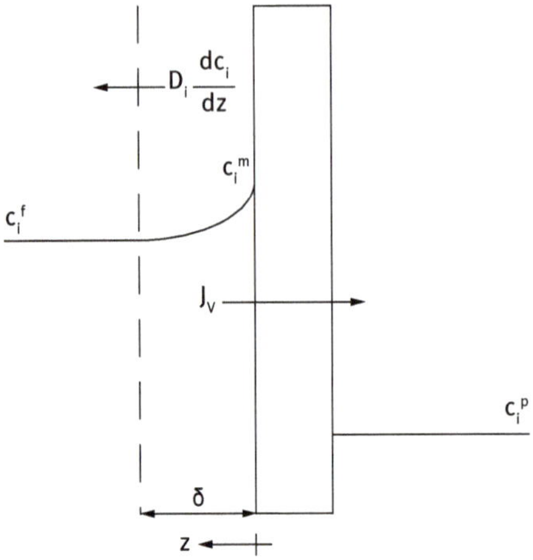

Fig. 2.9: Schematic illustration of the concentration polarization phenomenon in a cross-flow membrane system.

An intrinsic rejection coefficient R is defined as

$$R = 1 - \frac{c_i^p}{c_i^m}. \tag{2.51}$$

Combining equations (2.50) and (2.51), and integrating through the boundary layer having a thickness δ,

$$\int_0^\delta \frac{J_v}{D_i}\, dz = -\int_{c_i^m}^{c_i^f} \frac{dc_i}{[c_i - c_i^m(1-R)]}, \tag{2.52}$$

the following solution is obtained:

$$\frac{J_v \delta}{D_i} = \ln \frac{c_i^m R}{c_i^f - c_i^m(1-R)}. \tag{2.53}$$

Ultimately, the following equation correlates the concentration of solute in the bulk (c_i^f) to the concentration of solute at the membrane interface (c_i^m):

$$\frac{c_i^m}{c_i^f} = \frac{\exp(J_v \delta / D_i)}{R + (1-R)\exp(J_v \delta / D_i)}. \tag{2.54}$$

Assuming an ideally completed rejection ($R = 1$), equation (2.54) reduces to

$$\frac{c_i^m}{c_i^f} = \exp(J_v \delta / D_i). \tag{2.55}$$

The solute mass transfer coefficient $k = D_i^m/\delta$ is obtained through the following empirical correlations [11]:

– *laminar flow*

$$Sh = \begin{cases} 0.664\, Re^{0.5}\, Sc^{0.33} (d_h/L)^{0.33}, & L < 0.029\, d_h\, Re \quad \text{(Grober)} \\ 1.86\, Re^{0.33}\, Sc^{0.33} (d_h/L)^{0.33}, & L > 0.029\, d_h\, Re \quad \text{(Graetz–Leveque)} \end{cases} \tag{2.56}$$

and

– *turbulent flow*

$$Sh = \begin{cases} 0.023\, Re^{0.8}\, Sc^{0.33}, & Sc < 1 \quad \text{(Chilton–Colburn or Dittus–Boelter)} \\ 0.023\, Re^{0.875}\, Sc^{0.25}, & 1 < Sc < 1000 \quad \text{(Deissler)} \\ 0.0096\, Re^{0.91}\, Sc^{0.35}, & Sc > 1000 \quad \text{(Harriott–Hamilton)}, \end{cases} \tag{2.57}$$

with

$$Sh = \frac{k d_h}{D}, \quad Re = \frac{\rho u D}{\mu}, \quad Sc = \frac{\mu}{\rho D}. \tag{2.58}$$

In equations (2.56)–(2.58), Sh is the Sherwood number, Re is the Reynolds number, Sc is the Schmidt number, d_h is the hydraulic diameter, L is the length (of the membrane fiber), ρ is the density, μ the viscosity, D the diffusion coefficient, and u the fluid velocity.

2.5 Fouling phenomena

The performance of membrane processes can be severely affected by fouling: due to its great complexity, fouling is very difficult to describe theoretically and to manage practically. Fouling intensity depends on many physico-chemical parameters, including concentration of specific species, temperature, pH, ionic strength, and specific intermolecular interactions. Below, some simple and commonly used models are proposed for a preliminary estimate of flux decline in fouled membranes.

2.5.1 Cake layer model

This model assumes that the progressive accumulation of particles at the surface of the membrane determines a cake layer resulting in an additional resistance R_c in series with respect to the resistance R_m of the membrane (Figure 2.10).

Following analogies with electric circuits, the Darcy's law is

$$J_v = \frac{\Delta P}{\mu R}, \tag{2.59}$$

where the total resistance R is given by

$$R = R_m + R_c.$$

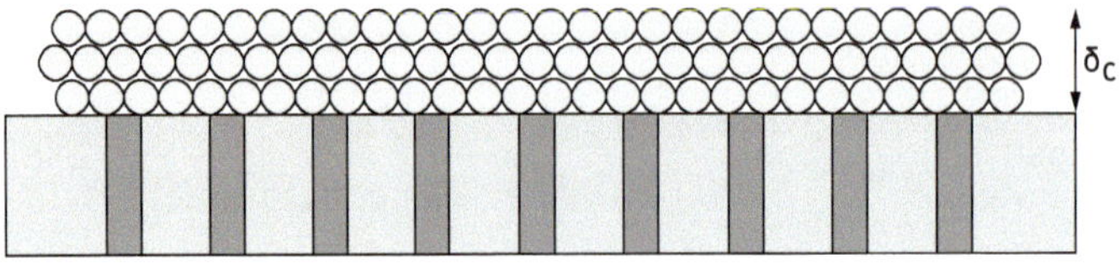

Fig. 2.10: Cake layer model for particulate fouling.

The value of R_c is provided by the Kozen–Carmany equation

$$R_c = \frac{180(1-\varepsilon_c)^2\delta_c}{d_s^2\varepsilon_c^3}, \tag{2.60}$$

with ε_c being the porosity of the cake, δ_c its thickness, and d_s the diameter of the single particle forming the cake.

2.5.2 Blocking models

Depending on the way one evaluates R in equation (2.59), four basic mechanistic models describing the effects of the deposition of particles on the permeability of porous membranes have been developed [12, 13].

In the *standard blocking model*, the original cylindrical shape of clean membrane pores is progressively reduced by the accumulation of solid particles at the pore walls. The resistance R is expressed as a function of time:

$$R = R_0\left(1+\frac{K_sJ_0t}{2}\right)^2, \tag{2.61}$$

where R_0 is the initial resistance of the membrane and J_0 the initial transmembrane flux (virgin membrane). The characteristic constant K_s (expressed in m^{-1}) is evaluated by filtration experiments linearly correlating the inverse of volume V to the inverse of filtration time t:

$$\frac{1}{V} = \frac{1}{J_0t} + \frac{K_s}{2}. \tag{2.62}$$

In the *intermediate* and *complete blocking models*, a portion of the pores becomes unavailable for flow. In particular, the available membrane area A reduces with the permeate volume V according to equation (2.63) for the intermediate blocking model,

$$\frac{A}{A_0} = e^{-K_iV}, \tag{2.63}$$

and to equation (2.64) for complete blocking model,

$$\frac{A}{A_0} = 1 - \frac{K_b}{J_0}V. \tag{2.64}$$

Constants K_i and K_b have the dimension of m^{-1} and s^{-1}, respectively. For constant transmembrane pressure operations, it is possible to derive the following equations for permeate volume V versus time t:

$$
\begin{aligned}
V &= \frac{1}{K_i}\ln(1 + K_i J_0 t) \quad \text{for intermediate pore blocking,} \\
V &= \frac{J_0}{K_b}\ln(1 - e^{-K_b t}) \quad \text{for complete pore blocking.}
\end{aligned}
\tag{2.65}
$$

In the *cake filtration model*, the transport resistance is increased by the presence of a cake layer as previously discussed in Section 2.5.1. The evolution of R in time is

$$\frac{R}{R_0} = \sqrt{(1 + 2K_c J_0^2 t)}, \tag{2.66}$$

where the R_0 is the initial membrane resistance; the constant K_c has units of s/m^2. The permeate volume as a function of time is

$$V = \frac{1}{K_c J_0}\left(\sqrt{1 + 2K_c J_0^2 t} - 1\right). \tag{2.67}$$

2.6 The transport of gases through porous membranes

Mass transport of a gaseous i-th species through porous media occurs under a partial pressure gradient ∇p_i (related, through the Dalton's law, to molar fraction x_i and total pressure P) with additional contribution of surface diffusion. However, surface diffusion is observed to provide a nonnegligible role only for very specific systems and operational conditions [14]; furthermore, transport is usually assumed under isothermal conditions.

As a result, mass transfer is limited by viscous resistance (resulting from the momentum transferred to the membrane by transported species), Knudsen diffusion resistance (related to collisions between molecules and pore wall), and ordinary diffusion resistance (due to collisions between molecules); the electrical analogy of the aforementioned transport resistances is illustrated in Figure 2.11.

Predominance or coexistence between Knudsen and ordinary diffusion is discriminated by the Knudsen number (Kn), defined as the ratio between the free mean path λ of diffusing molecules to the mean pore size of the pores (d_p):

$$Kn = \frac{\lambda}{d_p}. \tag{2.68}$$

The kinetic theory of ideal gases provides the formula for evaluating λ:

$$\lambda = \frac{k_B T}{P\sqrt{2}\pi\sigma^2}, \tag{2.69}$$

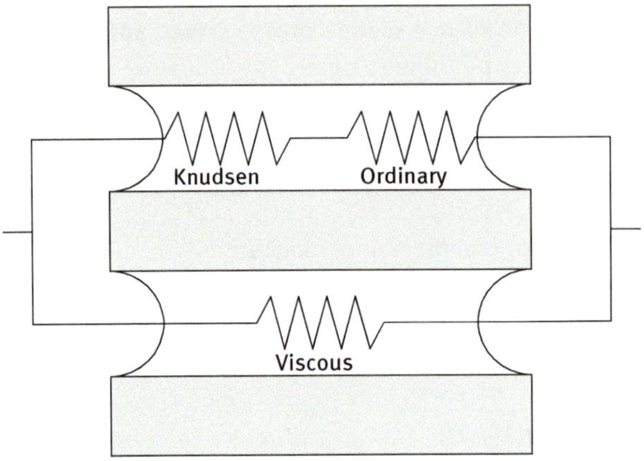

Fig. 2.11: Most frequent typologies of mass transfer resistances in porous media.

Tab. 2.3: Collision diameter (σ) and free mean path (λ) of gases at 0 °C and 1 atm [15].

Gaseous species	λ [10^{-9} m]	σ [Å]
Carbon dioxide	39.0	3.90
Carbon monoxide	58.6	3.71
Chloroform	161.0	5.43
Nitrogen	454	3.75
Nitrous oxide	362	3.88
Oxygen	425	3.54
Sulphur dioxide	300	4.29

where k_B is the Boltzmann constant (1.380×10^{-23} J/K), T the absolute temperature, P is the pressure, and σ the collision diameter of the molecule. In Table 2.3 collision diameter and free mean path of different gases are reported.

For small pores, molecule wall interactions are expected to dominate ($Kn \gg 1$), and Knudsen diffusion will be the controlling mechanism; vice versa, in large pores, the frequency of such collisions is expected to be lower than molecule-molecule collisions, and ordinary (or molecular) diffusion will be predominant ($Kn \ll 1$).

In many practical cases, λ is comparable to the pore size of microporous membranes, and both diffusional mechanisms need to be considered; in this transition region, the Bosanquet equation is used to estimate an overall effective diffusion coefficient of the i-th component D_i^{e} [16]:

$$D_i^{\mathrm{e}} = \left(\frac{1}{D_{ij,\mathrm{m}}^{\mathrm{e}}} + \frac{1}{D_{i,\mathrm{k}}^{\mathrm{e}}} \right)^{-1}, \tag{2.70}$$

where subscripts m and k refer to molecular and Knudsen mechanism, respectively.

Specifically,

$$D^{\mathrm{e}}_{i,\mathrm{k}} = \frac{2\varepsilon r_{\mathrm{p}}}{3\tau}\sqrt{\frac{8RT}{\pi M_i}},$$
$$D^{\mathrm{e}}_{ij,\mathrm{m}} = \frac{\varepsilon}{\tau} P D_{ij,\mathrm{m}}. \tag{2.71}$$

In equations (2.71), ε is the porosity, τ the tortuosity, and M_i the molecular weight of the i-th component.

Diffusion coefficient D^{e}_i can be used in Fick's law, the simplest model describing the diffusional transport of gases through porous media under a partial pressure gradient:

$$J_i = -\frac{D^{\mathrm{e}}_i}{RT}\nabla p_i, \tag{2.72}$$

where J_i is the flux of the i-th species.

In presence of a viscous flux, Fick's law is extended by introducing Darcy's law, obtaining the following formula:

$$J_i = -\left(\frac{1}{RT}\nabla p_i + \frac{B^0}{\mu} p_i \nabla P\right), \tag{2.73}$$

where ∇p_i is the partial pressure gradient, ∇P is the pressure gradient, B^0 is the membrane permeability, and μ the viscosity. Assuming that the membrane has uniform cylindrical pores with radius r_{p},

$$B^0 = \frac{\varepsilon r_{\mathrm{p}}^2}{8\tau}. \tag{2.74}$$

2.7 Transport of gases through dense membranes

Dense polymeric, inorganic, and mixed-matrix membranes are also widely used for gas separation on the basis of the selective permeability of different molecules from a gaseous mixture. In dense polymeric membranes, the most commonly used in biomedical and tissue engineering applications for biocompatibility reasons, the main transport mechanism is the solution-diffusion that includes three main steps: sorption of gas molecules at the membrane surface, diffusion through the polymeric matrix, and desorption on the permeate side. For glassy polymers, the sorption mechanism is described by a combination of Henry's law and Langmuir isotherm kinetics ("dual mode sorption theory"). Briefly, the penetrant concentration c is mathematically expressed as the sum of two contributions: c_{D}, which is the penetrant concentration by Henry's law mode, and c_{L}, which is the penetrant concentration by Langmuir mode [17]:

$$c = c_{\mathrm{D}} + c_{\mathrm{L}} = k_{\mathrm{D}} p + \frac{c'_{\mathrm{L}} b p}{1 + bp}, \tag{2.75}$$

where k_{D} is the Henry's law dissolution constant, p the (partial) pressure, b the microvoid affinity constant, and c'_{L} the Langmuir saturation constant.

First Fick's law mathematically describes the transport of a gas through the membrane:

$$J = -D\frac{\mathrm{d}c}{\mathrm{d}x}, \tag{2.76}$$

where J is the transmembrane flux of gas, D the diffusion coefficient, c the concentration, and x the spatial coordinate. Assuming a constant concentration profile within the membrane, the integration of equation (2.76) gives

$$J = D\frac{c^{\mathrm{feed}} - c^{\mathrm{perm}}}{\delta}, \tag{2.77}$$

where δ is the membrane thickness.

At low pressure, Henry's law is often adequate to express the concentration of the gas solubilized in the membrane phase

$$c = S\,p, \tag{2.78}$$

where S is the solubility constant.

By substituting equation (2.78) in (2.77),

$$J = DS\frac{p^{\mathrm{feed}} - p^{\mathrm{perm}}}{\delta} = \bar{P}\frac{p^{\mathrm{feed}} - p^{\mathrm{perm}}}{\delta}. \tag{2.79}$$

The permeability $\bar{P}$ is therefore defined as a product of the diffusivity and solubility coefficients of the gas species:

$$\bar{P} = DS. \tag{2.80}$$

Gas diffusion through membranes is a thermally activated process, and dependence of diffusion coefficient D on temperature is described by an Arrhenius-type relationship:

$$D_i = D_0 \exp\left(\frac{E_{\mathrm{D}}}{RT}\right), \tag{2.81}$$

where D_0 is the preexponential factor and E_{D} the activation energy to the diffusional transport.

The dependence of solubility on the temperature is analytically given by the van 't Hoff relationship

$$S_i = S_0 \exp\left(-\frac{\Delta H_{\mathrm{S}}}{RT}\right), \tag{2.82}$$

where S_0 is the preexponential factor and ΔH_{S} the partial molar enthalpy of sorption. In comparison to diffusion, solubility is less sensitive to temperature; therefore, permeability generally is enhanced at increasing temperature.

Selectivity α is defined as the ratio of individual gas permeabilities:

$$\alpha_{A/B} = \frac{\bar{P}_A}{\bar{P}_B} = \frac{D_A S_A}{D_B S_B} \tag{2.83}$$

and is therefore promoted by differences in diffusivity and/or solubility of the two gases.

Enrichment of oxygen from air (O_2/N_2), a process of industrial relevance also for biomedical applications, is generally a diffusivity-based gas separation, taking place due to the preferential permeation of the smaller – and more mobile – molecule.

2.8 Fluid oxygenation

Blood oxygenators are medical devices used to supplement the respiratory function of the lung by oxygenating the blood and removing carbon dioxide. Current blood oxygenators are manufactured in the form of microporous hollow fibers membranes (outer diameter typically within 200–400 µm, porosity ≥ 50 %). Membranes are made from hydrophobic polymers in order to prevent that pores be filled by liquid, thus facilitating the diffusion of the gas across them. They are used both as temporary extracorporeal devices to support short-term respiratory failure, and as a cardiopulmonary bypass during open-heart surgery. In the most common module configuration, oxygen flows in the lumen of the hollow fibers, while blood is fed through the interstitial spaces of the shell side.

The total molar flux of oxygen J transferred to a fluid is correlated to the gas concentration gradient Δc:

$$J = K\,\Delta c. \tag{2.84}$$

From a mass balance over the liquid phase, the overall mass transfer coefficient K is expressed as

$$K = \frac{Q}{A}\int_{c_{in}}^{c_{out}} \frac{\mathrm{d}c}{c^* - c}, \tag{2.85}$$

where Q is the fluid flowrate, A is the membrane surface area, c_{in}, c_{out}, and c^* are the inlet, outlet, and equilibrium concentrations of oxygen. The transport coefficient K results from the contribution of three individual mass transfer coefficients describing the transfer of oxygen: (1) from the bulk of the gas phase to the membrane interface (the concentration boundary layer on the gas side is generally negligible, and null if pure oxygen is supplied to the system); (2) across the membrane (since the membrane is hydrophobic, pores are gas-filled and the membrane resistance can be neglected); and (3) from the membrane interface to the bulk of the liquid phase (liquid-side concentration boundary layer is the controlling resistance to mass transport).

The shell-side mass transfer coefficient k depends on both the fluid dynamics and the physico-chemical parameters. For Newtonian liquids, the following general correlation is usually adopted:

$$Sh = a\,Re^b Sc^c, \tag{2.86}$$

where Sh is the Sherwood number, Re is the Reynolds number, and Sc is the Schmidt number. Values of empirical constants a, b, and c suggested by Wickramasinghe et al. are 0.8, 0.59, and 0.33, respectively [18].

For blood, exhibiting a non-Newtonian behavior, Hewitt et al. [19] developed the following correlation:

$$k = a\alpha d_{\mathrm{h}}^{-1} D^{0.66} \nu^{0.33} Re_{\mathrm{h}}^{b}, \tag{2.87}$$

with: $a = 0.524$, $b = 0.523$ [20], α is the solubility of oxygen in the blood, D its diffusivity, and ν is the kinematic viscosity of the blood.

If ε is the fiber bundle porosity, the hydraulic diameter d_{h} is assumed equal to the diameter of the fiber d_0 for $\varepsilon > 0.5$, while $d_{\mathrm{h}} = \varepsilon d_0/(1-\varepsilon)$ for $\varepsilon \leq 0.5$.

Some relevant physico-chemical parameters of the bovine blood are reported in Table 2.4.

Tab. 2.4: Some relevant physico-chemical parameters of the bovine blood.

Bovine blood parameter	Value	Reference
Density	1050 kg/m^3	[20]
Viscosity	2.9×10^{-3} kg/ms	[20]
Hemoglobin concentration	0.166 m^3 Hb/m^3 blood	[20]
Solubility of oxygen in the blood	3.0×10^{-5} m^3 O_2/m^3 blood/mmHg	[19]
Diffusion coefficient of oxygen in blood	1.8×10^{-9} m^2/s	[19]

In order to evaluate the pressure drop ΔP along the hollow fiber module, the following equation can be adopted:

$$\Delta P = \frac{4fL(1/2\rho u^2)}{d_{\mathrm{h}}}, \tag{2.88}$$

where f is the friction factor, L is the length of the module, ρ is the fluid density, and u is the fluid velocity.

The friction factor f can be estimated by the following experimental correlations [18]:

$$f = \begin{cases} 260\,Re^{-1.1}, & 0.1 < Re < 5, \\ 100\,Re^{-0.5}, & 5 < Re < 100. \end{cases} \tag{2.89}$$

For the oxygen uptake by hemoglobin, blood oxygenation modelling has to take into account a globular protein composed of four polypeptide chains each containing a hemo group.

The oxygen-binding curve is empirically described by an equilibrium reaction between hemoglobin (Hb) and oxygen (O_2):

$$\mathrm{Hb} + n\mathrm{O}_2 \overset{k_1,k_{-1}}{\longleftrightarrow} [\mathrm{Hb(O_2)}_n]. \tag{2.90}$$

For bovine blood, the value for n (a measure of the cooperativity between haeme units) is 2.85; for human blood, $n = 2.7$.

The apparent equilibrium constant K_e for reaction (2.90) is defined as

$$K_e = \frac{k_1}{k_{-1}} = \frac{[Hb(O_2)_n]}{[Hb][O_2]^n}, \tag{2.91}$$

where k_1 and k_{-1} are the forward and backward rate constants, and $[Hb(O_2)_n]$ is the concentration of the oxygenated hemoglobin in the plasma.

The degree of oxygen saturation S, given by the between oxygenated hemoglobin concentration and total hemoglobin concentration, is often described by the Hill's equation [21]

$$S = \frac{[Hb(O_2)_n]}{[Hb] + [Hb(O_2)_n]} = \frac{(P_{O_2}/P_{50})^n}{1 + (P_{O_2}/P_{50})^n}, \tag{2.92}$$

where P_{O_2} and P_{50} are the oxygen partial pressure and the oxygen partial pressure at 50 % hemoglobin saturation. At pH 7.4 and 37 °C, P_{50} is 3870 and 3550 Pa for bovine blood and human blood, respectively [22].

Comparison between equations (2.91) and (2.92) gives

$$\frac{1}{K_e} = \frac{1}{(\alpha P_{50})^n}, \tag{2.93}$$

when considering that at equilibrium $[O_2] = \alpha$.

The reaction between oxygen and hemoglobin leads to an enhancement of the mass transfer coefficient k':

$$k' = kE, \tag{2.94}$$

where k is the mass transfer coefficient in absence of chemical reaction, and E the enhancement factor. For blood oxygenation, an expression for E was derived by Wickramasinghe and Han [23]:

$$E = 1 + \frac{nD_{Hb(O_2)_n}}{D_{O_2}} \cdot \frac{1}{1 + (D_{Hb(O_2)_n}/D_{Hb})K_e\alpha^n} \cdot \frac{\alpha^n - c^n_{O_2,bulk}}{\alpha - c_{O_2,bulk}} K_e c_{Hb,bulk}, \tag{2.95}$$

where D is the diffusion coefficient, and $c_{O_2,bulk}$ and $c_{Hb,bulk}$ are the oxygen and hemoglobin concentrations in the bulk of the liquid phase, respectively.

2.9 Transport through charged membranes

2.9.1 The Donnan equilibrium

If a solution contains charged components (ions) and the membrane contains fixed charges, the equilibrium between the membrane and the adjacent solution will be established if the electrochemical potential of all ions in the membrane and in the solution are equal:

$$\bar{\mu}_i^m = \mu_i + z_i F\phi^m = \bar{\mu}_i^s = \mu_i^s + z_i F\phi^s, \tag{2.96}$$

where z is the valence, F the Faraday constant, and ϕ the electrical potential; superscripts m and s refer to membrane and solution, respectively. The electrical potential difference between membrane and solution is called the "Donnan potential" ϕ_{Don}:

$$\phi_{\mathrm{Don}} = \phi^{\mathrm{m}} - \phi^{\mathrm{s}} = -\frac{1}{z_i F}(\mu_i^{\mathrm{m}} - \mu_i^{\mathrm{s}}) = -\frac{1}{z_i F}\left[RT \ln \frac{a_i^{\mathrm{m}}}{a_i^{\mathrm{s}}} + \bar{V}_i(p_i^{\mathrm{m}} - p_i^{\mathrm{s}})\right]. \tag{2.97}$$

Here, a_i is the activity and $\bar{V}_i$ the partial molar volume of the i-th ion. The term $(p_i^{\mathrm{m}} - p_i^{\mathrm{s}})$ represents the swelling pressure that is the opposite of the osmotic pressure:

$$\bar{V}_i(p_i^{\mathrm{m}} - p_i^{\mathrm{s}}) = -\bar{V}_i\left(RT \ln \frac{a_{\mathrm{w}}^{\mathrm{m}}}{a_{\mathrm{w}}^{\mathrm{s}}}\right), \tag{2.98}$$

where subscript w refers to water.

According to the principle of electroneutrality, the electrical potentials for cations and anions of a dissociated electrolyte must be equal, and therefore

$$-\frac{RT}{z_{\mathrm{c}} F}\left[\ln \frac{a_{\mathrm{c}}^{\mathrm{m}}}{a_{\mathrm{c}}^{\mathrm{s}}} - \frac{\bar{V}_{\mathrm{c}}}{V} \ln \frac{a_{\mathrm{w}}^{\mathrm{m}}}{a_{\mathrm{w}}^{\mathrm{s}}}\right] = -\frac{RT}{z_{\mathrm{a}} F}\left[\ln \frac{a_{\mathrm{a}}^{\mathrm{m}}}{a_{\mathrm{a}}^{\mathrm{s}}} - \frac{\bar{V}_{\mathrm{a}}}{V} \ln \frac{a_{\mathrm{w}}^{\mathrm{m}}}{a_{\mathrm{w}}^{\mathrm{s}}}\right], \tag{2.99}$$

where V is the total volume ($V = \sum_i \bar{V}_i$); subscripts a and c refer to anion and cation, respectively.

Considering that

$$z_{\mathrm{a}} \nu_{\mathrm{a}} = -z_{\mathrm{c}} \nu_{\mathrm{c}}, \tag{2.100}$$

where ν is the stoichiometric coefficient, a rearrangement of equation (2.99) gives

$$\left(\frac{a_{\mathrm{c}}^{\mathrm{m}}}{a_{\mathrm{c}}^{\mathrm{s}}}\right)^{\nu_{\mathrm{c}}} \left(\frac{a_{\mathrm{a}}^{\mathrm{m}}}{a_{\mathrm{a}}^{\mathrm{s}}}\right)^{\nu_{\mathrm{a}}} = \left(\frac{a_{\mathrm{w}}^{\mathrm{m}}}{a_{\mathrm{w}}^{\mathrm{s}}}\right)^{\frac{\nu_{\mathrm{c}} \bar{V}_{\mathrm{c}} + \nu_{\mathrm{a}} \bar{V}_{\mathrm{a}}}{V}}. \tag{2.101}$$

For diluted solutions $V \gg \nu_{\mathrm{c}} \bar{V}_{\mathrm{c}} + \nu_{\mathrm{a}} \bar{V}_{\mathrm{a}}$, and the right-side term of equation (2.101) reduces to 1.

The activity of an ion can be expressed as the product of concentration and activity coefficient $\gamma_{\pm}$:

$$a_i = c_i \gamma_{\pm}. \tag{2.102}$$

For a concentration below 0.1 M, $\gamma_{\pm}$ can be estimated by the Debye–Huckel theory.

For monovalent salts such as NaCl, $\nu_{\mathrm{a}} = \nu_{\mathrm{c}} = 1$, and

$$c_{\mathrm{co}}^{\mathrm{m}} = \frac{(c_{\mathrm{salt}}^{\mathrm{s}})^2}{c_{\mathrm{fix}}}. \tag{2.103}$$

Here the term c with subscripts "co", "salt", and "fix" refer to the concentration of co-ions in the membrane, the concentration of electrolyte in solution, and the concentration of fixed charges in the membrane, respectively.

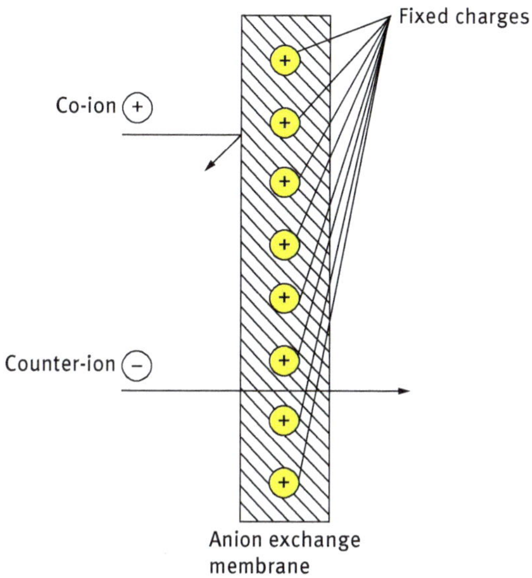

Fig. 2.12: Schematic drawing of an anion exchange membrane.

2.9.2 Nanofiltration

Nanofiltration (NF) is usually classified as a pressure-driven membrane process between ultrafiltration (UF) and reverse osmosis (RO). The main driving force to the selective transport of species is represented, in fact, by a hydrostatic pressure (about one order of magnitude higher than in UF) applied at the feed side. The main difference between UF and NF is that, while UF membranes exhibit a molecular weight cutoff roughly spanning between 10^3 and 10^6 Da with pores having a diameter in the interval of 2–10 nm, NF membranes have a molecular weight cutoff of about 400 ± 100 Da and pore size between 0.5 and 2 nm. Consequently, while in NF there is assumed a sieving-type filtration mechanism, in RO the selective permeation through dense membranes takes place according to a solution-diffusion mechanism.

An additional difference comes from the surface chemistry of membranes: while both UF and RO membranes are neutral, NF membranes often carry positive or negative fixed charges, thus affecting the permeability of dissociated electrolytes by Donnan exclusion mechanism.

In NF the total volumetric transmembrane flux J_v is given by the sum of the molar fluxes of each individual i-th component (J_i) multiplied by its partial molar volume ($\bar{V}_i$):

$$J_v = \sum_i J_i \bar{V}_i. \tag{2.104}$$

In general, the total volumetric flux is largely attributed to the flux of the solvent J_W; under this assumption,

$$J_V \cong J_W = \bar{V}_W L_W \frac{d}{dx}(\bar{V}_W p + RT \ln a_W) + L_V \frac{dp}{dx}. \tag{2.105}$$

Here L_W is a phenomenological coefficient related to the diffusion of the solvent through the membrane pores, L_V is the hydraulic permeability of the membrane related to the viscous flow through pores, a_W is the activity of the solvent, x the coordinate, and p the pressure.

Expressing the term related to the activity as a function of osmotic pressure $\Delta\pi$ (see equation (2.14)), then equation (2.105), after integration over the pore length Δx, becomes

$$J_V \cong \left(\frac{D_W \bar{V}_W}{RT} + L_V\right)\frac{\Delta p}{\Delta x} - \frac{D_W \bar{V}_W}{RT}\frac{\Delta\pi}{\Delta x}, \tag{2.106}$$

having expressed the coefficient L_W as

$$L_W = \frac{D_W}{\bar{V}_W RT}. \tag{2.107}$$

As previously mentioned, the Donnan potential arising from a nanofiltration membrane with fixed charges affects the separation. Since the concentration of species at the feed and permeate sides are different, an electrical potential is established across the membrane, which influences the flux J_i of a charged i-th component:

$$J_i = D_i^m \left[\left(\frac{\bar{V}_i k_i c_i}{RT}\frac{dp}{dx} + k_i \frac{dc_i}{dx}\right) + \frac{z_i F k_i c_i}{RT}\frac{d\phi_{Don}}{dx}\right] + L_V k_i c_i \frac{dp}{dx}, \tag{2.108}$$

where D^m is the diffusion coefficient within the membrane phase, $\bar{V}$ is the partial molar volume, k is the partition coefficient (depending on both size exclusion and Donnan exclusion), c the concentration in solution, p the pressure, F the Faraday constant, z_i the valence of the ion, ϕ_{Don} the Donnan potential, and x the spatial coordinate.

The Donnan potential between an ion-exchange membrane and a diluted electrolytic solution is given, to a first approximation, by

$$\phi_{Don} = \sum_i \frac{1}{z_i F}\left[RT \ln \frac{c_i}{{}^m c_i}\right], \tag{2.109}$$

where ${}^m c_i$ is the concentration if the i-th ion in the membrane (Donnan exclusion).

The typical trend of rejection, solute, and solvent fluxes as a function of the driving force ($\Delta p - \Delta\pi$) is shown in Figure 2.13 for NF operations carried out at constant feed concentration.

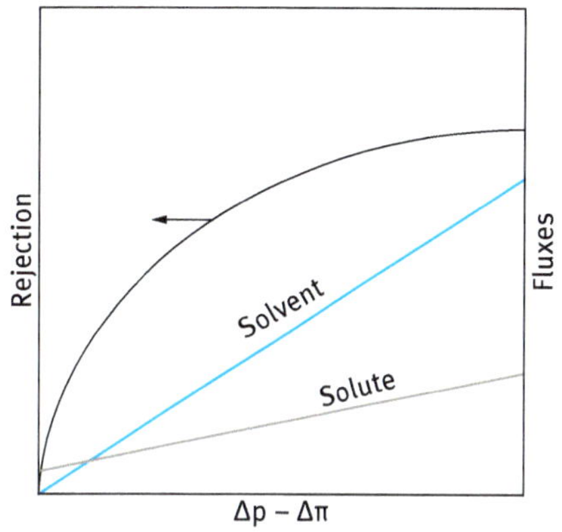

Fig. 2.13: Qualitative trends of rejection and trans-membrane fluxes through a charged NF membrane.

2.10 Hemodialysis

With more than two million patients affected by end-stage kidney disease, hemodialysis is a huge medical treatment that reached the industrial level.

Mathematical models used to predict and optimize this membrane operation span in a wide range of sophistication. As a first step, nonequilibrium thermodynamics (Section 2.2) is the most widely used approach to describe the phenomenological behavior of fluid and solute transport phenomena within membranes.

In hemodialysis, blood flows through an extracorporeal circuit and is cleaned in dialyzers made of polymeric permselective membrane that separates blood from dialysis fluid.

The clearance (K), a key parameter describing the solute removal from the body, is a proportionality constant that relates the rate of extraction of the solute from the blood and the concentration of solute in the blood at the outlet of the dialyzer. K is fixed by the design criteria of the hemodialyzer and operative conditions. Ultrafiltration enhances the clearance of the hemodialyzer by adding a convective transport to solute diffusion.

A simple two-compartment model is shown in Figure 2.14. The body and its fluids are represented by the block labelled “1”, while blood is represented by compartment “2” recirculated through box “1” at constant volumetric flow rate Q. Both compartments are assumed perfectly mixed, so that solute concentration within compartments “1” and “2” are uniformly set at c_1 and c_2, respectively. Within the body, the solute is being formed at constant rate G, while it is extracted from the blood by the dialyzer at rate Kc_2.

Despite its simplicity, the model is able to describe both the decay of solute (typically urea and creatinine) concentration during dialysis, as well as the phenomenon of solute rebound (i.e. the rapid increase of solute concentration after dialysis).

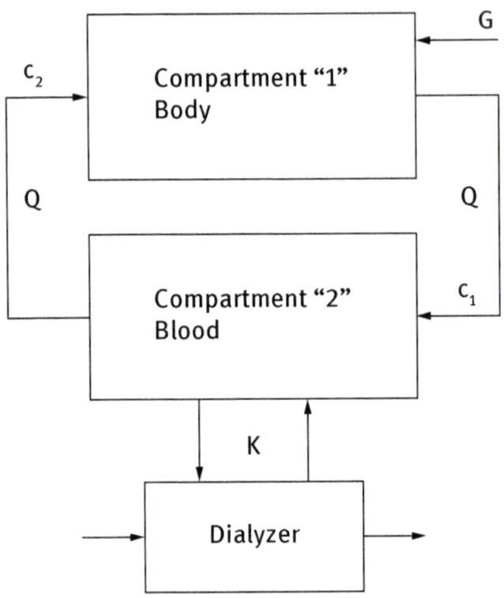

Fig. 2.14: Simplified two-compartments representation of the hemodialysis process.

Mass balance of solute within compartments "1" and "2" gives the following set of ordinary differential equations:

$$\begin{aligned} V_1\frac{dc_1}{dt} &= -Q(c_1 - c_2) + G, \\ V_2\frac{dc_2}{dt} &= Q(c_1 - c_2) - Kc_2. \end{aligned} \tag{2.110}$$

Adding equations (2.110) and solving for the time derivative of c_1,

$$\frac{dc_1}{dt} = \frac{G}{V_1} - \frac{Kc_2}{V_1} - \frac{V_2}{V_1}\frac{dc_2}{dt}. \tag{2.111}$$

By deriving equation (2.111) with respect to time,

$$V_2\frac{d^2c_2}{dt^2} = Q\frac{dc_1}{dt} - (Q + K)\frac{dc_2}{dt}, \tag{2.112}$$

and finally, combining equations (2.111) and (2.112),

$$\frac{d^2c_2}{dt^2} + \left(\frac{Q}{V_1} + \frac{Q}{V_2} + \frac{K}{V_2}\right)\frac{dc_2}{dt} + \frac{QK}{V_1V_2}c_2 = \frac{QG}{V_1V_2}. \tag{2.113}$$

The following initial conditions are required to solve equation (2.113):

$$\text{at } t = 0, \quad c_2 = c_0 \quad \text{and} \quad \frac{dc_2}{dt} = -\frac{Kc_0}{V_2} \quad \text{[from equation (2.110)]}$$

For recovery period after dialysis, the clearance K is 0 assuming that the patient completely lost his/her kidney function, and equation (2.113) reduces to

$$\frac{d^2c_2'}{dt^2} + Q\left(\frac{V_1 + V_2}{V_1V_2}\right)\frac{dc_2'}{dt} = \frac{QG}{V_1V_2}. \tag{2.114}$$

Bird et al. [24] provided a mathematical solution to the time evolution of creatinine for the following specific parameter values: $V_1 = 43\,\text{l}$, $V_2 = 4.54\,\text{l}$, $Q = 5.4\,\text{l/min}$, $K = 0.3\,\text{l/min}$, $G = 0.0024\,\text{g/min}$, $c_0 = 0.140\,\text{g/l}$. The time-variant profile of creatinine for a hemodialysis operation protracted for 50 min and a recovery period of 2 h is shown in Figure 2.15.

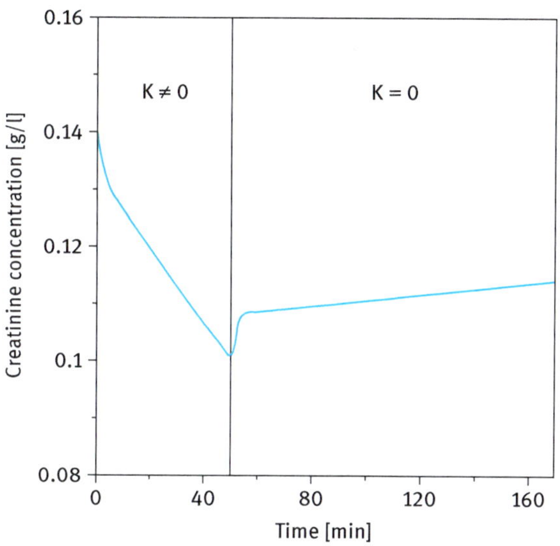

Fig. 2.15: Model prediction of creatinine concentration in time during and after hemodialysis operation.

One important aspect in hemodialysis is the regulation of ion content in the body and the control of hydration in order to limit adverse reactions in patients such as nausea, muscle cramps, headache, arterial hypertension, dialysis-associated hypotension, etc. [25]. Since sodium is a prominent osmotic regulator of water distribution within the body, a special attention is paid to its modelling.

In general, for a system that includes a membrane without fixed ions and a solution with one or more electrolytes, the balance of charges requires

$$\sum_a |z_a| c_a = \sum_c z_c c_c. \tag{2.115}$$

Here z is the valence and c the concentration; subscripts a and c refer to anion and cation, respectively. For NaCl dissociating in Na^+ and Cl^-, $z_a = -1$ and $z_c = +1$.

The concentration of the single ion is related to the concentration of the salt c_s by

$$\begin{aligned} c_a &= i \nu_a c_s, \\ c_c &= i \nu_c c_s, \end{aligned} \tag{2.116}$$

where i is the van 't Hoff coefficient, defining the degree of dissociation of the salt (for NaCl, $i \approx 1$), and ν is the stoichiometric coefficient ($NaCl \rightarrow Na^+ + Cl^-$, $\nu_a = 1$, $\nu_c = 1$).

If there is no electrical current, the conservation of charge requires that the flux of ions will occur in the same direction:

$$\sum_{a} |z_a| J_a = \sum_{c} z_c J_c. \tag{2.117}$$

When transported in solution, the different diffusivities of cations and anions will cause a local charge imbalance and, ultimately, an electrical potential gradient. Although it might be small, this diffusion potential will be able to slow down the fast diffusion ion and speed up the slower ion, so that all ions of the dissociated electrolyte will diffuse with the same rate:

$$J_s = J_a = J_c = -D_s \frac{dc_s}{dx} = -\bar{D}_a |z_a| \frac{dc_a}{dx} = -\bar{D}_c |z_c| \frac{dc_c}{dx}, \tag{2.118}$$

where J is the flux (subscript s refers to salt), c the concentration, D the diffusion coefficient, $\bar{D}$ an average diffusion coefficient, z the valence of the ion, and x the coordinate.

The diffusion coefficient of the salt D_s can be expressed as a function of the average diffusion coefficients of cations and anions:

$$D_s = \frac{\bar{D}_a \bar{D}_c (|z_a| + z_c)}{\bar{D}_a |z_a| + \bar{D}_c z_c}. \tag{2.119}$$

If the solution contains more than one electrolyte, and if the membrane is charged, the complexity of the system increases due to the possibility that certain ions are transported against their concentration gradient (uphill transport).

2.11 References

[1] Onsager L. Reciprocal Relations in Irreversible Processes. I Phys Rev. 1931; 37: 405–426.
[2] Spiegler KS, Kedem O. Thermodynamics of hyperfiltration (reverse osmosis): criteria for efficient membranes. Desalination. 1966; 1: 311–326.
[3] Erickson HP. Size and shape of protein molecules at the anometer level determined by sedimentation, gel filtration, and electron microscopy. Biol proceed Online 2009; 11: 32–51.
[4] Pan S, Nguyen DA, Sridar T, Sunthar P, Prakash JR. Universal solvent quality crossover of the zero shear rate viscosity of semidilute DNA solutions. J Rehol. 2014; 58: 339–68.
[5] Armstrong JK, Wenby RB, Meiselman HJ, Fisher TC. The hydrodynamic radii of macromolecules and their effect on red blood cell aggregation. Biophys J. 2004; 87: 4259–4270.
[6] Sakai K. Determination of pore size and pore size distribution. 2. Dialysis membranes. J Membrane Science. 1994; 96: 91–130.
[7] Curcio E, De Bartolo L, Barbieri G, Rende M, Giorno L, Morelli S, Drioli E. Diffusive and convective transport through hollow fiber membranes for liver cell culture. J Biotechnology. 2005; 117: 309–321.
[8] Cussler EL. Diffusion: mass transfer in fluid systems. New York: Cambridge University Press; 1984.
[9] Daley J. Packings and approximate packings of spheres. National Instuitute of Statistical Sciences, Research Triangle Park, NC, Technical Report; 2000.

[10] Karode SK. Laminar flow in channels with porous walls, revisited, J Membrane Science. 2001; 191: 237–241.
[11] van den Berg GB, Racz IG, Smolders CA. Mass Transfer Coefficients in Cross-Flow Ultrafiltration. J Membrane Science. 1989; 47: 25–51.
[12] Hlavacek M., Bouchet F. Constant flowrate blocking laws and an example of their application to dead-end microfiltration of protein solutions. J Membrane Science. 1993; 82: 285–295.
[13] Bolton G, LaCasse D, Kuriyel R. Combined models of membrane fouling: Development and application to microfiltration and ultrafiltration of biological fluids. J Membrane Science. 2006; 277: 75–84.
[14] Smolders HJ, Sloot CA, van Swaaij WPM, Versteeg GF. Surface diffusion of hydrogen sulfide and sulfur dioxide in alumina membranes in the continuum regime. J Membrane Science. 1992; 74: 263–278.
[15] Hirschfelder JO, Curtiss CF, Bird RB. Molecular Theory of Gases and Liquids. New York: Wiley; 1954.
[16] Zalc JM, Reyes SC, Iglesia E. The effects of diffusion mechanism and void structure on transport rates and tortuosity factors in complex porous structures. Chemical Engineering Science. 2004; 59: 2947–2960.
[17] Feng H. Modeling of vapor sorption in glassy polymers using a new dual mode sorption model based on multilayer sorption theory. Polymer. 2007; 48: 2988–3002.
[18] Wickramasinghe SR, Garcia JD, Han B. Mass and momentum transfer in hollow fiber blood oxygenators. J Membrane Science. 2002; 208: 247–256.
[19] Hewitt TJ, Hattler BG, Federspiel WJ. A mathematical model of gas exchange in an intravenous membrane oxygenator. Ann Biomed Eng. 1998; 26: 166–178.
[20] Zhang J, Nolan TDC, Zhang T, Griffith BP, Wu ZJ. Characterization of membrane blood oxygenation devices using computational fluid dynamics. J Membrane Science. 2007; 288: 268–279.
[21] Hill AV. The possible effects of the aggregation of the molecules of hemoglobin on its dissociation curves, J Physiology. 1910; 40: 4–7.
[22] Mockros LF, Leonard R. Compact cross-flow tubular oxygenators. Trans Am Soc Artif Intern Organs. 1985; 31: 628–633.
[23] Wickramasinghe SR, Han B. Designing microporous hollow fibre blood oxygenators. Chemical Engineering Research and Design. 2005; 83: 256–267.
[24] Bird RB, Stewart WE, Lightfoot EN. Transport phenomena, 2nd edition. New York: John Wiley & Sons; 2007.
[25] Wizemann V, Wabel P, Chamney P, Zaluska W, Moissl U, Rode C, Malecka-Masalska T, Marcelli D. The mortality risk of overhydration in haemodialysis patients. Nephrol Dial Transplant. 2009; 24: 1574–1579.

3 Artificial organs

3.1 Membrane artificial organs

A membrane artificial organ is an artificial membrane device designed to replace the organ functions. The growing need for these technologies is substantial. Improved health care has resulted in an increased life span for the general population and, when coupled with a growing shortage of donor organs, it is thus clear that organ assistance and substitution devices will play a larger role in managing patients with end-stage diseases by providing a bridge to recovery or transplantation. In US the annual need for organ replacement therapies increases by about 10 percent per year.

These devices can be extracorporeally connected to the patient body through blood circulation or implanted inside the body. They generally are used to temporarily support patients until the recovery of a failing organ, or as bridge until transplantation. There are few cases in which the membrane device permanently supports the patient. In artificial organs membranes act as a selective barrier for removal of endogenous and exogenous toxins from the blood of a patient with renal disease (e.g. hemodialysis, hemofiltration, hemodiafiltration) or with hematological or autoimmune disorders (e.g. plasmapheresis, plasmatherapy). Membrane-based processes are also used for gas exchange with blood, allowing oxygenation of the blood and removal of CO_2 [1, 2]. All these therapeutic membrane processes simulate the function of organ *in vivo* thanks to the highly selective properties of separation and interaction. Thus, the types of membranes, devices, and processs are designed on the basis of the organ functions to be replaced. Scientists make membranes and membrane devices which imitate nature in order to realize the same *in vivo* organ functions.

3.2 Hemodialysis

3.2.1 Kidney functions: What hemodialysis must replace

The kidney performs several functions, including excretion of wastes, acid-base homeostasis, osmolality regulation, blood pressure regulation, and hormone secretion. The kidney functions are based on mechanisms of filtration, reabsorption, and secretion which take place in the nephron. The nephron is the structural and functional unit of the kidney, which consists of a long tubule (about 30–55 mm), the Bowman's capsule, which encloses a cluster of microscopic blood vessels – capillaries – that form the glomerulus (Figure 3.1). Blood flows into and away from the glomerulus through the inner wall of the capsule and into the nephron tubule. The glomerular filter is freely permeable to water, mineral ions (Na^+, K^+, Ca^{++}), and small

DOI 10.1515/9783110268010-004

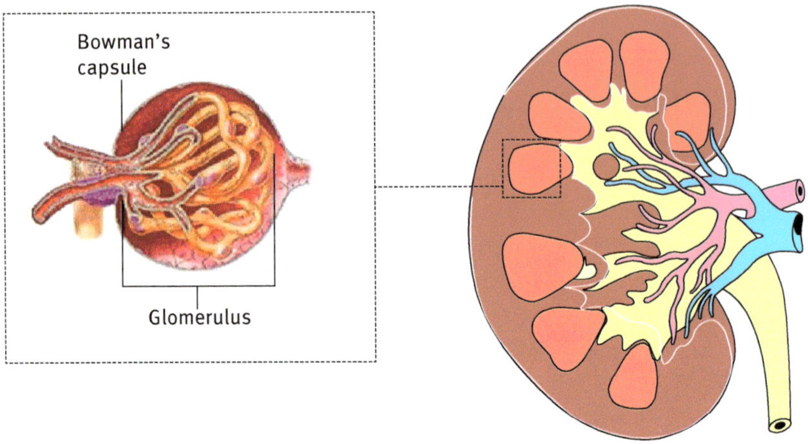

Fig. 3.1: Structural and functional unit of the kidney.

organic molecules such as glucose. A substantial fraction of high molecular weight proteins (with MW up to 20 000 Da) are also filtered, but proteins with MW greater than 40 000 Da are filtered only in trace amounts. Glomerular capillaries are about 10 times more permeable than other capillaries in the body. As the filtrate passes through the tubule, its composition is altered by the secretion of certain substances and by the selective reabsorption of water and other constituents. The filtration process produces an ultrafiltrate that becomes urine, which is conveyed through the collecting tubules into the renal pelvis. 1.2 million nephrons per kidney are involved in the filtration and removal of wastes as well as in the reabsorption of water, proteins, other essentials into the blood. About 180 liters of blood enter the kidneys every day while reabsorbing a large percentage, allowing for the generation of only approximately two liters of urine.

The kidneys eliminate a variety of waste products produced by metabolism, including urea derived from protein catabolism, as well as uric acid, from nucleic acid metabolism (Table 3.1). These nitrogenous wastes are eliminated through the formation of urine, which occurs by a counter-current multiplication system. This requires water and ion permeability in the descending limb of the loop, water impermeability in the ascending loop, and active ion transport out of most of the ascending loop of the tubules. The kidneys contribute to maintaining acid-base homeostasis, which is the maintenance of pH around a relatively stable value. This function is performed by reabsorbing bicarbonate from urine and excreting hydrogen ions into urine.

The kidney performs water reabsorption and creates an increase in urine concentration, maintaining plasma osmolality to its normal levels. An increase in osmolality causes the secretion of antidiuretic hormone (ADH), which binds to the principal cells in the collecting duct that translocate aquaporin to the membrane, allowing water to leave the normally impermeable membrane and be reabsorbed into the body, thus increasing the plasma volume of the body.

Tab. 3.1: Molecules excreted, secreted and reabsorbed by kidneys.

Excreted molecules		Secreted substances	Reabsorbed substances
Urea	30 g/day	– Hydroxybenzoates	– Glucose
Creatinine	2 g/day	– Hippurates	– Amino acids
Salt	15 g/day	– Neutrotransmitters (dopamine)	– Phosphate
Uric acid	0.7 g/day		– Sulfate
Water	1500 ml/day	– Bile pigments	– Lactate
		– Uric acid	– Succinate
		– Antibiotics	– Citrate
		– Morphine	
		– Saccharin	

Kidneys are also involved in the long-term regulation of blood pressure. This primarily occurs through the maintenance of the extracellular fluid compartment, the size of which depends on the plasma sodium concentration. Renin is the first chemical messenger that makes up the renin-angiotensin system. Changes in renin ultimately principally alter the hormones angiotensin II and aldosterone, which both increase the kidney's absorption of sodium chloride, thereby expanding the extracellular fluid compartment and raising blood pressure. Elevated levels of renin produce an increase of angiotensin II and aldosterone concentrations, leading to increased sodium chloride reabsorption, expansion of the extracellular fluid compartment, which increases the blood pressure. On the other hand, low renin levels produce a decrease of angiotensin II and aldosterone levels, resulting in a decrease of blood pressure. A variety of hormones are secreted by the kidneys, including erythropoietin and the enzyme renin. When levels of oxygen at the tissue level are low, erythropoietin is released in response to hypoxia in the renal circulation. It stimulates the production of red blood cells in the bone marrow.

When kidneys fail, it is necessary to replace the kidney functions with a dialysis treatment that involves the diffusion of solutes through a membrane driven by a concentration gradient. For the treatment of uremic patients an artificial membrane (hemodialysis) or a biological membrane (peritoneal dialysis) can be utilized.

3.2.2 Hemodialysis treatment

The global rate of end-stage renal disease (ESRD) was estimated to be 2 786 000 at the end of 2011 and, with a 6–7 % growth rate, continues to increase at a significantly higher rate than the world population. Approximately 2 164 000 people were undergoing hemodialysis (HD) or peritoneal dialysis (PD), and around 622 000 people were living with kidney transplants. Therefore, more than three-quarters of all ESRD patients were treated by dialysis.

If we look at the global geographical distribution we can see that most dialysis patients are in the US (19 %), Europe (15 %), and Japan (14 %). More than 50 % of the global dialysis patient population is treated in the US, Japan, China, Brazil, and Germany and 38 % in other countries [3] (Figure 3.2).

In hemodialysis, a selective artificial membrane simulates in the dialyzer the function of nephron. The patient is connected to an extracorporeal circuit by a venoarterious shunt. The patient blood flows outside of the body by means of plastic tubes (blood lines) into a dialyzer. The dialyzer membranes work as an artificial kidney by separating waste products and excess water from the blood. Arterial blood flows at 200–300 ml/min on one side of a membrane (0.5–2 m^2 surface area), while a buffered isotonic dialysate solution flows on the other membrane side in a counter-current mode. Low MW waste metabolites diffuse across the membrane in response to a transmembrane concentration difference. The hemodialysis machine pumps blood and anticoagulants, regulates the purification process, and controls the mixing of dialysis solution and the rate of its flow through the system.

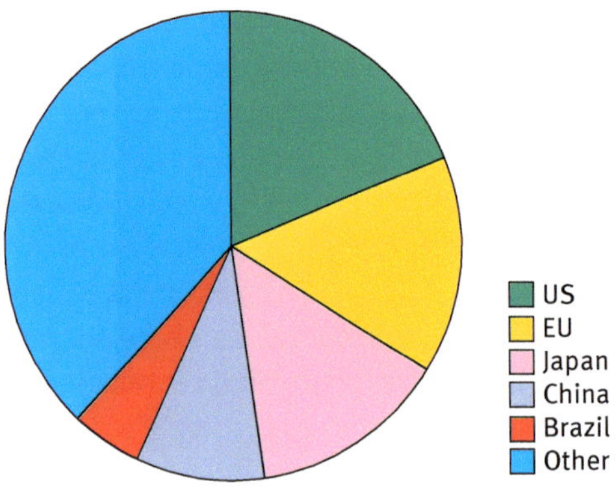

Fig. 3.2: Global distribution of ESRD patients treated by dialysis process (data from [3]).

Membrane hollow fibers and glomerular capillaries have many similarities: both employ a cylindrical cross-section that allows minimization of the perimembrane boundary layers and maximization of transport efficiency; they have the same ratio of wall thickness to the cross-sectional diameter; they are composed of linear hydrophilic polymers that rely on van der Waals forces, and islands of cristallinity to retain their integrity; they allow the passage of small solutes and retain proteins. Differences concern the size: the glomerule has a diameter of 4–8 μm, whereas synthetic hollow fiber membranes have diameters in the range of 200 μm. Furthermore there are other differences regarding the operating time, which in the case of nephron is 168 h/week, whereas in the case of dialyzer is 12 h/week. The cutoff of the dialysis membrane is lower than that of the glomerule. Chemical species with a molecular weight between 5000 and 12 000 Da, eliminated by natural kidneys, are not removed by hemodialysis treatment, and their accumulation might be responsible in some hemodialysis complications (anemia, bone and joint pain, neuropathy, itching). For these reasons the development of new dialysis membranes are devoted to realize a "high-flux dialyzer" and to prepare highly permeable membranes, such as new polymethylmetacrylate (BK-F) [4] which, compared to the conventional PMMA and cellulose acetate, have demonstrated a satisfactory dialytic removal of solutes including beta-2 microglobulin, which plays an important role in the complication of dialysis failure.

3.2.3 History of hemodialysis

Dialysis was first described by Thomas Graham in 1854 [5]. He used a membrane created from an ox bladder to cover the wide open end of a bell-shaped vessel filled with urine. Graham suspended the vessel inside a larger container filled with distilled water. After several hours, the bell-shaped vessel was removed (Figure 3.3). The larger container was heated so that the fluid inside boiled to dryness. Graham showed that the residue in the larger container consisted mainly of sodium chloride and urea, the principal components of urine. This proved that urea had passed through the membrane.

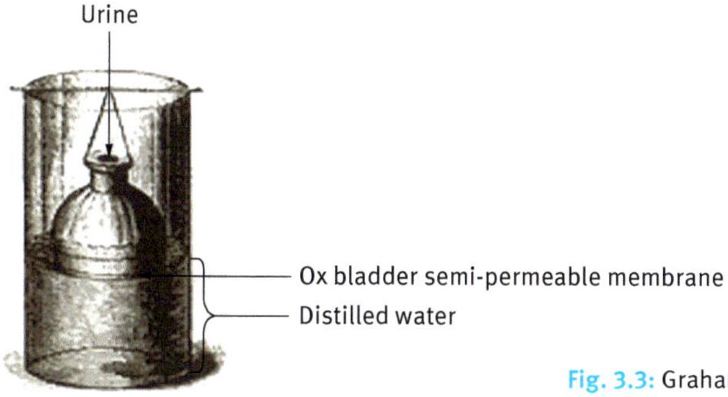

Fig. 3.3: Graham dialysis system [5].

In 1914 John Abel et al. developed and tested the first efficient dialysis system, consisting of a filtering device made of cellulose trinitrate (collodion) tubes and an attached burette containing hirudin solution obtained from leech heads used as the anticoagulant (Figure 3.4) [6].

The first human hemodialysis was performed in a uremic patient by Haas in 1924 by using a tubular device made of collodion. Only in 1937, the first flat hemodialysis membrane made of cellophane was produced (Figure 3.5) [7].

The first practical artificial kidney was developed by Willem Kolff [8].The Kolff kidney used a 20 m long tube made of cellophane sausage casing as a dialyzing membrane. The tube was wrapped around a slatted wooden drum (Figure 3.6). Powered by an electric motor, the drum revolved in a tank filled with dialyzing solution. The

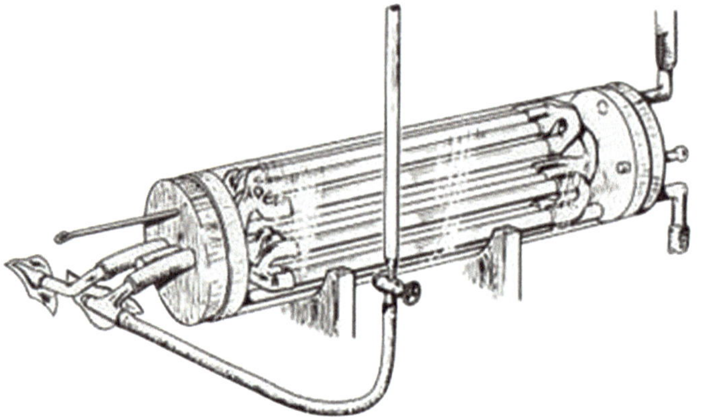

Fig. 3.4: Dialysis device developed by John Abel.

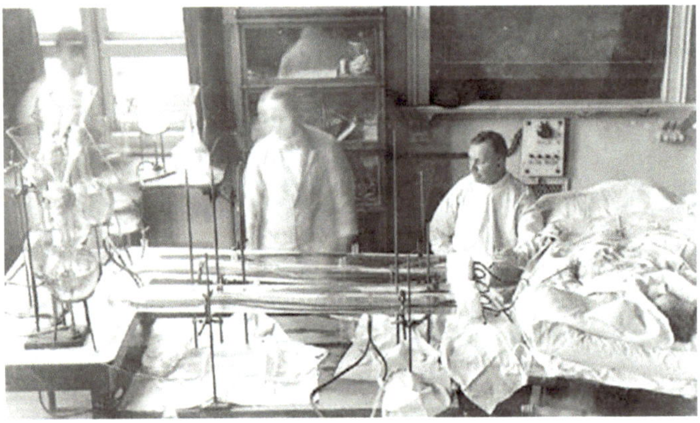

Fig. 3.5: Haas dialyzer (image from the Institut für Geschichte der Medizin, Gießen).

patient's blood was drawn through the cellophane tubing by gravity as the drum revolved. Toxic molecules in the blood diffused through the tubing into the dialyzing solution. Complete dialysis took about six hours. The Kolff kidney effectively removed toxins from the blood, but because it operated at low pressure, it was unable to remove excess fluid from the patient's blood. Modern dialysis machines are designed to filter out excess fluid while cleansing the blood of wastes.

Improvement of the membrane geometry led to the development of new hemodialyzers by using tubes, coils (Alwall dialyzer), flat sheets (Kiil dialyzer), and hollow fiber membranes (Stewart) [9, 10].

A major step forward was the development of the hollow-fiber dialyzer by Richard Stewart in 1964 [9] (Figure 3.7), who explored the medical applications of cellulose acetate capillary fibers. His group demonstrated that as in a kidney substances can be

Fig. 3.6: Kolff artificial kidney (www.edren.org/pages/history/early-development.php).

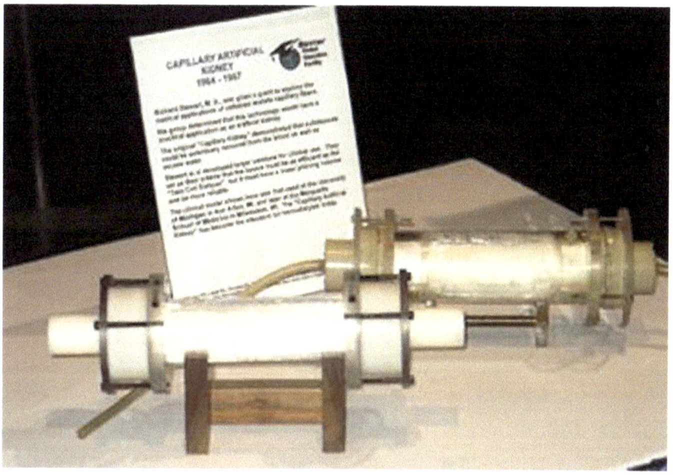

Fig. 3.7: Hollow-fiber dialyzer by Richard Stewart in 1964 (http://homedialysis.org/home-dialysis-basics/machines-and-supplies/dialysis-museum).

selectively removed from the blood as well as excess water. This technology replaced the traditional membranous tubes and flat membranes of the day with a number of capillary-sized hollow membranes. This procedure allowed the realization of dialyzers with a surface area large enough to fulfil the demands of efficient dialysis treatment.

The first capillary dialyzer used capillaries that were made of thin-wall (16 micron) cuprophan fibers (Figure 3.8). The size of dialyzers decreased over time, and the apparatus utilized for treatment has become less elaborate.

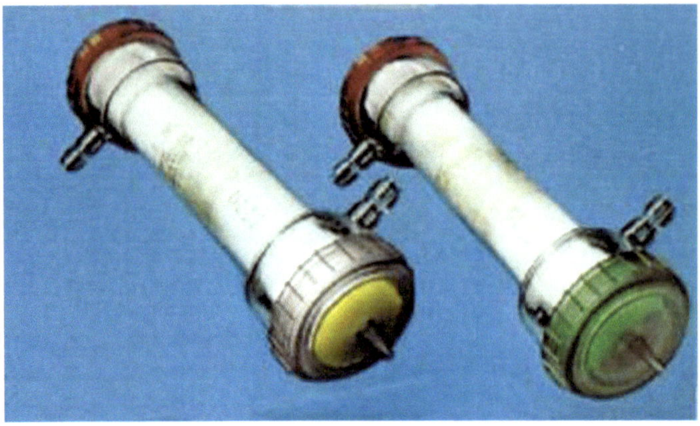

Fig. 3.8: Cuprophan hollow fiber dialyzer (http://homedialysis.org/home-dialysis-basics/machines-and-supplies/dialysis-museum).

3.2.4 Concepts of hemodialysis

In dialytic therapy, diffusion and convection have been combined in order to replace the renal function.

Diffusive transport allows the passage of solutes from one side of the membrane to the other side according to their concentration gradients. The diffusive transport (J_S) through a membrane depends on the temperature (T), the surface area (A), and it is inversely proportional to the membrane thickness (δ):

$$J_S = D_S A T \frac{\Delta c_S}{\delta}. \tag{3.1}$$

As a result, separation between solutes is obtained as a difference in the diffusion rates across the membrane arising from differences in molecular size. In hemodialysis the primary mechanism of solute transport is diffusion, although there is also a convective contribution. The membrane separates blood from the dialysate and small solutes diffuse across the membrane from the blood to the dialysate compartment according to their concentration gradient, passing from the higher to the lower concentration side [11]. Blood and dialysate flow are in countercurrent mode in order to maintain optimized concentration gradients over the whole length of the dialyzer (Figure 3.9).

High flux can be obtained using membranes as thin as possible. Considering that the diffusion coefficient is dependent on the molecular weight of the chemical species, the performance of the dialysis process is more efficient than ultrafiltration for species in a wide molecular weight range.

In the convective transport the driving force for the passage of solutes through a membrane is the transmembrane pressure gradient. Therefore, the convective flux of a solute (J_C) depends on the ultrafiltration rate (Q_{UF}), the solute concentration in

Fig. 3.9: Scheme of diffusive transport of solutes across membrane in hemodialysis.

plasma water (c_p), and the solute sieving coefficient (S) [12]:

$$J_c = Q_{UF} c_p S. \tag{3.2}$$

The sieving coefficient is the ratio from solute filtrate concentration c_f to the respective solute plasma concentration c_p:

$$S = \frac{c_f}{c_p}. \tag{3.3}$$

From a theoretical point of view the sieving coefficient value can be considered in ideal conditions as $S = 1 - \sigma$, where σ is the reflection coefficient of the membrane, which is a measure of the selectivity of a membrane and has a value between 0 (no selectivity) and 1 (no solute transport). With porous membranes the major contribution for the retention of a given solute is its molecular size in relation to that of the membrane pores [11]. Therefore a relationship exists between the reflection coefficient and the solute sizes. The reflection coefficient increases with increasing solute size, and the membrane becomes more selective.

The dialyzer performance can be evaluated by (a) the ultrafiltration coefficient or flux, (b) the ability to remove urea and other molecules with low MW, (c) proficiency to expel molecules with middle MW, and (d) the capacity to retain albumin and larger molecules. One of the roles of the dialyzer is the removal of water excess from the patient, which can be achieved by using a membrane highly permeable to water. The ultrafiltration coefficient (K_{UF}) describes the water permeability of the membrane dialyzer and depends on the hydraulic permeability (L_h) and the surface area of the membrane (A), as reported in equation (3.4):

$$K_{UF} = L_h A. \tag{3.4}$$

Thus, at a given transmembrane pressure gradient ΔP the ultrafiltration rate (Q_{UF}) is dependent of the ultrafiltration coefficient:

$$Q_{\mathrm{UF}} = K_{\mathrm{UF}}\,\Delta P. \tag{3.5}$$

The dialyzer performance is measured through the clearance, which is the ratio of removal rate to blood concentration [13]:

$$\text{Clearance} = Q_{\mathrm{b}}\left(\frac{c_{\mathrm{in}}^{\mathrm{b}} - c_{\mathrm{out}}^{\mathrm{b}}}{c_{\mathrm{in}}^{\mathrm{b}}}\right) + Q_{\mathrm{f}}\frac{c_{\mathrm{out}}^{\mathrm{b}}}{c_{\mathrm{in}}^{\mathrm{b}}}, \tag{3.6}$$

where $c_{\mathrm{in}}^{\mathrm{b}}$ and $c_{\mathrm{out}}^{\mathrm{b}}$ are the inflow and outflow blood concentrations, respectively.

The diffusive term is described in the first part of equation (3.6), whereas the convective term is described by the second part.

Solute clearance increases as the dialyzer blood and dialysate flow rates increase.

The dialysance accounts for the accumulation of solute in the dialysate compartment:

$$\text{Dialysance} = Q_{\mathrm{b}}\left(\frac{c_{\mathrm{in}}^{\mathrm{b}} - c_{\mathrm{out}}^{\mathrm{b}}}{c_{\mathrm{in}}^{\mathrm{b}} - c_{\mathrm{in}}^{\mathrm{d}}}\right) + Q_{\mathrm{UF}}\left(\frac{c_{\mathrm{out}}^{\mathrm{b}}}{c_{\mathrm{in}}^{\mathrm{b}} - c_{\mathrm{in}}^{\mathrm{d}}}\right), \tag{3.7}$$

where $c_{\mathrm{in}}^{\mathrm{d}}$ is the inflow concentration of a given solute in the dialysate.

The removal rate can be measured instantaneously by sampling blood on either side of the dialyzer and multiplying the difference by the inflow rate:

$$K_{\mathrm{d}} = Q_{\mathrm{b}}\frac{c_{\mathrm{in}}^{\mathrm{b}} - c_{\mathrm{out}}^{\mathrm{b}}}{c_{\mathrm{in}}^{\mathrm{b}}}, \tag{3.8}$$

where Q_{b} is blood flow, c_{in} and c_{out} are the inflow and outflow concentration, respectively.

The performance of the whole treatment is best measured from the fall in solute concentration from before dialysis (c_0) to after dialysis (c). The clearance is a measure of a solute's fractional removal rate, which for convection or diffusion tends to remain constant, despite the marked decrease in concentration during dialysis [14]. A constant fractional removal rate means that the absolute removal rate is proportional to the concentration, as illustrated in Figure 3.10.

Generally low molecular weight solutes like urea, creatinine, phosphate, and vitamins are used as marker to evaluate the performance of hemodialysis, which reflects the removal through a diffusive mechanism. Larger solutes (e.g. inulin, β2 microglobulin, and albumin) can be removed through a convective transport, although the diffusive clearance of these solutes does not provide reliable information. The sieving coefficient is an important parameter to be considered in this case. The larger solutes are used as marker molecules in the hemofilters and hemodiafilters. However, the total clearance (equation (3.6)) provides information concerning the performance of high-flux dialyzers.

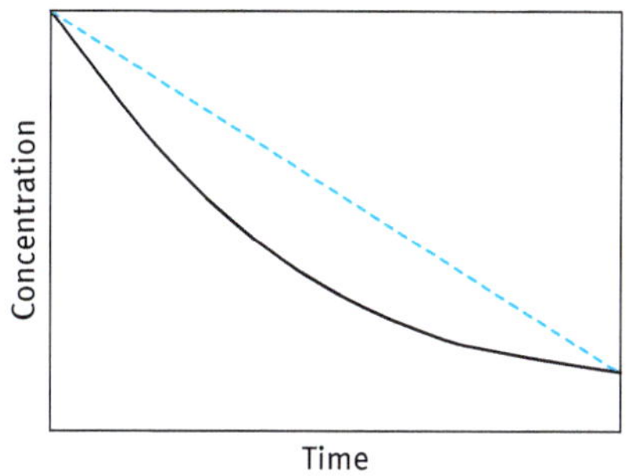

Fig. 3.10: Decrease of solute concentration with time in hemodialysis, a straight line when plotted on a logarithmic scale (adapted from [14]).

The clearance of molecules is affected by such factors as the blood flow rate and the molecular weight of molecules, as depicted in Figure 3.11. For small molecules the clearance rate increases significantly with increasing blood flow reaching plateau values. For molecules with middle molecular weight the clearance is affected to a small extent by the blood flow rate.

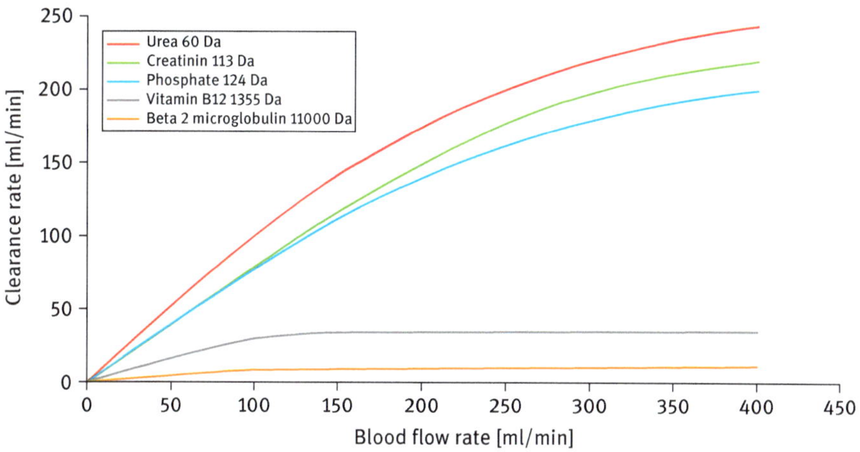

Fig. 3.11: Schematic diagram of the clearance rate of molecules with small and middle MW as a function of the blood flow rate.

Other factors which influence mainly small solute clearance on the basis of equation (3.1) are also the membrane thickness and surface area. At a given blood flow rate small solute clearance increases linearly with surface area up to 1.2–1.4 m^2 and then flattens off [13]. The influence of the membrane surface becomes significant with larger solutes which are removed by convection. The diffusive clearance can be improved by using thinner membranes.

The maximum clearance achievable for a particular solute at infinite blood and dialysate flow rates is described by the dialyzer mass transfer area coefficient (K_0A) [15, 16]:

$$K_0A = \frac{Q_b Q_d}{Q_b - Q_d} \ln \frac{Q_d(Q_b - K_d)}{Q_b(Q_d - K_d)}, \tag{3.9}$$

where K_0 is the mass transfer coefficient, A is the surface area of the membrane, Q_b and Q_d are the blood flow rate and dialysate flow rate, respectively, K_d is the mean of blood and dialysate side solute clearance.

K_0A is directly proportional to the dialyzer membrane surface area, which depends on the length and number of fibers in a hollow-fiber dialyzer. This parameter is inversely proportional to the diffusion pathway represented by the average length of the membrane pores.

Therefore, the mass transfer area coefficient is the intrinsic clearance of the dialyzer for the measured solute (e.g., urea), which is a property of the solute and of the dialyzer, and is independent on flow rates and concentrations. For this reason it is often used to compare different dialyzers and dialyzer models. On the basis of the K_0A parameter, hemodialyzers are distinguished as low or high efficiency [17, 18]. In Figure 3.12 we can see the comparison of urea clearance rates between low and high efficiency hemodialysis. At low blood flow the clearance rate is similar in both hemodialyzers. As blood and dialysate flow rates increase, the clearance increases in a curvilinear fashion that reaches a plateau (K_0A) which is higher ($K_0A > 600$ ml/min) for the high efficiency hemodialyzer and low ($K_0A < 500$ ml/min) for the low efficiency hemodialyzer [19].

High-flux dialyzers are more effective in removing larger molecules as well as water and electrolytes.

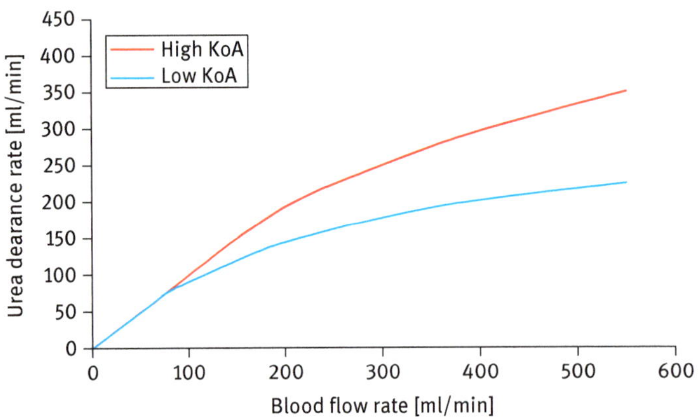

Fig. 3.12: Comparison of urea clearance rates between high and low efficiency hemodialyzers.

Another property used to evaluate the performance of the hemodialyzer is the permeability, which is a measure of the clearance of the middle molecular weight molecule (e.g. β2-microglobulin). Hemodialyzers that allow β2-microglobulin clearance < 10 ml/min have been defined as low permeability, whereas hemodialyzers which allow β2-microglobulin clearance > 20 ml/min have been defined as high permeability hemodialyzers [19, 20].

Solute removal by hemodialysis is greatly influenced by the molecular size of the solute and its diffusion in the blood compartment. A decrease in the efficacy of solute removal could be due to large molecular size, high protein binding, and slow diffusion of the solutes to the blood compartment.

The physico-chemical properties of membrane (hydrophobic/hydrophilic domains, ionic charges, etc.) and solute (e.g. electrical charge) may affect the diffusion across the hemodialysis membrane. Associated with the membrane are the layers at the interface of membrane; depending on the fluid the mixing may change the concentration at the interface of solutes and consequently the concentration gradient representing the driving force of the diffusive transport.

The presence of proteins in the blood can affect the diffusive transport of solutes by binding to proteins or by a Gibbs–Donann effect. Solutes may bind to the blood proteins or other constituents forming complexes which are not permeable to the membrane.

Negatively charged blood proteins are impermeable to the membrane and tend to accumulate at the membrane surface. Consequently in order to ensure electroneutrality a certain number of cations (e.g. sodium, calcium) must be retained by the blood, leading to an imbalance in the concentration of ions across the membrane.

3.3 Hemofiltration

Membrane-based therapies for the treatment of patients with acute renal failure are not limited to dialysis, but also include other modalities. Hemofiltration (HF) has been developed to overcome the reduced efficacy of diffusion for molecules with middle and large MW. In hemofiltration, similarly to hemodialysis, the blood passes through a hemofilter in an extracorporeal circuit. However, the mechanisms by which the molecules are removed by blood differ markedly. In this process the driving force is a hydrostatic pressure gradient that induces a large flux of water across the membrane from the blood side to the filtrate side (Figure 3.13). As a result the rate of solute removal is proportional to the applied pressure. Because the water flux drags solutes across the membrane, the selectivity of the process is determined exclusively by the sieving properties of the membrane [21].

Fig. 3.13: Solute transport in hemofiltration.

The clearance for a given solute is expressed in terms of its sieving coefficient and the total water flux:

$$\text{Clearance} = Q_{UF}S. \tag{3.10}$$

Factors that influence the effective convection are the hydraulic permeability and ultrafiltration coefficient of the membrane, the solute permeability in the useful molecular weight range, and the pressure gradient. Therefore, the convective transport depends strongly on pore size and pore size distribution of the membrane, molecular shape and geometry of solutes and their molecular weight, as well as on the physicochemical properties of membranes and solutes.

As reported above, the solute sieving coefficient (S) is the ratio of the solute filtrate concentration c_f to the respective solute plasma concentration c_p in the absence of a diffusion gradient. However, in the blood compartment, solute distribution is not homogeneous: solute tends to accumulate at the blood–membrane interface when filtration increases, thus creating gradients for diffusion towards the bulk region inside the hollow fiber. As a consequence, a contribution to the solute transport is also given by diffusion. The sieving coefficient could be overestimated, because the concentration in the bulk region (which is the measured value) is generally lower than that at the blood–membrane interface [11]. The sieving coefficient is also dependent on the interfacial membrane interactions with plasma proteins. For example the protein adsorption on the membrane surface reduces the sieving coefficient slightly or strongly depending on the formation of thin or thick protein layer and additional phenomenon of polarization (Figure 3.14).

The separation properties of membranes used in hemofilters are closer to those of renal glomerule: membrane pore size ranges from 40 to 50 kDa. In this way, middle molecular weight toxins can be eliminated, thus avoiding their accumulation in the blood and consequently complications. In hemofiltration, an excess of water can

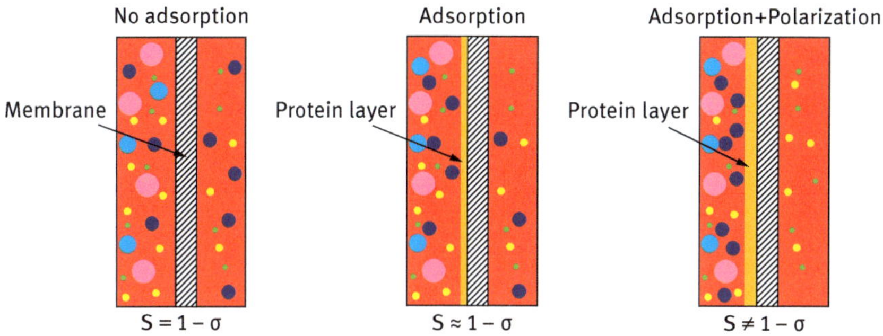

Fig. 3.14: Effects of interactions between the membrane and plasma proteins on solute sieving coefficient.

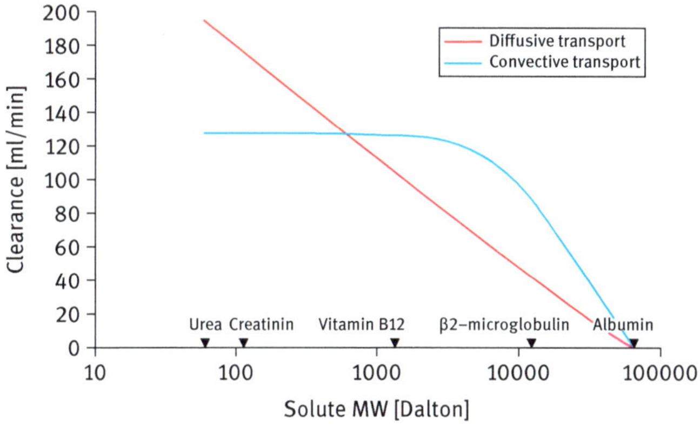

Fig. 3.15: Comparison of convective and diffusive transport of solutes with different MW.

be removed, and due to the convective mass transfer, solute transport is independent on the molecular size in contrast to the diffusion process, dominant in conventional dialysis, which favors the elimination of small solutes. In fact, uremic middle molecules are more effectively eliminated during hemofiltration than during dialysis. During hemofiltration, in contrast to hemodialysis, urea, creatinine, and phosphate are cleared at lower rates, but convective transport is favorable for larger MW solutes (Figure 3.15). Overall, the efficacy is reduced for small molecular weight substances (urea, creatinine, electrolytes) and improved for higher molecular weight solutes (small proteins, mediators, etc.).

Larger molecules, which are cleared from the blood only in negligible quantities by a dialyzer, are cleared efficiently by the hemofilter. Although the properties of the membrane in hemofiltration are closer to the those of renal glomerule, in this process the filtrate is not selectively reabsorbed, as occurs in the kidney. During hemofiltration the filtrate is discarded, and the patient receives infusions of a substitution solution.

Substitution fluid, typically a buffered electrolyte solution close to a plasma-water composition, can be administered pre- or post-filter. For this reason generally this type of therapy is used for patients who show fluid retention to avoid any fluid substitution, or for patients with acute renal failure in intensive care units (where most patients have multiorgan failure) who receive more than 2 l of fluid a day, many of which is nutrition [22–24].

On the other hand, it is recognized that in hemofiltration the use of membranes with higher cutoff involves a loss of selectivity of the membrane: small solutes are removed less effectively than during dialysis, thus limiting the reduction of treatment time. To match HF small MW transport with HD performance, large amounts of exchange volume are needed.

However, hemofiltration inevitably leads to an increase in the concentration of both red cells and plasma protein in the blood of the extracorporeal circuit. The plasma component of the blood represents the sole source of the salt and water that are the principal constituents of the filtrate [25]. Therefore, there is thus a tendency to produce viscous blood. This can be reduced by inducing a filtration rate that is lower than 30 % of the blood flow rate, or by using as predilution by administering the replacement fluid proximal to the hemofilter.

Several treatment formats are currently in use, including continuous arterio-venous hemofiltration (CAVH) and continuous veno-venous hemofiltration (CVVH).

3.4 Hemodiafiltration

Hemodiafiltration (HDF) originated from work done in the late 1970's by Leber et al. and is the combination of hemodialysis and hemofiltration processes. The high rate of ultrafiltration is combined with the efficient diffusion, overcoming the disadvantages related to each single treatment. Due to a combination of convective mass transfer and diffusion, the clearance values of both small and larger molecules are significantly higher than during hemofiltration and hemodialysis alone with the same membrane. Hemodiafiltration allows the removal of middle- and large-size molecules, which influence the incidence and progress of ESRD complications such as amyloidosis that is due to the accumulation of β2-microglobulin. In contrast to the membranes used in HD, which have a hydraulic permeability of 5–6 ml/h mmHg m^2, membranes used in HDF have higher hydraulic permeability of one order of magnitude (30–40 ml/h mmHg m^2). Considering that the ultrafiltration coefficient is related to the MW of the solutes, the removal of solutes with MW of 5000 Da requires an ultrafiltration coefficient about 15 ml/h mmHg (Figure 3.16). The ultrafiltration coefficient for HDF membranes is greater than 10 ml/h mmHg. This process requires volume substitution (pre- or post-filter) and dialysate fluid.

HDF, like HD, is influenced by the blood and dialysate flow rate and red cell concentration. The solute transport in the case of HDF is also influenced by the dilution

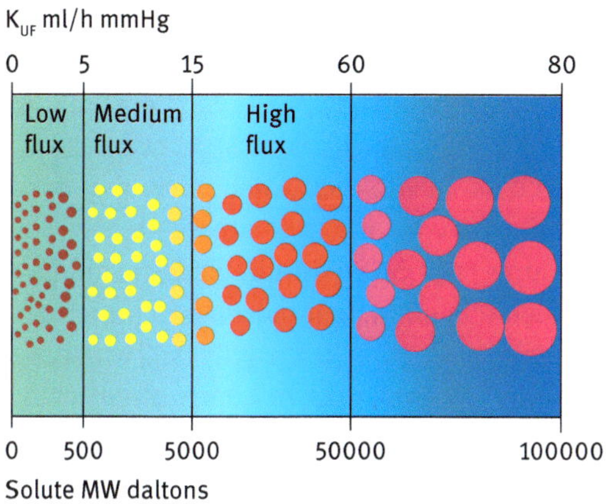

Fig. 3.16: Ultrafiltration coefficient of membranes related to the MW of solutes to be removed.

mode if the dilution of the blood occurs before to enter in the filter or post filter. The dilution of blood before entering into the filter ensures better rheological conditions and higher convective transport of large molecules, but clearances of small MW solutes are slightly reduced because of the decreased concentration gradients across the membrane. In the post-dilution mode, the efficacy of molecules with small MW is slightly increased, but clearances of high MW solutes may reach the three-fold level of HD. For this reason post-dilution is the best compromise between solute removal and substitution fluid consumption.

Many works have shown the effect of HDF in comparison to HD and HF, underlying the efficacy of HDF in the removal of high MW solutes (small proteins, mediators, etc.) and small MW substances (urea, creatinine, electrolytes, buffer base).

If we compare the clearance of low and high MW molecules among HD, HF, and HDF, we can see that the elimination potential of HDF approaches that of the natural kidney, ensuring an efficient removal in a wide range of MW (Figure 3.17).

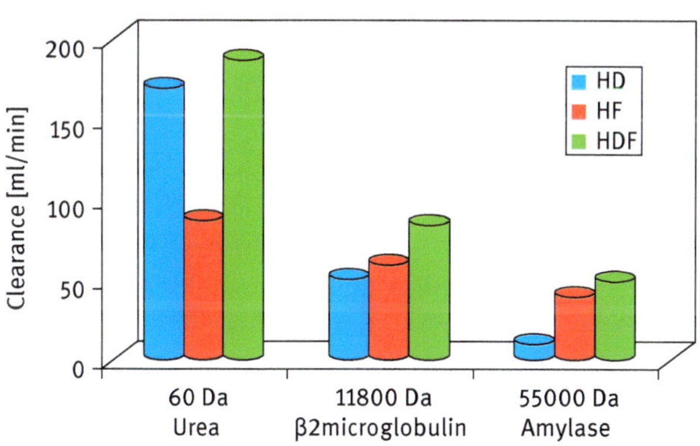

Fig. 3.17: Comparison of clearances for small and high MW solutes among HD, HF, and HDF processes.

3.5 Membranes used in hemodialyzers/hemofilters

Membranes used in dialysis are of cellulosic or synthetic nature. The early membranes used in hemodialysis were manufactured from regenerated cellulose or cellulose acetate by using dry-wet and dry-spinning process. Due to their small thickness, these hydrophilic membranes exhibited good diffusivity for solutes with small MW and could be sterilized by the common procedures (e.g. EtO, γ-irradiation, steam). One of the problems of these membranes has been related to their biocompatibility properties due to the presence of the reactive hydroxyl groups in the polymer, which are responsible for complement activation. In order to improve the biocompatibility cellulose has been modified by substituting hydroxyl groups with N,N,-diethylaminoethyl (DEAE) [27]. These DEAE groups create hydrophobic regions in hydrophilic chain and sterically hinder the interaction with complement factors, thus improving membrane biocompatibility. This membrane has been commercialized under the name Hemophan® by Membrana. A similar approach consisted in replacing less than 1 % of hydroxyl groups with benzyl groups [28]. The membrane biocompatibility of cellulose was also improved by grafting the cellulose backbone with polyethylene glycol (PEG), which creates a layer between the membrane surface and blood hindering the contact with plasma proteins that lead to complement activation (AM-BIO, Asahi) [29]. With the attempt to enhance the biocompatibility of the cellulose membranes a coating with vitamin E has been used. The most common method to modify cellulose was to substitute the hydroxyl groups with acetyl groups (cellulose acetate) and on the basis of the number of hydroxyl groups substituted with acetyl groups the polymer becomes cellulose diacetate or triacetate. These membranes are produced by Gambro, Toyobo, Baxter, Tejin, and Helbio.

Most of the synthetic polymers used to prepare membranes are hydrophobic in nature (except EVAL) and have to be made more hydrophilic during their production process. PAN is made hydrophilic by copolymerization with sodium methallyl sulfonate (AN69® Hospal) and with methylmethacrylate and the addition of acrylic acid (PA, Asahi) [30].

These membranes have high hydraulic permeabilities and are therefore suitable for high-flux dialysis, hemodiafiltration, and hemofiltration.

In order to overcome problems with anaphylactic reactions, PEI is sprayed onto the membrane to mask the negatively charged sulfone groups of the AN69®.

PMMA membranes are modified by copolymerization with small amounts of p-styrene sodium sulfonate. These membranes are commercialized by Toray [31]. In the polyamide membranes used in hemofilters (Gambro), PA is blended with PVP to make the membrane more hydrophilic. Membranes of polysulfone/poly(aryl)ethersulfone family meet most of the requirements made on a membrane. Polysulfone membranes can be sterilized by all the common methods, are biocompatible, and are of high physical strength and chemical resistance. Low- and high-flow versions exhibit good performance characteristics. Most of various polysulfone membranes currently available on the market differ slightly in their basic polymer. This membrane family is manufactured by Fresenius, Asahi Medical, Toray Industries, Saxonia, Membrana, Hospal, Gambro, Minntech, Nikkisio, and Kimal [13].

3.6 Plasmapheresis

Membrane plasmapheresis exchange therapy (from the Greek "apheresis" – to suppress) is a simple method, without side effects, for healing or achieving homeostasis of the body. The procedure consists of separating the blood from the cellular elements (red cells, leukocytes, platelets, etc.) from the plasma, the fluid part of the blood which stores degraded metabolic components, chemical components, drug toxins, environmental components – heavy metals, for example – and immunological components characteristic of humoral immunity and of its autoimmune informational content. In summary, all the fractions that cause the disease.

Membrane-based plasma therapy is an important treatment for attending to some immunologic and dismetabolic diseases due to the accumulation of macromolecules such as IgG, IgM, cryoglobulins immune complexes, and LDL cholesterol [16]. Patients with polyneuropathy, neuropathy, systemic glomerulonephritis, and familial hypercholesterolemia are treated with plasma filters for the removal of specific factors. In membrane plasmapheresis, blood is pumped through a plasma filter which consists of a membrane and ultrafiltration solution. The plasma filter separates blood into a plasma and a cellular component. The cellular component is reinserted into the blood stream along with the plasma component. Membrane plasmapheresis processes work in much the same way as the hemodialysis process. Blood is generally passed through a number of membranes: usually one but possibly two in more selective applications. Secondary plasma filtration allows for a more selective elimination of dangerous macromolecules while retaining important components to be recombined with plasma. The primary membrane is used to separate the cellular and plasma components. The secondary filter is then used to separate the plasma component into two solutions, one which is discarded, and another which is recombined, which are selectively re-added to the cellular component.

Once plasma is separated from the cellular components, it is easy to remove certain toxic substances. It is possible to selectively substitute molecules at the level of a few million Daltons molecular weight, e.g. immune complexes, or a few hundred thousand Daltons molecular weight, e.g. gammaglobulin fractions, or below 100 000 Da molecular weight, e.g. albumin fractions, by using a cascade of membrane technology [33] (Figure 3.18).

Thus, this procedure involves the filtration of plasma through membrane filters with different pore size to remove toxins and retain clotting factors and proteins. Smaller pore size plasmafilters are used to remove molecules with a molecular weight ≥ 180 kDa, such as IgG, immune complexes, and cryoglobulins. Larger pore filters are

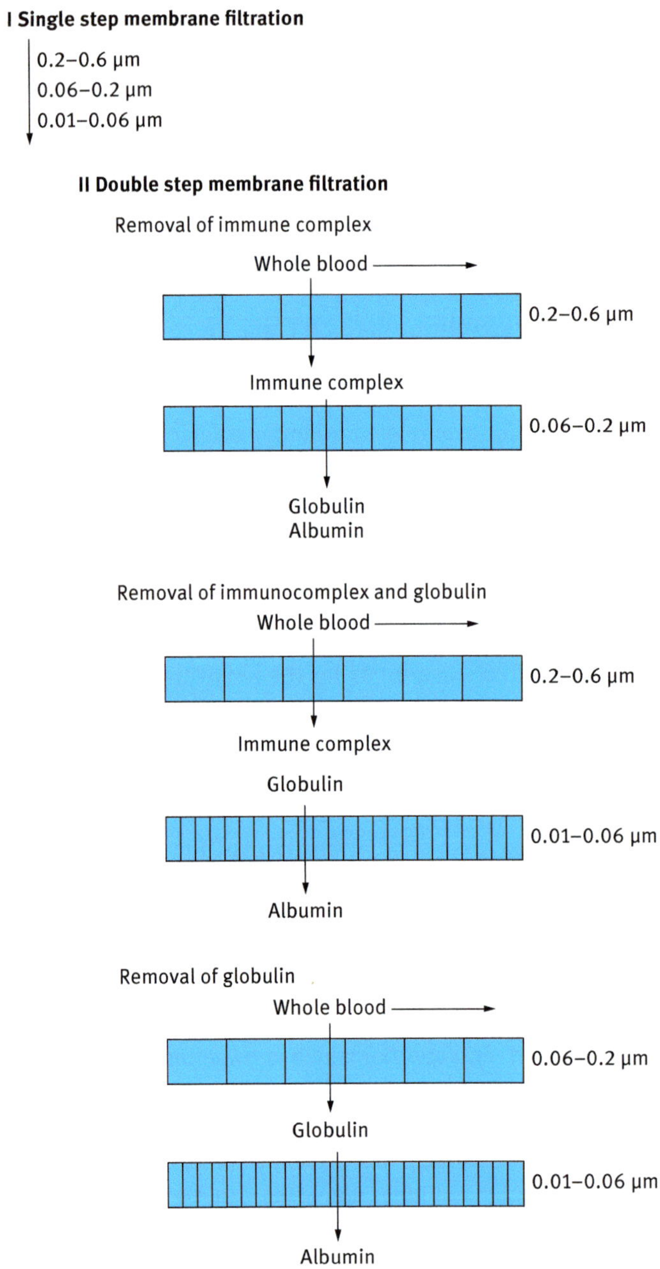

Fig. 3.18: Cascade of membrane technology for selective removal of macromolecules (adapted from [33]).

able to remove macromolecules with molecular weight ≥900 kDa, such as IgM and LDL-cholesterol. In plasmapheresis microfiltration membranes are used (e.g. polysulfone, polypropylene, cellulose diacetate) with pore size ranging from 0.1 to 0.7 μm and membrane surface area 0.2–0.5 m^2. The membranes are available in hollow fiber and flat configurations and are assembled in devices produced by several companies (Table 3.2).

Tab. 3.2: Commercial membrane plasma filters (adapted from [32]).

Membrane type	Pore size [μm]	Surface area [m^2]	Filter
cellulose diacetate (hollow fiber)	0.20 0.20	0.50 0.65	Plasmaflo-AP05H (Asahi Medical, Tokyo, Japan) Plasmaflo-AP05 (Asahi Medical)
polypropylene (hollow fiber)	0.55 0.20 0.65	0.17 0.50 0.12	Fenwal CPS-10 (Baxter, Deerfield, IL, USA) Hemaplex-BT900 (Sorin Group Italia srl) Curesis M82 (Organon Teknika, Durham, NC, USA)
polypropylene (flat plate)	0.60	0.13	Centry TPE (Cobe, Lakewood, CO, USA)
polysulfone (hollow fiber)	0.20	0.50	Sulflux-FS (Kaneka, New York, NY, USA)
polymethylmeth-acrylate (hollow fiber)	0.50	0.50	Plasmax-PS05 (Toray, Tokyo, Japan)

The efficacy of plasmafilters in removing toxins is comparable with that of plasma exchange, but the requirement for albumin is greatly reduced. This procedure is used in patients with Guillain–Barre syndrome, chronic idiopathic demyelinating polyneuropathy. In this case the use of plasmafilters permits an effective removal of pathogenic high molecular weight substances.

The plasma treatment with sorbents is an alternative technique used for the removal of toxins from patients' blood. The application of this technique consists of the separation of plasma from the whole blood by filtration through microporous membranes, the removal of toxins from the plasma through sorption, and the reinfusion of the purified plasma into the blood through filters that prevent the passage of the sorbent. Multisorbent systems contain anion exchange resin and activated charcoal for the removal of middle molecular weight solutes and molecules.

Other therapeutic apheresis in which membrane devices are also used include erythrocytapheresis, which consists of the separation of erythrocytes from whole blood; the plateletpheresis (thrombopheresis, thrombocytapheresis), consisting of the collection of platelets by apheresis, while returning the red blood cells, white blood cells and plasma component; leukapheresis consisting of the removal of peripheral monocytes, basophils, eosinophils for transfusion into patients whose peripheral monocytes are ineffective or where traditional therapy has failed.

The primary physiologic effects of therapeutic apheresis include citrate anticoagulation, fluid balance, plasma dilution, and a potential for cellular loss. The most common side effects include vasovagal reaction, citrate toxicity (hypocalcemia), and transfusion reactions.

3.7 Artificial lung

3.7.1 Lung functions

Lungs are a pair of spongy, air-filled organs located on either side of the chest (thorax). The trachea (windpipe) conducts inhaled air into the lungs through its tubular branches, called bronchi. The bronchi then divide into smaller and smaller branches (bronchioles), finally becoming microscopic.

The bronchioles eventually end in clusters of microscopic air sacs called alveoli where the gas exchange of carbon dioxide and oxygen takes place.

In the alveoli, oxygen from the air is absorbed into the blood. Carbon dioxide, a waste product of metabolism, travels from the blood to the alveoli, where it can be exhaled. Between the alveoli is a thin layer of cells called the interstitium, which contains blood vessels and cells that help support the alveoli.

Each alveol can be compared to a small sac delimited by a very thin layer of tissue surrounded by blood capillaries. The thin layer of tissue and the capillaries form the respiratory membrane which is responsible for gas exchange: the oxygen passes from the lungs to the blood and carbon dioxide from the blood to the lungs. The oxygen and carbon dioxide concentrations are different at both sides of the respiratory membrane: the oxygen concentration is higher in the alveoli, whereas the carbon dioxide concentration is higher on the blood side (Figure 3.19). The transport of both gases is due to the diffusive mechanism: the oxygen from the alveoli to the blood red cells and the carbon dioxide from the blood red cells to the alveoli. The thin thickness of the respiratory membrane (0.1 micron) ensures a fast diffusion of oxygen and carbon dioxide. The carbon anhydrase enzyme localized in the endothelium of pulmonary capillaries facilitates the removal of carbon dioxide from the blood. This enzyme catalyses the rapid conversion of carbon dioxide to bicarbonate and protons.

In patients with impaired lung functions a membrane oxygenator is able to replace the lung functions and to support the patient for a limited time as bridge.

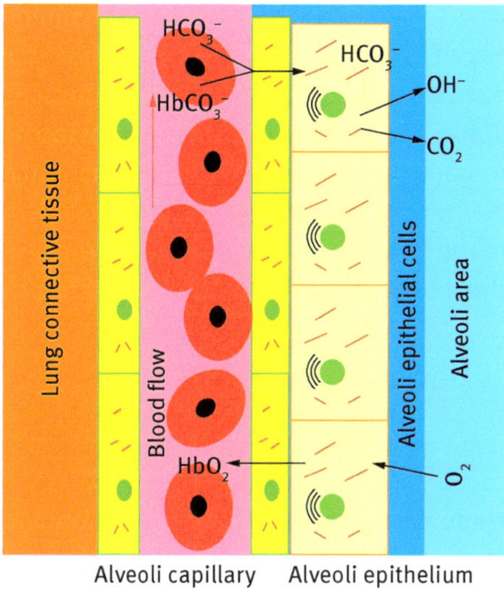

Fig. 3.19: Transport of oxygen and carbon dioxide between the alveolar epithelium and capillaries.

3.7.2 Membrane oxygenator

Membranes are used in blood oxygenation for cardiopulmonary bypass and/or respiratory insufficiency by substituting the lung function for oxygen supply to the blood and removal of CO_2 from the blood. Today the extracorporeal membrane oxygenation system is an accepted technique for temporary lung support.

Acute respiratory failure occurs with an incidence of 80–90 cases/100 000 inhabitants per year [34, 35]. The more severe forms of acute respiratory failure, such as acute lung injury and acute respiratory distress syndrome (ARDS), occur with incidences of 20–70 [36] and 1.5–13.5/100 000 inhabitants per year [36, 37, 39], respectively. Patients with ARDS are treated with advanced methods, which include, but are not limited to, various forms of mechanical ventilation with positive end-expiratory pressure (PEEP) and permissive hypercapnia, positional manoeuvres, sophisticated fluid regimens, and inhalational pulmonary vasodilators [40–42]. However, a number of ARDS patients whose pulmonary gas exchange cannot be sufficiently improved by the above mentioned methods, and extracorporeal membrane oxygenation (ECMO) represents an additional therapeutic option during the acute phase. Furthermore, ECMO is widely used in heart surgery. ECMO is most commonly used to support mature newborn infants with acute respiratory failure. In neonates, quite promising survival rates of 56 % have been observed since ECMO was introduced, whereas in adults comparable good survival rates have not been shown. Bartlett et al. [43] attributed the success of ECMO in newborns to the fact that in neonatal respiratory failure the lungs require only a short time for recovery, and extracorporeal techniques in the 1970s could be carried out safely for a few days. In the 1980s, the technology was adapted to pediatric and, later on, in the 1990s, to adult acute respiratory failure and cardiac failure.

In the neonate, persistent pulmonary hypertension (PPHN) is the most frequent cause of acute respiratory failure. The etiology of PPHN can either be idiopathic or a secondary consequence of a variety of diseases that include, but are not limited to, sepsis, pneumonia, meconium aspiration, diaphragmatic hernia, and hyaline membrane disease. PPHN is pathophysiologically characterized by the distinct pulmonary hypertension, pronounced hypoxia, and the extrapulmonary monary right-to-left shunt across the patent ductus arteriosus and the foramen oval. The course of the disease may be complicated by the effects of direct lung damage, leading to additional intrapulmonary right-to-left shunt.

3.7.3 History of ECMO

The first device to oxygenate blood extracorporeally for perfusion of isolated organs was developed in 1885 by von Frey and Gruber [44]. In this device gas exchange was achieved by conducting a continuous flow of oxygen through an inclined rotating cylinder with the inner surface covered with a thin film of blood (Figure 3.20).

In 1937 Gibbon [45] developed the heart-lung machine in order to allow open heart surgery. In this system anticoagulated blood was directly exposed to oxygen ("film" or "bubble oxygenators"), but the use of this device was limited to a few hours [46] due to severe hemolysis, thrombocytopenia, hemorrhage, and organ failure caused by the direct contact between blood and the gaseous phase (Figure 3.21).

In 1956, Clowes and Neville [47] developed the first membrane artificial lung that separated the gaseous from the liquid phase by a membrane. They used a teflon membrane with an area of 25 m^2, but it was bulky and difficult to sterilize. In 1964 a new disposable oxygenator using silicon-rubber membrane was developed, and only in 1963 hollow-fiber membrane oxygenators were proposed. The membrane oxygenator

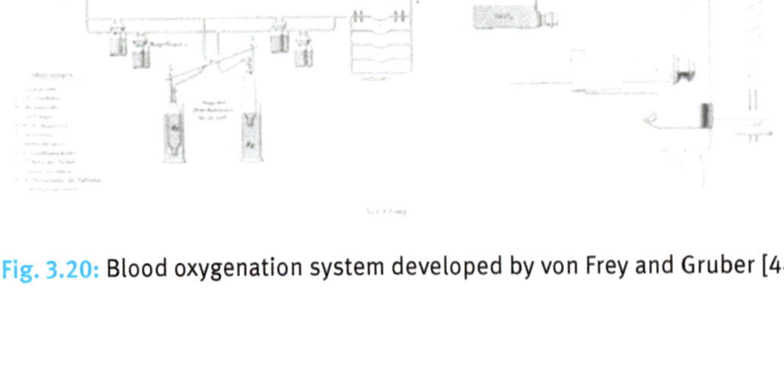

Fig. 3.20: Blood oxygenation system developed by von Frey and Gruber [44].

Fig. 3.21: Heart-lung machine used by John Gibbon successfully on an 18-year-old patient [45].

provided faster and more efficient blood oxygenation with fewer complications than the film or bubble oxygenators, and were practical for cardiopulmonary bypass that lasted longer than a few hours [48]. In 1972, clinical application of ECMO in respiratory failure of newborns and adults was attempted. In that year, Hill et al. [49] reported the survival of a 24 year old patient with ARDS who had been treated with ECMO during the acute phase of the disease, and four years later, Bartlett et al. [43] reported the first newborn treated with ECMO who survived.

Bleeding due to complete anticoagulation has been reported as the major complication during extracorporeal respiratory support [50, 51]. In 1983, Larm et al. [52] developed a technique in which the heparin molecule is covalently attached to synthetic surfaces, which allowed heparinization of all surfaces of the extracorporeal gas exchange device that come into contact with blood. In 1987, Bindslev et al. [53] reported the first long-term application of a surface-heparinized extracorporeal circuit in a 44-year-old woman with severe ARDS. Since that time, nearly all European ECMO centers have switched to the surface-heparinized extracorporeal circulation technique, with no or minimal systemic heparinization [41, 54].

3.7.4 Extracorporeal membrane oxygenation (ECMO) system

The term extracorporeal membrane oxygenation (ECMO) was initially used to describe long-term extracorporeal support that focused on the function of oxygenation. Subsequently, in some patients, the emphasis shifted to carbon dioxide removal, and the term extracorporeal carbon dioxide removal was coined. Extracorporeal support was later used for postoperative support in patients following cardiac surgery. Other variations of its capabilities have been tested and used over the last few years, making it an important tool in the armamentarium of life and organ support measures for clinicians. It works as an artificial heart and lung by providing propulsion pressure for blood flow through a circuit and an interface for oxygen and carbon dioxide exchange. ECMO can be categorized according to the organ supported (respiratory, cardiac, or cardiorespiratory), or according to the circuit used (veno-arterial and veno-venous). Veno-arterial ECMO provides both gas exchange and circulatory support, while veno-venous ECMO allows only gas exchange. With all of these uses for extracorporeal circuitry, a new term, extracorporeal life support (ECLS), has come into vogue to describe this technology. There are several differences between ECMO and cardiopulmonary bypass. ECMO is frequently instituted using only cervical cannulation, which can be performed under local anesthesia; standard cardiopulmonary bypass is usually instituted by transthoracic cannulation under general anesthesia. Unlike standard cardiopulmonary bypass, which is used for short-term support measured in hours, ECMO is used for longer-term support ranging from 3 to 10 days. The purpose of ECMO is to allow time for intrinsic recovery of the lungs and heart; a standard cardiopulmonary bypass provides support during various types of cardiac surgical procedures.

In the ECMO, the blood from a systemic vein (e.g. cava vein) flows by a pump to the membrane device for oxygenation, and the oxygenated blood comes back to the aorta branch in systemic circulation (Figure 3.22). In membrane devices for blood oxygenation, the membrane is used in order to have a suitably wide surface area (2–10 m^2) to gas exchange between the blood and the oxygen enriched stream. The membranes used in oxygenators are generally hydrophobic membranes, permeable only to gas and not to liquids (e.g. polypropylene, teflon, silicone). In this way, the resistance to

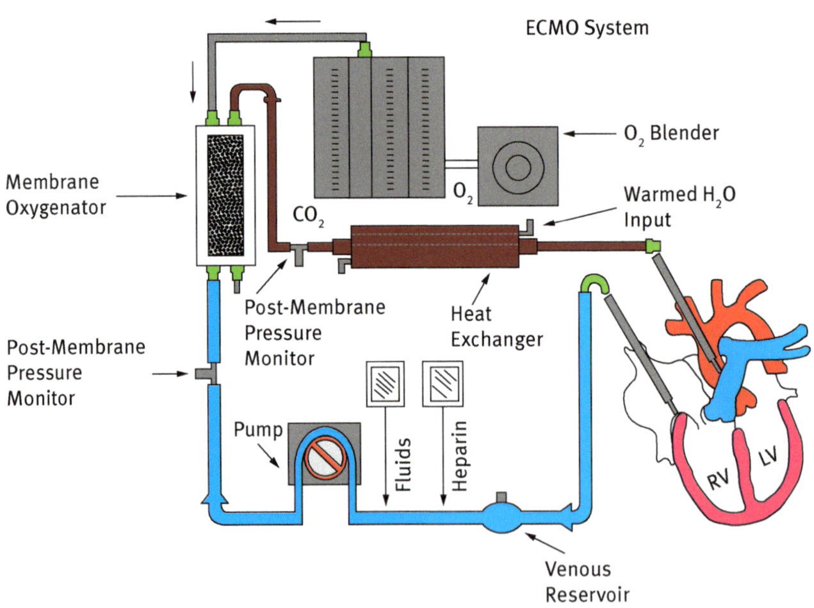

Fig. 3.22: Diagram of extracorporeal membrane oxygenation (www.perfusione.net).

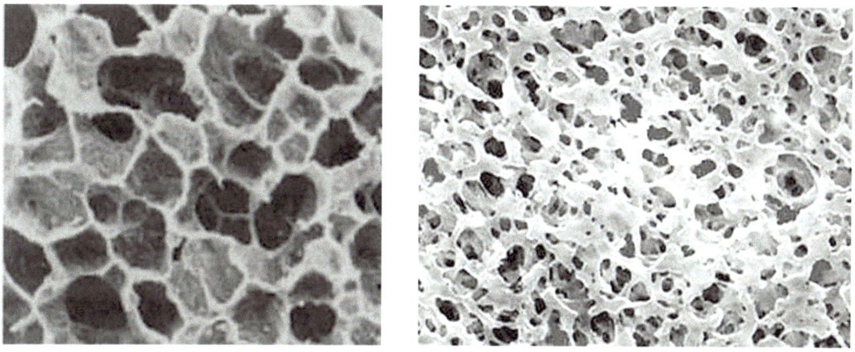

(a) (b)

Fig. 3.23: Comparison between alveoli (a) and synthetic membranes of polypropylene (b) used in membrane oxygenators.

gas transport through the membrane is reduced, and consequently also the surface of the membrane necessary to obtain transmembrane sufficient gas flux. The first type of membranes used in oxygenator was silicone dense membrane. The introduction of polypropylene porous membranes successfully permitted the reduction of the resistance to oxygen transport.

The performance of the membrane oxygenator closely approximates that of the natural lung. Synthetic and natural membranes are similar: both the saccular form of the alveoli and the very thin alveoli septa as well as the polypropylene membrane used in oxygenators for the massive reticulated surface deployed as fine, porous, open-cell foams (Figure 3.23).

In the natural lung the blood flows through fine capillaries in the alveoli and exchanges O_2 and CO_2 across the capillary endothelial cells and a very thin mucopolysaccaride hydrogel film. In the oxygenator the blood flows on the outside of a large fiber with diameter of approximately of 300 μm and exchanges respiratory gases across a meniscus that forms across the membrane pore at the interface of the blood and the gas. In the natural lung the gas exchange surface (50–100 m^2) is one order of magnitude greater than artificial lung; therefore to sature the blood a higher residence time in the device and the use of a gas current more enriched of oxygen [55] are necessary (Table 3.3).

First, red blood cells pass through pulmonary capillaries, making the distance for O_2 diffusion much shorter than in an artificial lung. The rate of oxygen transfer in natural lungs is not limited by diffusion, except in the case of severe lung disease or extreme exercise. Indeed, the difference between gas tensions measured in the natural alveoli and in the blood is mostly due to ventilation-perfusion mismatching. In contrast, in artificial lungs operating under normal conditions, a significant difference in partial pressures occurs between the gas and blood phases.

Tab. 3.3: Comparison of physical characteristics of a membrane artificial lung and natural lung (adapted from [56]).

Characteristic	Membrane lung	Natural lung
Surface area [m^2]	0.5–4	70
Blood path width [µm]	200	8
Blood path length [µm]	250 000	200
Membrane thickness [µm]	150	0.5
Maximum O_2 transfer [ml/min]	400–600	2000

A second disadvantage of the artificial lung is the significantly smaller surface area over which to exchange gases (typically less than 10 % of the natural lung's area). Current membrane lungs compensate for these shortcomings by increasing the blood path length (the distance that the blood travels past the gas exchange surface), thereby increasing the time available for blood exposure to the gas exchange surface. In addition, secondary flows are induced in artificial lungs to promote mixing and bring deoxygenated blood closer to the exchange surface. Artificial lungs can be ventilated with 100 % O_2 to maximize the driving pressure difference for O_2 diffusion, without the toxic effects that would occur in the natural lung, and the artificial lung can be ventilated with a high flow of fresh gas to keep the CO_2 fraction in the gas phase low.

A typical membrane oxygenator has a gas surface exchange of 1.7 m^2 and operates with a maximum blood flow rate of 800 ml/min, and it uses a priming volume of 260 ml. Both oxygen and carbon dioxide flow rates increase with increasing blood flow rate (Figure 3.24) approaching to saturation values at 800 ml/min.

3.7.5 Intravenous membrane oxygenator (IMO)

An intravenous membrane oxygenator (IMO) was developed at the University of Pittsburgh in order to provide temporary and portable respiratory support to military and civilian personnel whose lungs are acutely damaged and impaired [57]. This device consists of several hundred hollow fiber membranes (HFMs) manifolded to gas supply lines for O_2 delivery, CO_2 removal, and helium supply to a balloon integer located within the fiber bundle. The IMO device is several feet in length and consists of about 1000 hollow fiber membranes. Oxygen enters these membranes through an external tube, and flows through the fibers under vacuum pressure. Oxygen within each fiber diffuses through tiny pores in the fiber wall into the blood, and is exchanged for carbon dioxide, which diffuses into the fibers and exits the device through a second tube. A key element is a central balloon within the bundle of hollow fiber membranes that pulsates to move the fibers and mix blood (Figure 3.25). The inflation and deflation of the balloon draws blood directly across the fiber membranes and greatly facilitates the exchange of oxygen and carbon dioxide. Since the balloon essentially pumps the

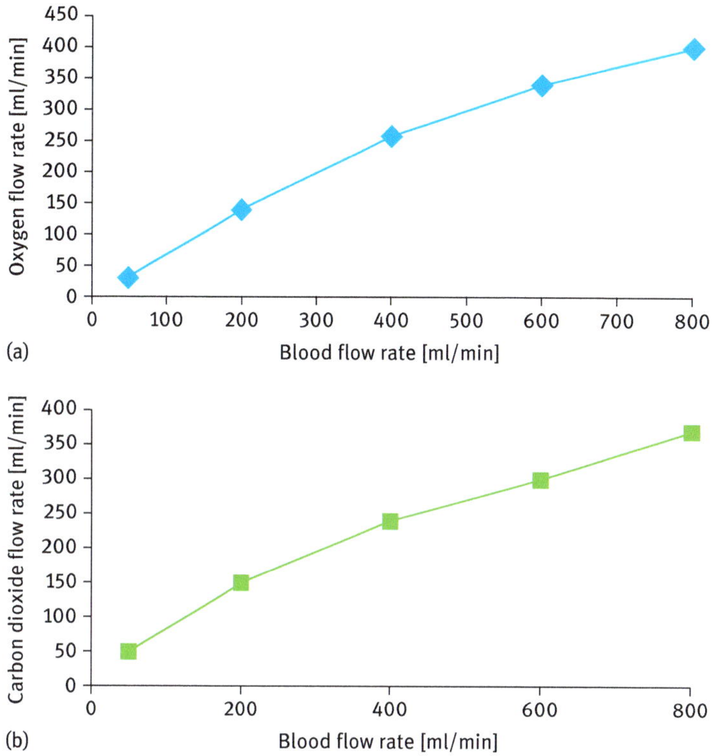

Fig. 3.24: Relationship between gas flow rate and blood flow rate: (a) oxygen and (b) carbon dioxide.

blood over the fibers, the device can be designed to offer little impediment to blood flow returning to the heart. Rapid pulsation of the balloon generates additional convective flow of blood across the HFMs and enhances the rate of O_2 delivery and CO_2 removal. The IMO prototypes exchanged O_2 and CO_2 at rates equal to or exceeding our design target for gas transfer per fiber surface area. This target is based on attaining 50 % of the normal baseline metabolic requirements for O_2 supply and CO_2 removal with an IMO device of 0.4–0.5 m^2 fiber surface area. The next phase of IMO development has already begun and involves scaling up to full-size IMO devices intended for human implantation [57].

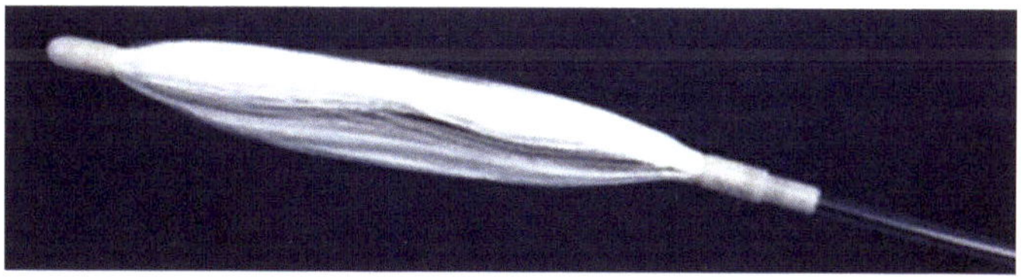

Fig. 3.25: IMO device (www.pitt.edu).

The IMO is used for patients who have a chance to reverse their respiratory problems. Because the prolonged use of respirators could eventually cause damage to the patient's lungs over a period of days or weeks, the artificial lung may eventually replace existing mechanical ventilation in patients with respiratory failure.

3.7.6 Gas transfer in a membrane oxygenator

In an oxygenator, gas transfer from the gas to the liquid phase (or the opposite direction) is driven by diffusion according to the partial pressure difference of the gas on the basis of Fick's law. The rate of diffusion is proportional to the partial pressure gradient of the gas in the direction of diffusion (i.e. the change in the partial pressure of the gas per unit distance). Mathematically, the rate of diffusion per unit area, $J\delta$, along the diffusion path can be described as

$$J = -D\frac{\Delta P}{\delta}, \tag{3.11}$$

where D is the diffusivity constant (a characteristic of the material and gas), and P is the partial pressure of the gas at any particular location, δ. [58].

Several inferences can be made from this relationship. The rate of total gas transfer can be increased by increasing the partial pressure difference (represented by ΔP) or the surface area available for diffusion. The rate of gas transfer can also be increased by decreasing the distance δ through which the gas must diffuse.

The diffusivity, D, is constant for a particular gas and diffusion barrier material at a constant temperature. Kinetic theory dictates that the diffusivity is related to the molecular speed of the gas molecules and, according to Graham's law, is inversely proportional to the square root of the molecular weight of the gas. The diffusivity of a gas is related to the solubility of the gas, because increasing solubility enhances movement of the gas through a solid or liquid. This latter factor complicates the analysis of gas transfer [59, 60]. Gas diffusion in blood, particularly oxygen diffusion, is somewhat more complicated than predicted by Fick's law. In addition to simple diffusion through the blood plasma, the absorption of oxygen by red blood cells must be considered. It is not difficult to imagine that the O_2 concentration within a volume of blood will increase as the exposure time of blood to either a bubble or membrane interface increases.

The gas exchange of the oxygenator can rapidly be determined using either blood or gas phase measurements. In the blood phase, gas transfer can be calculated by application of conservation of mass (Fick's principle). This assumes that the transfer of O_2 across the membrane or bubble interface causes the difference between the oxygen content flowing into and emerging from the oxygenator. Thus, the oxygen content of the arterial and venous perfusion tubes is determined by continuous in-line monitors or blood gas determinations, and the difference is multiplied by the flow rate of blood (pump flow rate) as shown in equation (3.12):

$$V_{O_2} = Q\,(c_a - c_v). \qquad (3.12)$$

where V_{O_2} is the oxygen transfer rate, Q is the blood flow rate, and c is the oxygen content, with a and v representing the arterial and venous values, respectively.

3.7.7 Membrane configuration

Membrane lungs attempt to achieve separation between the blood and gas in a manner analogous to the natural lung. Membrane lungs provide a complete barrier between the gas and blood phases, so that gas transfer depends totally on the diffusion of gas through the membrane material. Most membrane lungs today have micropores, as for example the Kolobow spiral coil membrane lung [61] (Avecor Cardiovascular, Inc., Minneapolis, MN). This lung is primarily used in extracorporeal membrane oxygenation because of its ability to maintain stable CO_2 and O_2 for long periods of time (weeks) without the decrease in gas transfer that is commonly seen with microporous membrane lungs. The Kolobow membrane lung consists of a silicone membrane in the shape of an envelope that is coiled on it. Blood flows through the integral heat exchanger and then past the membrane (Figure 3.26 (a)). This oxygenator is available in gas exchange surface area in the range of 0.5–4.5 m^2 to provide extracorporeal membrane oxygenation to patients ranging from neonates with congenital lung disease to adults with adult respiratory distress syndrome.

Microporous membranes allow at least transient direct blood-gas interfacing at the initiation of cardiopulmonary bypass. After a short time, a protein coating of the membrane and gas interface takes place, and no further direct blood and gas contact exists. Typically, the surface tension of the blood prevents large amounts of fluid from traversing the small micropores during cardiopulmonary bypass. The micropores provide conduits through the polypropylene membrane that give sufficient diffusion capability to the membrane for both oxygen and carbon dioxide exchange.

However, over several hours of use, the functional capacity of microporous membrane oxygenators decreases because of evaporation and subsequent condensation of serum that leaks through the micropores [62–64]. It has been suggested that this transfer may be reduced by heating the membrane lung and the gas entering it [62], the premise being that the condensation within the fibers can be minimized by maintaining the ventilating gas temperature above blood temperature to minimize condensation. Although currently unconfirmed, the initial results of this approach appear promising. Blood surface tension prevents gas leakage into the blood (provided excessive gas compartment pressures do not occur).

Two primary membrane configurations are currently being used: the hollow fiber design originally described by Bodell et al. [65] and the folded envelope design. Hollow fiber membranes of polypropylene with micropores less than 1 µm are required to inhibit both gas and serum leakage across the membrane. The widespread current use of membrane lungs depended on the development of the microporous membrane. Before this innovation, available materials with the necessary structural integrity (teflon, cellulose) were incapable of sufficient gas exchange without excess surface areas. The microporous membrane provides the necessary gas transfer capability via the micropores, where there is a direct blood-gas interface with minimal resistance to diffusion.

Two basic types of hollow fiber membrane lungs have been made (Figure 3.24 (b) and (c)): those with the blood phase on the inside or on the outside of the fibers. In the former, blood flows within the lumen of a bundle of parallel connected hollow fibers typically 200 µm internal diameter and 25 nm wall thickness. Gas flows in the fiber shell counter current to the blood (Figure 3.24 (b)). The blood side transfer resistance is high, owing to the laminar rectilinear blood flow. This could be reduced by decreasing the diffusion path length when using smaller fiber bores. However, it is impractical to reduce the inner diameter, d, below the present level of 200 µm for several reasons, such as the hydrostatic pressure drop over the bundle and the potential for fiber occlusion by cellular aggregates. Decreased oxygenator function from thrombosis within the fibers has occurred with the former design [66]. However, satisfactory clinical performance continues to be achieved with both blood flow patterns. For oxygenators with blood flow outside the fiber, the blood flows either perpendicular

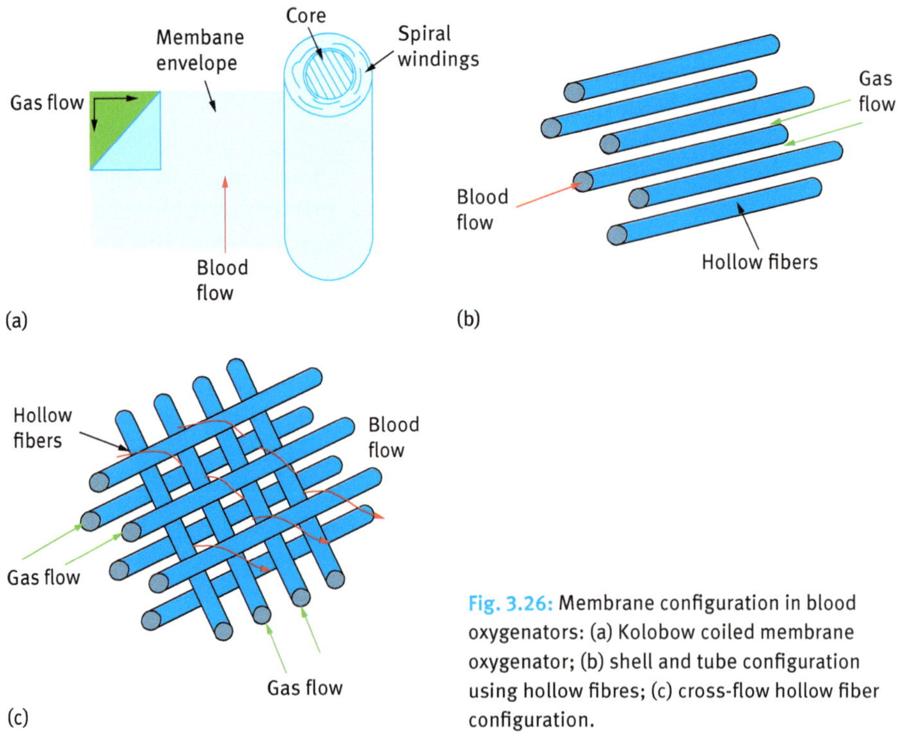

Fig. 3.26: Membrane configuration in blood oxygenators: (a) Kolobow coiled membrane oxygenator; (b) shell and tube configuration using hollow fibres; (c) cross-flow hollow fiber configuration.

to the fiber bundle (cross-current) or in the direction of the fibers (Figure 3.24 (c)). In the latter case, blood usually flows in the opposite direction to the gas flow (countercurrent). Cross-current blood flow offers the advantage of naturally induced secondary flow generation. The fibers tend to "trip" the flow, inducing eddies downstream of each fiber. This flow alteration reduces the diffusion boundary layer of downstream fibers, thereby enhancing gas exchange. Another advantage to routing blood outside the hollow fibers is reduced pressure drop across the blood compartment. The generation and strength of the secondary flows is dependent on the blood velocity and viscosity, the fiber external diameter, and the inter fiber spacing. For similar O_2 transfer rates, the membrane area required for a cross-flow design is about 2–2.5 times smaller than that for a shell and tube design.

In both parallel plate/screen and cross-flow configurations, the primary blood flow generates convective mixing in combination with the flow path geometry. The gas transfer may be greatly enhanced by active mixing units, in which an additional external energy input is required, e.g. the rotation of a membrane-lined disc [67]. Although many active mixing oxygenators have been tested, very few have reached clinical application because of their greater complexity, higher production cost, and the need for ancillary drives. The pulsed vortex design was marketed as the interpulse membrane oxygenator [68]. In this oxygenator each blood channel is composed of two opposing sheets of microporous membrane supported over plastic plates to form longitudinal furrows perpendicular to the blood flow direction. As the blood is pulsed across the membrane, vortices are cyclically formed within and ejected from the furrows. This method of mixing is highly effective and reduces by a factor of six the membrane area needed by a shell and tube unit.

Nowadays membrane oxygenators are extensively used in North America and Western Europe because of their suitability for long-term perfusion in cardiac surgery, although they present some problems concerning red cell damage and granulocyte and platelet activation. However, because technological advances have reduced the cost difference between membrane and bubble oxygenators, routine use of membrane oxygenators has become the norm in most developed countries. Benefits may be realized from using membrane oxygenators in long perfusions in adults, and possibly in all perfusions in infants and small children.

3.8 References

[1] Drioli E, De Bartolo L. Membrane bioreactor for cell tissues and organoids. Artificial Organs. 2006; 30: 793–802.

[2] De Bartolo L, Drioli E. Membranes in artificial organs In: Haris PI, Chapman D (eds.). New Biomedical Materials – Basic and Applied Studies. Biomedical and Health Research vol. 16. Amsterdam/Berlin/Tokjo/Washington: IOS Press; 1998. pp. 167–181.

[3] Fresenius Medical Care. ESRD Patients in 2011 – A Global Perspective. 2011; 1–10.

[4] Bonomini M, Fiederling B, Bucciarelli T, Manfrini V, Di Ilio C, Albertazzi A. A new polymethyl-methacrylate membrane for hemodialysis. Int J Artif Organs. 1996; 19: 232–239.
[5] Graham T. The Bakerian lecture: Osmotic force. Philos Trans R Soc Lond. 1854; 144: 117–128.
[6] Abel J, Roundtree L, Turner B. On the removal of diffusible substances from the circulating blood of living animals by dialysis. J Pharmacol Exp Ther. 1914; 5: 275–316.
[7] Benedum J. Pioneer of dialysis, George Haas (1886–1971). Med Hist. 1979; 14: 196–217.
[8] Kolf, WJ, Berk HTJ. The artificial kidney: A dialyzer with a great area. Acta Med Scand. 1944; 117: 121–134.
[9] Stewart RD, Lipps BJ, Baretta ED, Piering WR, Roth DA, Sargent JA. Short-term hemodialysis with the capillary kidney. Trans Am Soc Artif Intern Organs. 1968; 14: 121–125.
[10] Alwall N. On the artificial kidney. I. Apparatus for dialysis of blood in vivo. Acta Med Scand. 1947; 128: 317–321.
[11] Ronco C, Ghezzi PM, Brendolan A, Crepaldi C, La Greca G. The haemodialysis system: basic mechanisms of water and solute transport in extracorporeal renal replacement therapies. Nephrol Dial Transplant. 1998; 13: 3–9
[12] Ronco C, Canaud B, Alijama P. Hemodiafiltration Contribution to Nephrology, vol. 158. Basel: Karger AG; 2007.
[13] Grassmann A, Uhlenbusch-Körwer I, Bonnie-Schorn E, Vienken, J. Understanding membranes and dialysers. Lengerich, Berlin: PABST Science Publishers; 2004.
[14] Depner TA. Hemodialysis adequacy: Basic essentials and practical points for the nephrologist in training. Hemodialysis International. 2005; 9: 241–254.
[15] Michaels AS. Operating parameters and performance criteria for hemodialyzers and other membrane-separation devices. Trans Am Soc Artif Intern Organs. 1966; 12: 387–392.
[16] Yeun JY, Depner TA. Principles of hemodialysis. In: Pereira BJ, Sayegh MH, Blake P (eds.). Chronic Kidney Disease, Dialysis, and Transplantation. Philadelphia: Saunders; 2005. pp. 307–340.
[17] Cheung AK, Leypoldt JK. The hemodialysis membranes: a historical perspective, current state and future prospect. Sem Nephrol. 1997; 17: 196–213.
[18] Leypoldt JK, Cheung AK, Agodoa LY, Daugirdas JT, Greene T, Keshaviah PR. Hemodialyzer mass transfer–area coefficients for urea increase at high dialysate flow rates. Kidney Int. 1997; 51: 2013–2017.
[19] Collins AJ. High-flux, high-efficiency procedures. In: Henrich Norwalk W (ed.). Principles and Practice of Hemodialysis, CT: Appleton & Large; 1996. pp. 76–88.
[20] Ambalavanan S, Rabetoy G, Cheung A. High efficiency and high flux hemodialysis. In: Schrier RW (ed.). Atlas of Diseases of the Kidney. Philadelphia: Current Medicine; 1999. Vol. 5, pp. 3.1–3.10.
[21] Forni LG, Hilton PJ. Continuous hemofiltration in the treatment of acute renal failure. The New England Journal of Medicine. 2003; 336: 1303–1309.
[22] Price CA. Continuous renal replacement therapy: the treatment of choice for acute renal failure. Anna J. 1991; 18: 239–244.
[23] Kierdorf H. Continuous versus intermittent treatment: clinical results in acute renal failure. Contrib Nephrol. 1991; 93: 1–12.
[24] Clark WR, Mueller BA, Alaka KJ, Macias WL. A comparison of metabolic control by continuous and intermittent therapies in acute renal failure. J Am Soc Nephrol. 1994; 4: 1413–1420.
[25] Ronco C. Continuous renal replacement therapies for the treatment of acute renal failure in intensive care patients. Clin Nephrol. 1993; 40: 187–198.
[26] Leber H-W, Wizemann V, Goubeaud G, Rawer P, Schutterle G. Hemodiafiltration: a new alternative to hemofiltration and conventional hemodialysis. Artif Organs. 1978; 2: 150–153.

[27] Akzo Faser AG. Hemophan, a DEAE modified cellulose for dialysis. US patent no. 4,668,396, European patent no. 172,497
[28] Bowry SK, Rintelen TH. Synthetically modified cellulose (SMC): A cellulosic hemodialysis membrane with minimized complement activation. ASAIO J. 1998; 44: M579–M583.
[29] Hoenich N, Woffindin C, Stamp S, Roberts SJ, Turnbull J. Synthetically modified cellulose: an alternative to synthetic membranes for use in Haemodialysis? Biomaterials. 1997; 18: 1299–1303.
[30] Gohl H, Buck R, Strathmann H. Basic features of the polyamide membranes. In: Berlyne GM, Giovannetti S (eds.). Polyamide – The evolution of a synthetic membrane for renal therapy. Basel: Karger, vol. 96; 1992. pp. 1–25.
[31] Sugaya H, Sakai Y. Polymethylmethacrylate: from polymer to dialyzer. In: Ronco C . Polymethylmethacrylate. A flexible membrane for a tailored dialysis, Contrib. Nephrol. 125. Basel: Karger; 1998. pp. 1–8.
[32] Siami G, Siami F. Membrane Plasmapheresis in the United States: A Review Over the Last 20 Years. Therapeutic Apheresis. 2001; 5: 315–20
[33] Nose Y, Malchesky PS. Therapeutic Membrane Plasmapheresis. Therapeutic Apheresis and Dialysis. 2000; 4: 3–9.
[34] Lewandowski K, Metz J, Deutschmann C, Preiss H, Kuhlen R, Artigas A, Falke KJ. Incidence, severity, and mortality of acute respiratory failure in Berlin, Germany. Am J Respir Crit Care Med. 1995; 151: 1121–1125.
[35] Luhr OR, Antonsen K, Karlsson M, Aardal S, Thorsteinsson A, Frostell CG, Bonde J. Incidence and mortality after acute respiratory failure and acute respiratory distress syndrome in Sweden, Denmark, and Iceland. Am J RespirCrit Care Med. 1999; 159: 1849–1861.
[36] Luce JM. Acute lung injury and the acute respiratory distress syndrome. Crit Care Med. 1998; 26: 369–376.
[37] Villar J, Slutsky AS. The incidence of the adult respiratory distress syndrome. Am Rev Respir Dis. 1989; 140: 814–816.
[38] Nolan S, Burgess K, Hopper L, Braude S. Acute respiratory distress syndrome in a community hospital ICU. Intens Care Med. 1997; 23: 530–538.
[39] Webster NR, Cohen AT, Nunn JF. Adult respiratory distress syndrome: how many cases in the UK? Anaesthesia. 1988; 43: 923–926.
[40] Amato MBP, Barbas CSV, Medeiros DM, Magaldi RB, Schettino GP, Lorenzi-Filho G, Kairalla RA, Deheinzelin D, Munoz C, Oliveira R, Takagaki TY, Carvalho CRR. Effect of a protective-ventilation strategy on mortality in the acute respiratory distress syndrome. N Engl J Med. 1998; 338: 347–354.
[41] Lewandowski K, Rossaint R, Pappert D, Gerlach H, Slama, KJ,Weidemann H, Frey DJ, Hoffmann O, Keske U, Falke KJ. High survival rate in 122 ARDS patients managed according to a clinical algorithm including extracorporeal membrane oxygenation. Intens Care Med. 1997; 23: 819–835.
[42] Rossaint R, Falke KJ, Lopez F, Slama K, Pison U, Zapol WM. Inhaled nitric oxide for the adult respiratory distress syndrome. N Engl J Med. 1993; 328: 399–405.
[43] Bartlett RH, Gazzaniga AB, Jefferies MR, Huxtable RF, Haiduc NJ, Fong SW. Extracorporeal membrane oxygenation (ECMO) cardiopulmonary support in infancy. Trans Am Soc Artif Intern Organs. 1976; 22: 80–93.
[44] Von Frey M, Gruber M. Studies on metabolism of isolated organs. A respiration-apparatus for isolated organs Virchows. Arch Physiol. 1885; 9: 519–532.
[45] Gibbon JH. Artificial maintenance of circulation during experimental occlusion of pulmonary artery. Arch Surg. 1937: 34: 1105–1131.

[46] Lee LH, Krumhaar D, Fonkolsrud EW, Schjeide OA, Maloney JV. Denaturation of plasma proteins as a cause of morbidity and death after intracardiac operations. Surgery. 1961; 50: 29–37.
[47] Clowes GH, Hopkins AL, Neville WE. An artificial lung dependent upon diffusion of oxygen and carbon dioxide through plastic membranes. J Thorac Cardiovasc Surg. 1956; 32: 630–637.
[48] Kolobow T, Bowman RL. Construction and evaluation of an alveolar membrane artificial heart-lung. Trans Am Soc Artif Intern Organs. 1963; 9: 238–243.
[49] Hill JD, O'Brien TG, Murray JJ, Dontigny L, Bramson ML, Osborn JJ, Gerbode F. Prolonged extracorporeal oxygenation for acute post-traumatic respiratory failure (shock-lung syndrome). N Engl J Med. 1972; 286: 629–634.
[50] Zapol WM, Snider MT, Hill JD, Fallat RJ, Bartlett RH, Edmunds LH, Morris AH, Peirce II C, Thomas AN, Proctor HJ, Drinker PA, Pratt PC, Bagniewski A, Miller RG. Extracorporeal membrane oxygenation in severe acute respiratory failure. A randomized prospective study. JAMA. 1979; 242: 2193–2196.
[51] Gattinoni L, Pesenti A, Mascheroni D, Marcolin R, Fumagalli R, Rossi F, Iapichino G, Romagnoli G, Uziel L, Agostoni A, Kolobow T, Damia G. Low-frequency positive pressure ventilation with extracorporeal CO2-removal in severe acute respiratory failure. JAMA. 1986; 256: 881–886.
[52] Larm O, Larsson R, Olsson P. A new thrombogenic surface prepared by selective covalent binding of heparin via a modified reducing terminal residue. Biomater Med Device Artif Organs. 1983; 2: 161–173.
[53] Bindslev L, Eklund J, Norlander O, Swedenborg J, Olsson P, Nilsson E, Larm O, Gouda I, Malmberg A, Scholander E. Treatment of acute respiratory failure by extracorporeal carbon dioxide elimination performed with a surface heparinized artificial lung. Anesthesiology. 1987; 67: 117–120.
[54] Bindslev L, Böhm C, Jolin A, Hambraeus Jonzon K, Olsson P, Ryniak S. Extracorporeal carbon dioxide removal performed with surface-heparinized equipment in patients with ARDS. Acta Anaesthesiol Scand. 1991; 35: 125–131.
[55] De Bartolo L, Drioli E. Membranes and membrane reactors in artificial organs. In: Drioli E, Giorno L (eds.). Biocatalytic Membrane Reactors. London/Philadelphia: Taylor and Francis Group; 1998. pp. 191–207.
[56] Chauhan S, Subin S. Extracorporeal membrane oxygenation, an anesthesiologist's perspective: Physiology and principles. Part 1. Ann Card Anaesth. 2011; 14: 218–29
[57] Hattler BG, Reeder GD, Sawzik PJ, Walters FR, Pham SM, Kormos RL, Keenan RJ, Griffith BP, Armitage JM, Hardesty RL, et al. Development of an intravenous membrane oxygenator: a new concept in mechanical support for the failing lung. J Heart Lung Transplant. 1994; 13: 1003–1008.
[58] Bird RB, Stewart WE, Lightfoor EN. Transport phenomena. New York: John Wiley & Sons; 1960.
[59] Marx TI, Snyder WE, St. John AD, Moeller CE. Diffusion of oxygen into a film of whole blood. J Appl Physiol. 1960; 15: 1123–1129.
[60] Zapol WM, Qvist J. Artificial lungs for acute respiratory failure. New York: Academic Press; 1976.
[61] Kolobow T, Gattinoni L, Tomlinson T, Pierce JE: Control of artificial breathing using an extracorporeal membrane lung. Anesthesiology. 1977: 46: 138–141.
[62] Murphy W, Trudell LA. Friedman LI, Kakvan M, Richardson PD, Karlson K,. Galletti PM. Laboratory and clinical experience with a microporous membrane oxygenator. Trans Am Soc Artif Intern Organs. 1974; 20A: 278–285.
[63] Mottaghy K, Oedekoven H, Starmans H, Müller B, Kashef A, Hoffmann B, Böhm S. Technical aspects of plasma leakage prevention in microporous membrane oxygenators. Trans Am Soc Artif Intern Organs. 1989; 35: 640–643.

[64] Gille JP, Trudell L, Snider MT, Borsanyi AS, Galletti PM. Capability of the microporous membrane-lined, capillary oxygenator in hypercapnic dogs. Trans Am Soc Artif Intern Organs. 1970; 16: 365–371.
[65] Bodell BR, Head JM, Head LR, Formolo AJ, Head JR. A capillary membrane oxygenator. J Thorac Cardiovasc Surg. 1963; 46: 639–650.
[66] Dutton RC, Edmunds LH Jr. Formation of platelet aggregate emboli in a prototype hollow-fiber membrane oxygenator. J Biomed Mater Res. 1974; 8: 163.
[67] Gaylor JDS. Membrane oxygenators: current developments in design and application. J Biomed Eng. 1988; 10: 541–547.
[68] Spratt EH, Melrosc D, Bellhouse B, Badolato A, Thompson R. Evaluation of a membrane oxygenator for clinical cardiopulmonary bypass. Trans Amer Soc Artif Internal Organs. 1981; 27: 285–288.

4 Blood-membrane interactions

4.1 Thrombogenicity of membranes in contact with blood

One of the problems when devices such as hemodialyzers, oxygenators, or artificial valves contact blood is the thrombosis and thromboembolism [1]. When blood comes in contact with foreign substances by extracorporeal circulation, a reaction to foreign bodies takes place in the blood. Thus, to evaluate the safety of the material, a reaction to foreign bodies is used as an indicator of biocompatibility. A biocompatible membrane is a membrane with more affinity to a living body and which reacts less to foreign bodies.

Materials in contact with blood often initiate coagulation processes, affect platelet morphology and function, and cause an immunologic answer of the body. The first event of a complicated sequence that leads to thrombus formation and immunoreactions is protein adsorption, which plays a mediating role in bioaccumulation, systemic foreign body reactions, and tissue regeneration, and is therefore crucial for the biocompatibility and performance of medical devices [2]. The protein adsorption occurs in a few seconds; then platelets adhere within the first minute, followed by platelet aggregation that leads to blood coagulation and thrombus formation [3] (Figure 4.1).

Thus, the development of biomedical polymers has to comprise the detailed investigation of protein adsorption. According to the Vroman effect, proteins adhere with the sequence albumin, IgG, fibrinogen, fibronectin, factor XII, and high MW kininogen following an adsorption and desorption process in which the desorption is due to the arrival of proteins at higher affinity [4]. The adsorption of thrombin may activate the clotting cascade [5].

The protein adsorption is dependent on physico-chemical properties of both membranes and proteins, and thus the adsorbed protein layer in terms of composition and thickness affects the blood-materials interactions. As a consequence this layer supports the adhesion of platelets that undergo morphological changes, release factors and aggregate. The coagulation cascade activation can occur through the adsorption of proteins and release of platelet factors. In Figure 4.2 the interrelations among the biocompatibility parameters are shown.

There are two main mechanisms by which the blood clotting process is initiated by blood-surface interactions: platelet activation through protein adhesion and coagulation activation. Both these mechanisms depend on the properties of the substrate material.

DOI 10.1515/9783110268010-005

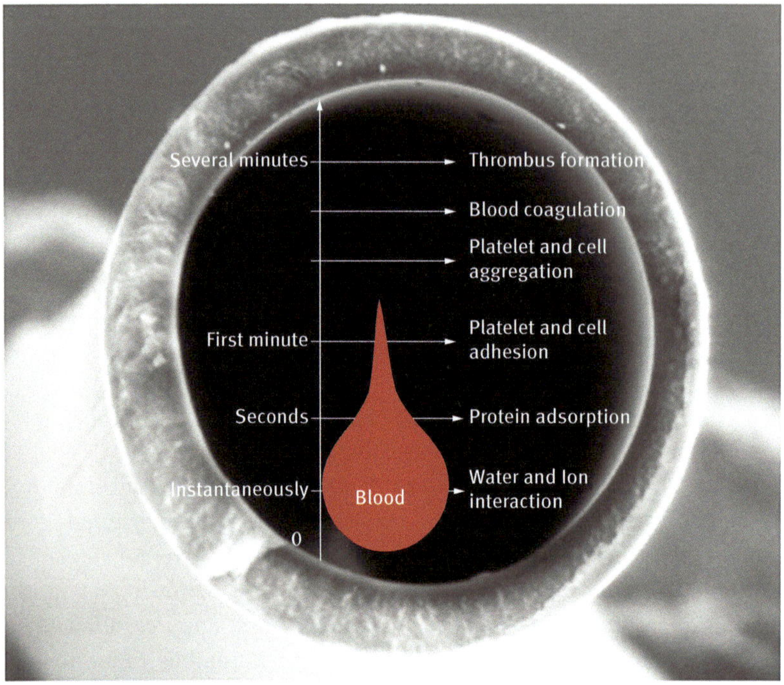

Fig. 4.1: Temporal sequence of thrombogenic events following membrane-blood interactions.

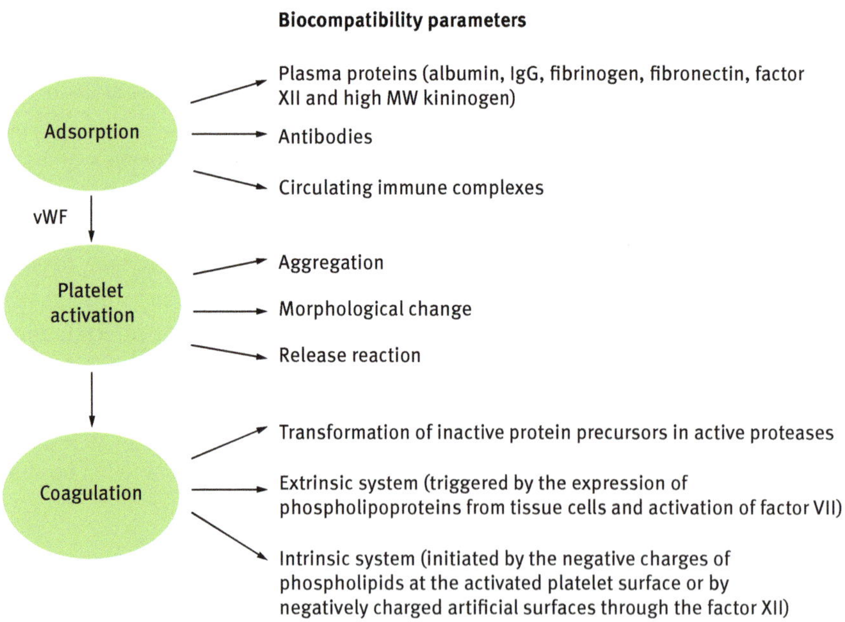

Fig. 4.2: Interrelations between the biocompatibility parameters.

4.2 Binding of platelets to the surface

The platelet interactions at the material interface play a major role in contributing to the thrombogenicity of foreign surfaces.

The surface chemistry (hydrophilicity, hydrophobicity, polarity, contact angle, critical surface tension) can have a profound impact on platelet reactivity. Platelets adhere to an artificial surface through interaction with surface adsorbed adhesion molecules like fibrinogen, von Willebrand factor, immunoglobulins, vitronectin, and fibronectin. When the adhesion receptors of platelets recognized the adsorbed proteins over the membrane, platelets undergo to the subsequent adhesion, spreading, and activation process.

For a wide range of surfaces the platelet adhesion is mediated by GPIIb/IIIa receptors. Once these receptors are activated they go through calcium-dependent conformational changes to express binding sites for the adhesive ligands such as fibrinogen, von Willebrand factor (VWF), fibronectin, and vitronectin, which support platelet aggregation and interactions with the subendothelium (Figure 4.2). Fibrinogen is the GPIIb/IIIa ligand involved in supporting platelet adhesion to artificial surfaces. Its adsorption has correlated well with the ability of many surfaces to promote platelet adhesion. In particular, fibrinogen has been shown to be the key protein in the adhesion process [6, 7], whereas fibronectin, VWF and immunoglobulins only show supporting effects that may be related to platelet activation. Furthermore, the amount of adsorbed fibrinogen required to facilitate platelet adhesion is extremely small, $\approx 7\,ng/cm^2$. Platelet adhesion to surface-bound fibrinogen is, however, intriguing, since the fibrinogen receptor integrin $\alpha_{IIb}\beta_3$ is required to be activated into a high affinity state by inside-out signalling to effectively bind fibrinogen in solution [8]. This mechanism is overridden as integrin $\alpha_{IIb}\beta_3$ in its "low affinity" state, evidently is capable of binding surface adsorbed fibrinogen. It is speculated that this is facilitated by a change in the conformation of fibrinogen when adsorbed to surfaces.

For the stable adhesion of platelets to the vessel wall or a growing thrombus *in vivo* are required integrins, which are noncovalent heterodimeric complexes of a transmembrane α- and β-chain. These are largely extracellular and contain short cytoplasmic tails [9]. The integrins need to be in an active conformation to bind ligand [10, 11]. Integrins on platelets, once activated, interact with several adhesive proteins of the extracellular matrix such as collagens, elastin, laminin, vitronectin, or in the blood plasma (VWF, fibrinogen, fibronectin) [12]. The integrin $\alpha_{IIb}\beta_3$ (GPIIb/IIIa) is the most abundant glycoprotein expressed by platelets (more than 80 000 copies expressed per platelet) and the most important adhesive receptor for platelet aggregation. In the active conformation it binds to several ligands, particularly fibrinogen, VWF, fibronectin, vitronectin, and CD40L. The activation of integrins occurs through several molecules that include the soluble agonists ADP, epinephrine, thromboxane A2, and thrombin, as well as the ligands of adhesive receptors like GPVI (collagen). Once activated the integrin $\alpha_{IIb}\beta_3$ also contributes to the platelet activation process.

The integrin $\alpha_2\beta_1$ (GPIa/IIa) is a less abundant platelet integrin, expressed at 1500–4000 copies per platelet [14]. In the active, extended conformation, it binds with increased affinity to collagens [14, 15]. Integrin $\alpha_2\beta_1$ sustains and enhances platelet adhesion via GPIb-V-IX and GPVI [16]. Therefore, platelet adhesion to artificial surfaces requires fibrinogen receptor (GPIIb/IIIa receptors) expression to bind fibrinogen adsorbed onto surfaces. Integrin $\alpha_v\beta_3$ (binding vitronectin and fibrinogen), integrin $\alpha_5\beta_1$ (binding fibronectin), and integrin $\alpha_6\beta_1$ (binding laminin) are moderately expressed on platelets (1000 copies per cell) [13]. They contribute to thrombus formation only in the absence of other functional integrins $\alpha_2\beta_1$ and $\alpha_{IIb}\beta_3$ [17]. The glycoprotein CD36 (GPIV, GPIIIb) is abundantly expressed on platelets (10 000–25 000 copies) as well as on mononuclear cells, macrophages, and endothelial cells. This protein acts as a receptor for thrombospondin-1, oxidized low-density lipoprotein, and oxidized lipids. VWF consisting of larger and smaller multimers of disulphide subunits unrolls from a globular to a filamental conformation, which becomes a high-affinity surface for the platelet GPIb-V-IX complex. The large multimers contain domains that support the interactions between platelets, endothelial cells, and subendothelial collagen. VWF together with specific glycoproteins GPIb/IX (CD42 a–d) expressed by platelets serve as bridge between the surface and activated platelets (Figure 4.3).

The adherent and activated platelets undergo morphological changes that lead to the formation of pseudopodia, to which other platelets attach through fibrinogen bound to GPIIb (CD41) and GPIIIa (CD61). Activated platelets release adenosine diphosphate and thromboxane A2 and other substances that accelerate the formation

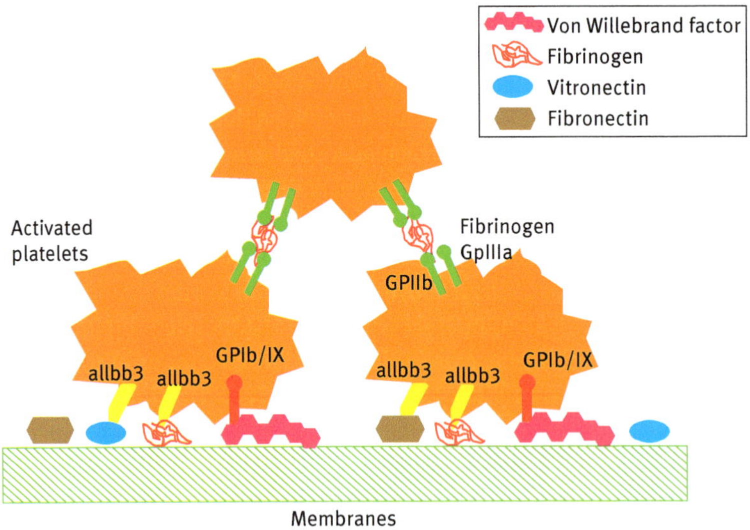

Fig. 4.3: Platelet activation and aggregation on membranes. The adhesion of platelets is mediated by vWF and the platelet receptors GPIb/IX (CD42 a–d).

of thrombin. The platelet stimulation via adhesive receptors and by soluble agonists (autocrine substances and coagulation products) is required for thrombus formation.

The kinetics of platelet adhesion to artificial surfaces have been revealed to be very rapid and initiation may take place in < 5 s for hydrophobic surfaces and < 30 s for hydrophilic surfaces [18, 19].

Hydrophobic surfaces are quickly fouled by fibrinogen adsorption, which causes the platelet adhesion and ultimately the formation of a clot. Negatively charged surfaces result in contact activation, where coagulation signalling factors present in the blood are activated on the substrate surface. This activation starts a signalling pathway that leads to the formation of thrombin and ultimately a blood clot. Both of these pathways are represented in Figure 4.4.

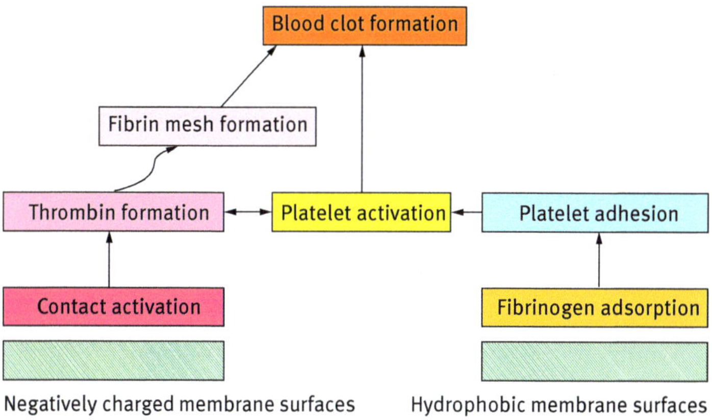

Fig. 4.4: Scheme of the two main mechanisms of blood clot formation caused by blood-material interaction.

4.3 Coagulation activation

The activation of coagulation can occur parallel to the activation of platelets [20]. The coagulation pathway consists of a series of reactions in which inactive precursors (clotting factors XII, XI, X, IX, VII, and II) are transformed in active proteases (factors XIIa, XIa, Xa, IXa, VIIa, and IIa). Factors VII, IX, and X and prothrombin require vitamin K for their synthesis and divalent cations (Ca^{2+}, Mg^{2+}). Factors V and VII are cofactors which accelerate the catalytic reactions. There are two pathways of activation: intrinsic and extrinsic pathway (Figure 4.5).

The extrinsic pathway initiates within seconds by the cell membrane-anchored tissue factor exposed during damage of the endothelium and activates factor VII. The intrinsic system is activated within minutes by artificial surfaces such as membranes through factor XII. Platelets express negatively charged phospholipids on

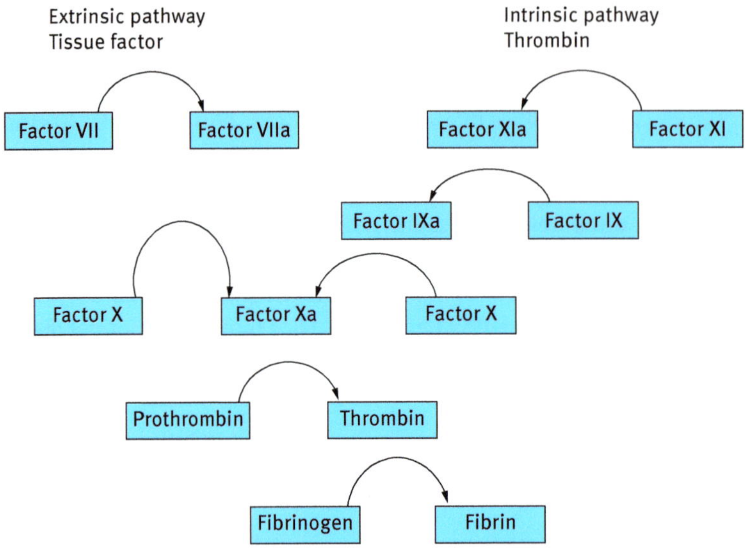

Fig. 4.5: Intrinsic and extrinsic pathways of coagulation activation.

their surface, and these either enhance the proteolitic activity of prekallikrein and high MW kininogen or trigger the clotting cascade by the factor XII. All factors activating platelets affect coagulation activation. The coagulation process involves three separate phases: (1) an initiation phase, in which low or of active coagulant factors are generated; (2) an amplification phase, in which the level of active coagulation factors is boosted; and (3) a propagation phase, in which coagulation factors bind to highly procoagulant membranes of activated platelets and fibrin clots are formed.

The initiation phase, classically referred to as the extrinsic pathway of coagulation, starts when the vasculature is disrupted, and subendothelial cells like smooth muscle cells and fibroblasts become exposed to the bloodstream. These cells expose a key initiator of the coagulation cascade, tissue factor, which binds coagulation factor VII. Tissue factor promotes proteolysis and activation to factor VIIa acting as a cofactor for factor VII. Upon endothelial damage, tissue factor is exposed to the bloodstream and binds factor VII, which is activated by factor VIIa. This complex enables subsequent activation of factor X and prothrombin, after which small amounts of thrombin activate the factor XI–IX feedback loop on the platelet surface. Factor IXa will then activate an additional factor X. Simultaneously, the trace amounts of thrombin will then activate factors VIII (cofactor to factor IX) and V (cofactor to factor X), which dramatically enhances catalytic activity of factors IX and X. Finally, thrombin (factor IIa) activation leads to fibrin deposition.

In parallel, local polyphosphate (polyP) release by activated platelets may additionally stimulate activation of factor XII, factor V, and factor XI and inhibit clot lysis. The intrinsic system is activated within minutes by artificial surfaces trough factor XII.

The clotting cascade in the case of injury is triggered via both intrinsic and extrinsic systems. The extrinsic system involves the expression of phospholipoproteins (tissue factor) from cells and activation of factor VII; on the other hand the intrinsic system starts with the negative charges of phospholipids expressed at the platelet surface or in the case of extracorporeal treatment by the charged membrane surfaces. For the contact phase activation are required: (1) negatively charged surfaces that induce a conformational change in (2) factor XII that becomes susceptible to the cleavage of (3) prekallikreinin and high MW kininogen. Activated factor XII triggers factor XI, which activates factor IX and which together with factor VII and calcium ions form a tenase complex. This complex also turns on factor X, which can also be activated by the tissue factor. The activated factor X forms together with factor V and prothrombin, which will be converted into thrombin. Thrombin is the final enzyme in the coagulation cascade and acts to cleave the plasma protein fibrinogen into fibrin monomers, which then polymerize and cross-link to form a fibrous mesh [21–23]. Finally erythrocytes and platelets are trapped in this fibrin mesh, forming the thrombus.

The membrane surface may also influence blood coagulation modifying the functions of the coagulation factors XII, XI, high-MW-kininogen and pre-kallikrein after adsorption. It is proposed that the adsorption of coagulation factor XII to negatively charged surfaces leads to subtle conformational changes in the enzyme that provoke autoactivation [21–23]. Therefore the density of negative charges on a membrane surface may influence the intensity of factor XII activation via positively charged amino acids in its heavy chain [23]. Through complexation with high-MW-kininogen, which binds to surfaces in a similar way, the factors XI and prekallikrein are brought into close vicinity with factor XII, triggering their reciprocal activation and leading to a strong amplification of contact activation [21].

4.4 Complement activation

As a component of the body's defence system, the complement can be activated in the case of invading microrganisms and when blood comes into contact with artificial surfaces [24]. In the case of extracorporeal treatments such as hemodialysis the complement activation, depending on the type of membranes, increases, reaching peak values within the first 15–30 min of treatment. This is due to the adsorption of complement proteins, which produce an acute inflammatory response. This proteolytic cascade, which consists of about 30 soluble and membrane-bound proteins, can be activated through three different pathways: (1) the classical pathway activated by the antigen-antibody complex; (2) the lectin pathway, which is not relevant in interactions with artificial surfaces; (3) the alternative pathway activated by binding the complement factor C3b to a foreign surface. The end-point of the classical and alternative pathways are the C3a, C4a, and C5a.

The classical pathway is activated when immune complexes (antigen-antibody) bind to the C1q component of the complement system. Subsequently there is the activation of C1r and C1s, both C1q-complexed proteases. Activated C1s cleaves C4 to form C4a and C4b, which bind to the cell surface; C2 is cleaved into C2a which binds to C4b to form the C4b2a complex (C3 convertase). C3 convertase cleaves C3 to form C3b, which binds to C4b2a (C3 convertase) to form C4b2a3b complex (C5 convertase). C5 convertase in turn cleaves C5 into C5a and C5b, and the latter triggers the formation of the membrane attack complex (MAC, C5b9), the final step in the complement cascade that leads to cell lysis. C4b and C3b promote opsonization, while C4a and C5a are anaphylatoxins with chemotactic properties and inflammatory response.

The activation of the lectin pathway is similar to the activation of the classical pathway. It is triggered when mannose-binding proteins (MBP) found in complexes of lectin-binding proteases (MASP) recognize the mannose on the surface of a pathogen. As these proteins bind to mannose, they are activated to cleave C4 and C2. The rest of the activation pathway is similar to the classical pathway.

The alternative pathway can be activated in conjunction with the classical and lectin pathways, but may be constitutively turned on at any time. The hydrolysis of C3 triggers its activation to form C3a and C3b. C3b binds to the cell surface and interacts with Factor B, which is cleaved by Factor D to generate fragment Bb. Fragment Bb is capable of binding to other C3b molecules on the cell surface to form C3bBb (C3 convertase). C3 convertase triggers an amplification loop to increase the hydrolysis of C3. The ensuing surplus of C3b binds to C3 convertase and promotes conversion to C5 convertase (C3b2Bb). C5 convertase then cleaves C5 to form C5a and C5b, and the latter factor (C5b) leads to the formation of MAC. C3a and C5a are anaphylatoxins.

A system of regulatory proteins, one of them being factor I, acts to prevent uncontrolled activation of the alternative pathway by inactivating C3b into C3bi (inactivated C3b). Factor H, the membrane cofactor protein (MCP, CD46), and thrombomodulin act as factor I cofactors to inactivate C3b (Figure 4.6). In the event that these regulatory proteins fail to perform their duties, the alternative complement pathway is over-activated, causing uncontrolled cell damage.

When membrane C3b coats artificial surface it attracts phagocytes and the terminal complement components C5b–9, which attack membranes and lyse microorganisms influencing also cell activation [25]. Although the nonspecific alternative pathway was long believed to be the relevant pathway in the extracorporeal treatment, immune complexes and the classical pathway are also involved. An important step of the complement activation is the cleavage of C5 in C5a e C5b. C5b initiates the assembly of other factors (C6, C7, C8, and C9) to form the complex that attack the membrane of foreign cells, leading to cell death. In the alternative pathway of complement activation C3 splits into C3a and C3b (initiation phase). C3b has normally a half-life and is inactivated, but in the presence of an artificial surface that exposes carbohydrates, C3b binds hydroxyl groups, or amino groups of the surface, and then cleaved (amplification phase). C3b attaches to the C3b component of convertase, with

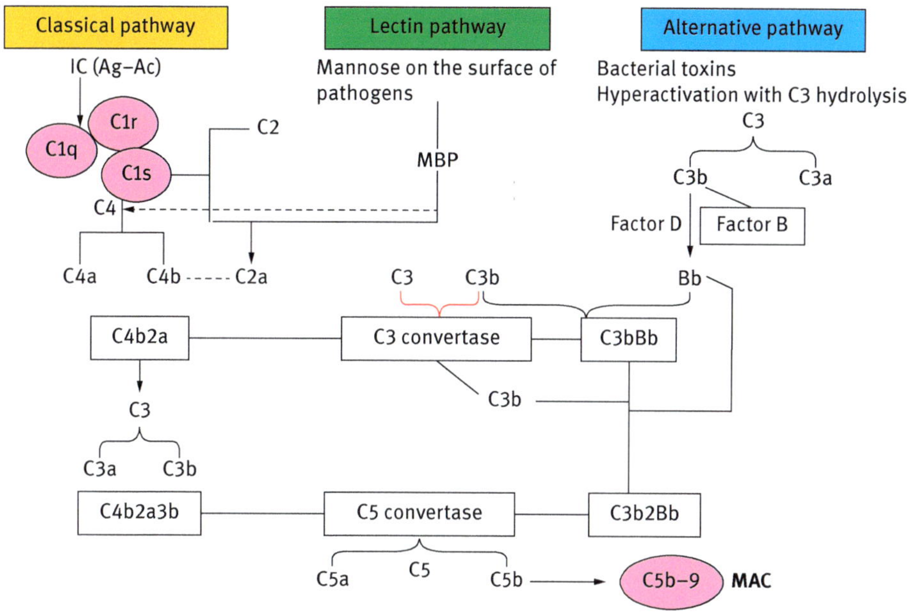

Fig. 4.6: Complement system activation pathways. IC: immune complexes (antigen-antibody); MBP: mannose-binding protein; C3bi: C3b inactivated; MCP: membrane cofactor protein; TM: thrombomodulin; MAC: membrane attack complex; amplification loop (red).

the consequent formation of complex C3b–C3bBb (C5 convertase), which cleaves C5 into C5a and C5b. The latter attaches to the surface, and together with C6 and C9 initiates the formation of the membrane attack complex.

4.5 Biocompatibility of dialysis membranes

Several studies indicate that platelet activation increases in a positively charged membrane and decreases in a membrane having a microdomain structure in which hydrophilic and hydrophobic groups coexist randomly as molecules. Also, the complement activation, which in the case of biomaterials proceeds via an alternative pathway, is mainly noted with cellulose membranes; free hydroxyl groups on the membrane surface are bonded with C3b and further with factor B that promotes the activation [26, 27].

Regenerated cellulose, which in the past was the main membrane material, is a strong activator of the complement owing to the considerable number of factor C3b binding sites such as hydroxyl groups that are responsible for activating complement systems through the alternative pathway [28]. C3 can associate with the adsorbed proteins, also playing a crucial role in the complement activation. Because the complement system is activated by the interaction with the –OH groups in cellulose, this polymer was modified, substituting some –OH groups with less reactive chemical groups. The first modification was performed replacing some hydroxyl groups with N,N,-diethylaminoethyl (DEAE) [29]. In the polymer, bulk tertiary amino groups replace only 1.5 % of all hydroxyl groups through ether bonds. These groups constitute a hydrophobic region on a hydrophilic surface and sterically hinder the interaction with the complement factor. Other modifications consist of the replacement of some hydroxyl groups of the glucose monomer (less than 1 %) with hydrophobic benzyl groups through ether bonds, in order to create hydrophobic domains onto a hydrophilic surface. This membrane named synthetically modified cellulose (SMC®) is produced by Membrana (Wuppertal, Germany) and is available in several dialyzers (e.g. Polysynthane™ by Baxter and SMC® by Bellco/Sorin, Kawasumi, Braun) [30]. An increase in biocompatibility was also achieved by grafting the cellulosic backbone with a polyethylenglycol (AM-BIO membrane, Asahi) [31]. One of the strategies undertaken to modify a cellulosic membrane was the coating of vitamin E performed by using a hydrophilic acrylic polymer in order to reduce the oxidative stress [32]. This membrane is marketed under the name Excebrane® by Terumo. In cellulosic membranes some hydroxyl groups of the cellulosic glucose monomer were replaced with acetyl groups developing an acetate cellulose membrane. On the basis of the substitution degree (grade 2 or 3) cellulose diacetate or triacetate is formed by replacing two and three hydroxyl groups of glucose monomer, respectively. Cellulose acetate membranes with different substitution grade (2, 2.4, or 3) are on the market [33]. In addition, different synthetic polymers were developed for dialysis membranes currently in use. These polymers include PSf, PES, and polyester polymer alloy (PEPA®), all of which use PVP as a hydrophilizing agent; PMMA; ethylene-vinylalcohol copolymer (EVAL®); and acrylonitrile and sodium methallylsulfonate copolymer (AN69®).

However, membranes made from synthetic polymers which are negatively charged may activate the complement system. The complement components C3a and C5a [34], inflammatory mediators, and the complement factor D [35], a 24 kDa molecule needed for complement activation, are adsorbed by these membranes. The cytokines and the complement factor D are efficiently removed using highly permeable membranes or hemodiafiltration, which also suppress some biological responses resulting from blood-membrane interactions. Therefore, anionic membranes (e.g. PAN) could be more thrombogenic than more neutral polymers (e.g. polysulfone). It has been shown that high factor XII adsorption and activation to factor XIIa was observed on the AN69® membrane. A comparison of different commercial membranes AN69® (PAN), Cuprophan® (regenerated cellulose), and polysulfone in ex vivo recirculation shows the higher thrombin-antithrombin III (TAT) complex generation in AN69® than regenerated cellulose and polysulfone [36] (Figure 4.7). Membranes with hydroxyl groups are relatively inert regarding the activation of the intrinsic coagulation pathway. On the other hand positive surface charges can activate membrane thrombogenicity. Hemophan® (DEAE-modified cellulose membrane), which is a low-level

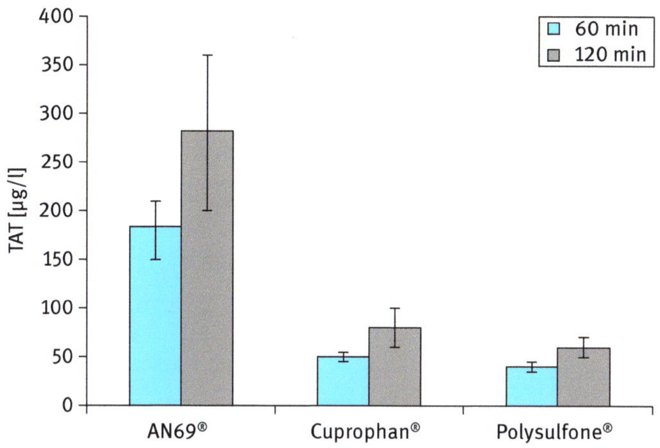

Fig. 4.7: Formation of TAT after 60 and 120 min of ex vivo recirculation by using commercial membrane dialyzers (adapted from [36]).

platelet activator, favors thrombin-antithrombin III complex generation rather than polyamide [37]. This is probably due to to the capacity of the DEAE positive groups to adsorb negatively charged heparin from the blood with the consequent reduction of heparin available for anticoagulation.

Unlike the coagulation factors, platelets adhere more to hydrophobic and cationic-charged membranes. A significant decrease in platelet count of about 9 % was observed by using cellulose acetate dialyzers (Duo-flux Cordis Dow) whereas on Hemophan® (GFS 120 Plus, Gambro), AN69® (Filtral 12, Hospal), polysulfone (F60, Fresenius) and cuprammonium rayon (AM-50-Bio Asahi) only an insignificant decrease was observed [38]. Striking differences have been found in the expression of the glycoprotein GMP-140 at the platelet surface. This protein is expressed by platelets after activation and is considered to be another important parameter of platelet activation. The analysis of samples from arterial and venous blood taken 5 min after beginning dialysis treatment evidenced the highest expression of GMP140 in cuprophan dialyzer (ST-15 Travenol), about 7 % of GMP140 expression increase was observed in cellulose diacetate (CA 110, Nissho) and PMMA dialyzer (Filtryzer B 1-1.6H, Toray), and the lowest expression was observed in polysulfone (Spiraflo HFT 10, Sorin) and AN69® (Filtral 10 Hospal) [39] (Figure 4.8).

Among dialysis membranes, regenerated cellulose membranes are the strongest complement activators, owing to the presence of hydroxyl groups that bind the complement factor C3b. Indeed the partial substitution of hydroxyl groups with acetyl, DEAE, benzyl groups, or the coating with PEG significantly reduces the activation of the complement. Figure 4.9 shows the improvement of the C5a generation modifying regenerated cellulose [6].

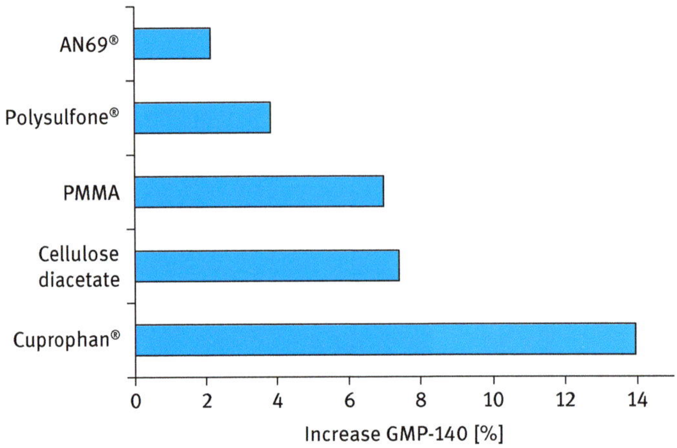

Fig. 4.8: Increase of GMP-140 expression in nine patients treated with different hemodialyzers (data from [39]).

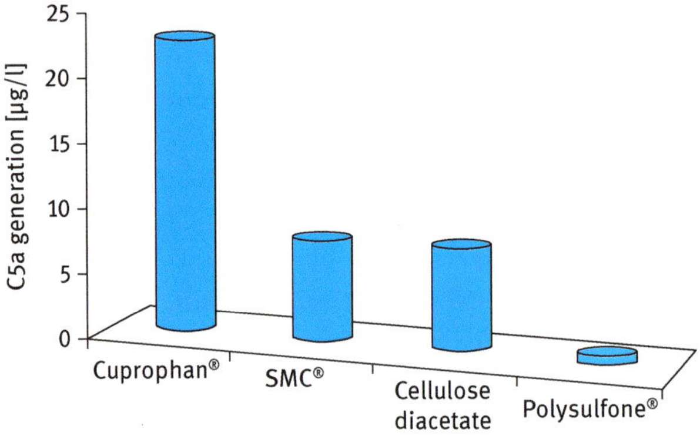

Fig. 4.9: C5a generation of cellulosic membranes native and modified compared to synthetic membranes [6].

However the degree of hydroxyl group substitution is not correlated to the degree of complement activation, because there are other factors involved. For this reason, synthetically modified cellulose – SMC® with less than 1 % of hydroxyl groups modified exhibits a complement activation similar to that of acetate cellulose membrane, where from 60 % (cellulose diacetate) to 90–100 % (cellulose triacetate) of hydroxyl groups are substituted [6]. Regulatory proteins such as factor D and factor H that downregulate the alternative pathway of the complement cascade are involved. Both factors bind the cellulose acetate membranes accelerating the degradation of surface-bound C3b [27]. Therefore a reduction of complement activation can be achieved preventing the interaction of C3 molecules with hydroxyl or amino groups.

Several types of synthetic polymeric membranes have a low incidence of bioincompatible responses. An example of bioincompatible response elicited by negatively charged membranes is the reaction that begins with coagulation factor XII activation and subsequently produces bradykinin [40]. Because bradykinin is broken down by angiotensin-converting enzyme (ACE), negatively charged membranes may cause anaphylactic- like shock in patients taking ACE inhibitors, due to the accumulation of bradykinin [41]. Therefore, dialysis treatment with AN69® membranes is contraindicated for patients using ACE inhibitors.

Membranes interacting with blood can also directly activate blood cells such as leukocytes, platelets, and red blood cells [28], or indirectly through the pathway that activates the complement system or coagulation factors, etc. (Figure 4.10).

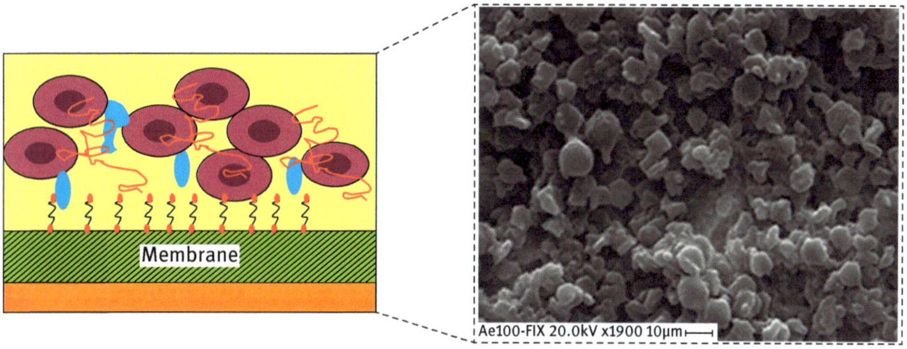

Fig. 4.10: Adhesion of blood cells and platelets on membranes in contact with blood.

Blood cells such as platelets and leukocytes regulate each other through activation and/or inhibition; therefore they directly or indirectly activate or inhibit each other [42–45]. The platelets that are activated upon exposure to a dialysis membrane adhere to and aggregate on the membrane, forming clots, and also bind to leukocytes, thereby activating them [46, 47]. The activated leukocytes release reactive oxygen species that also activate platelets [48]. Since platelet activation induces the activation and inhibition of various blood cells and biological reactions, the suppression of platelet activation is the key target to improving membrane biocompatibility. It has been reported that platelet adhesion in the case of cellulose-based membranes is associated with roughness [49], while on the other hand for PSf membranes the platelet adhesion is associated with albumin adsorption of the membrane surface [50]. In the case of membranes modified through coating, the surface modification alters not only the chemical composition of the membrane material, but also the physical properties of the surface such as pore size and roughness producing an effect on platelets and leukocytes adhesion.

Membrane dialyzers can activate leukocytes, which are determined by evaluating their disappearance from the blood 30 min after the starting of hemodialysis. The leukopenia is mainly induced by the overexpression of the leukocyte receptors CD11b/CD18, CD15s, which leads to adhesiveness and aggregation. These activation processes are mediated by the complement system. A very strong leukopenia is caused by regenerated cellulose membranes, while modified cellulosic membranes cause an intermediate to low drop in the number of leukocytes. On the other hand synthetic membranes induce a moderate decrease of leukocyte number. The increase of membrane permeability could remove the complement factors, and consequently could improve membrane biocompatibility, although the leukopenia is dependent on receptor expression.

4.6 The role of protein adsorption

Generally, when membranes come into contact with blood, plasma protein adsorption is the first event in the complex sequence that leads to thrombus formation and immunoreactions. The physical adsorption of protein molecules onto polymeric surfaces results in a modification of the interface properties of the membrane, so that in some cases the extent of subsequent cell adhesion increases, whereas in other cases the extent of adhesion decreases. Furthermore, the polymeric surface may induce conformational changes of the adsorbed proteins, with consequent influence on the biological activities of the molecules causing an immunoresponse.

Fibrinogen and immunoglobulin G favor the adhesion of platelets, whereas albumin causes a marked decrease in platelet adhesion. Materials that show preferential albumin adsorption demonstrate good blood compatibility, and materials that show preferential fibrinogen adsorption demonstrate bioincompatibility. Depending on the size, shape, overall charge, hydrophobicity, and internal stability of the molecule, proteins could adsorb onto hydrophobic or hydrophilic substratas [51, 52]. As a result, the interfacial properties of the polymer material are modified by the physical adsorption of protein molecules, with the consequent activation of platelets. Several studies have shown the wettability properties of a material surface to protein adsorption [51, 52]. It has been demonstrated that among human plasmatic proteins albumin, fibrinogen and immunoglobulin G, albumin adsorbs to a larger extent than the other proteins according to the following sequence: albumin → fibrinogen → IgG. The adsorption of proteins is affected by both physico-chemical properties of proteins and membranes. Indeed, proteins have different values for interfacial tension and their respective components. Protein adsorption depends on the acid-base parameter of the protein surface free energy (Table 4.1). Albumin is the protein with the highest value of total surface free energy γ and of acid-base parameter γ^{AB} with respect to fibrinogen and IgG; therefore it adsorbs to the greatest extent. Despite the different nature of the proteins, fibrinogen, albumin, and IgG follow the same trend: the lower

Tab. 4.1: Acid-base parameters of surface free energy of human plasma proteins.

Proteins	γ^{LW}	γ^{-}	γ^{+}	γ^{AB}	γ
	[mJ/m²]				
Albumin	36.1	49.9	2.2	21.0	57.1
Immunoglobulin G	37.8	37.4	0.2	6.2	44.0
Fibrinogen	43.5	11.4	3.3	12.3	55.8

the base parameter of the membrane surface free energy, the lower the amount of adsorbed protein.

Polymeric surfaces with a low base parameter of surface free energy exhibited low IgG, albumin, and fibrinogen adsorption. In particular, it has been shown that the amount of IgG at plasmatic concentration adsorbed on PEEK-WC membranes was reduced to 45 % of the CA value. The adsorption of plasmatic proteins observed on membranes with different physic-chemical properties (e.g. PP, PEEK-WC, CA, PU, PES) follows the same trend: it increases with increasing the base parameter of the membrane surface free energy with maximum adsorption on the membranes with $\gamma^{-} = 33.4\,\mathrm{mJ/m^2}$ [51, 52].

In recent years a great effort has been made to optimize both the chemical and physical properties of the membrane surface exposed to blood in order to improve biocompatibility.

4.7 Strategies for inhibiting immunoreactions

In hemodialysis, coagulation pathways are suppressed, due to the use of anticoagulants during the treatment, which cannot inhibit the activation of platelets. Currently, heparin and low molecular weight heparin derivatives are used as the main anticoagulants. Citrate and prostacyclin have also been investigated. Heparin is a glucosaminoglycan consisting of a mixture of sulphate polysaccharide chains with alternating residues of D-glucosamine and uronic acid. Heparin performs its activity through the activation of molecule antithrombin III. The anticoagulation activity is based on the binding of a specific pentasaccharide to antithrombin III, which causes a conformational change in the molecule. Thanks to this conformational change, antithrombin III binds and inactivates the clotting factors thrombin, Xa, IXa, XIa, and XIIa. In this way fibrin formation is avoided. The anticoagulant activity of heparin is related to its molecular size: decreasing the molecular weight increases the factor Xa inhibition and reduces the thrombin inhibition. Unfractionated heparin, which is used in hemodialysis, is a heterogeneous mixture of molecules, with a molecular weight ranging from 2000 to 40 000 Da. This also involves some side effects in chronic hemodialysis patients, such as hypersensitivity reactions, inhibition of platelet function, increased bleeding, thrombocytopenia, release of lipase, etc. Citrate chelates divalent cations such as calcium and magnesium are used as anticoagulant molecules. In hemodialysis blood is citrated only during the extracorporeal passage because of the interrelation of cations with the physiologic system. The application of citrate is restricted to patients suffering from bleeding, pericarditis, or to patients who react to heparin with severe thrombocytopenia. Indeed, citrate causes less bleeding and activation of coagulation than heparin. However, the use of citrate yielded conflicting results, and for this reason citrate anticoagulation is not yet considered to be suitable for standard hemodialysis treatment. Risks such as induction of hyper- or hypocalcemia, hypernatriemia, and metabolic alkalosis are associated with the use of citrate.

Prostacyclin is a strong inhibitor of platelet aggregation which *in vivo* is liberated by endothelial cells during injury. Synthetic prostacyclin was used in dialysis in the 1980s, but its use is currently restricted to patients with a high risk of bleeding, owing to side effects and high expense.

Recently, strategies aimed at avoiding the thrombogenicity of membranes involve the seeding of endothelial cells on membrane surfaces. These cells normally maintain hemostasis and regulate blood clotting, and thus it is expected to be a way to improve biocompatibility. Molecular characterization studies suggest that confluent endothelial cells exhibit a nonthrombogenic phenotype when cultured on a biomaterial [53]. The formation of a confluent monolayer of endothelial cells is dependent on the morphological and physico-chemical properties of the membrane surface.

Since the endothelization of biomaterials has been a topic for research, other applications would also benefit, such as vascular prosthesis, stents, prosthetic heart valves, and vena cava filters.

4.8 References

[1] Hanson SR. Device thrombosis and thromboembolism. Cardiovasc Pathol 1993; 2(Suppl): 157S–165S.

[2] Kokubo K, Kurihara Y, Kobayashi K, Tsukao H, Kobayashi H. Evaluation of the Biocompatibility of Dialysis Membranes. Blood Purif. 2015; 40: 293–297.

[3] Forbes CD. Blood-material interaction: importance of thrombosis. In: Falkenhagen D, Klinkmann H, Opatrny K. Blood-material interaction. INFA 1998. pp. 56–58.

[4] Vroman L, Adams AL, Fischer GC, Munoz PC. Interaction of high molecular weight kininogen, factor XII, and fibrinogen in plasma at interfaces. Blood- 1980; 55 (1): 156–159.

[5] Chuang HY, Mohammad SF, Sharma NC, Mason RG. Interaction of alpha-thrombin with artificial surfaces and reactivity of adsorbed alpha-thrombin. J Biomed Mater Res. 1980; 14: 467–476.

[6] Grassmann A, Uhlenbusch-Körwer I, Bonnie-Schorn E, Vienken J. Understanding membranes and dialysers. Lengerich, Berlin: PABST Science Publishers; 2004.

[7] Sperling C, Fischer M, Maitz MF, Werner C. Blood coagulation on biomaterials requires the combination of distinct activation processes. Biomaterials. 2009; 30: 4447–4456

[8] Versteeg HH, Heemskerk JWM, Levi M, Reitsma PH. New Fundamentals in hemostasis. Physiol Rev. 2013; 93: 327–358.
[9] Hemler ME. VLA proteins in the integrin family: structures, functions, and their role on leukocytes. Annu Rev Immunol. 1990; 8: 365–400.
[10] Plow EW, Pesho MM, Ma YQ. Integrin alpha IIb beta 3. In: Platelets. Burlington: Academic; 2007. pp. 165–184.
[11] Shattil SJ, Kim C, Ginsberg MH. The final steps of integrin activation: the end game. Nat Rev Mol Cell Biol. 2010; 11: 288–300.
[12] Hawiger J. Adhesive interactions of blood cells and the vascular wall in hemostasis and thrombosis. In: Colman JH, Marder VJ, Clowes AW, George JN (eds.). Hemostasis and Thrombosis, Basic Principles and Clinical Practice. Philadelphia, PA: Lippincott Williams & Wilkins; 2001. pp. 639–660.
[13] Clemetson KJ, Clemetson CJ. Platelet receptors. In: Michelson AD (ed.). Platelets. Burlington: Academic; 2007. pp. 117–143.
[14] Emsley J, Knight CG, Farndale RW, Barnes MJ, Liddington RC. Structural basis of collagen recognition by integrin alpha2beta1. Cell. 2000; 101: 47–56.
[15] Moroi M, Onitsuka I, Imaizumi T, Jung SM. Involvement of activated integrin alpha2beta1 in the firm adhesion of platelets onto a surface of immobilized collagen under flow conditions. Thromb Haemost. 2000; 83: 769–776.
[16] Gruner S, Prostredna M, Schulte V, Krieg T, Eckes B, Brakebusch C, Nieswandt B. Multiple integrin-ligand interactions synergize in shear-resistant platelet adhesion at sites of arterial injury in vivo. Blood. 2003; 102: 4021–4027.
[17] Siljander PR, Munnix IC, Smethurst PA, Deckmyn H, Lindhout T, Ouwehand WH, Farndale RW, Heemskerk JW. Platelet receptor interplay regulates collagen-induced thrombus formation in flowing human blood. Blood. 2004; 103: 1333–1341.
[18] Sivaraman B, Latour RA. The Relationship between Platelet Adhesion on Surfaces and the Structure versus the Amount of Adsorbed Fibrinogen. Biomaterials. 2010; 31(5): 832–839.
[19] Sperling C, Fischer M, Maitz MF, Werner C. Blood coagulation on biomaterials requires the combination of distinct activation processes. Biomaterials. 2009; 30: 4447–4456.
[20] Achneck HE, Sileshi B, Parikh A, Milano CA, Welsby IJ, Lawson JH. Pathophysiology of bleeding and clotting in the cardiac surgery patient: from vascular endothelium to circulatory assist device surface. Circulation. 2010; 122: 2068–2077.
[21] Kozin F, Cochrane CG. The contact activation system of plasma – biochemistry and pathophysiology. In: Gallin JI, Goldstein IM, Snyderman R (eds.). Inflammation: basic principles and clinical correlates, 2nd ed. New York: Raven Press; 1992. pp. 103–122.
[22] Chen X, Wang J, Paszti Z, Wang F, Schrauben JN, Tarabara VV, et al. Ordered adsorption of coagulation factor XII on negatively charged polymer surfaces probed by sum frequency generation vibrational spectroscopy. Anal Bioanal Chem. 2007; 388: 65–72.
[23] Colman RW, Schmaier AH. Contact system: a vascular biology modulator with anticoagulant, profibrinolytic, antiadhesive, and proinflammatory attributes. Blood. 1997; 90: 3819–3843.
[24] Mims C, Playfair J, Rottl, Wakelin D, Williams R. The innate defences of the body. In: Crowe L (ed.). Medical Microbiology, 2nd ed. London: Mosby International Limited; 1998. pp. 47–61.
[25] Johnson RJ. Complement activation during extracorporeal therapy: Biochemistry, cell biology and clinical relevance. Nephrol Dial Transplant. 1994; 9: 36–45.
[26] Wettero J, Askendal A, Bengtsson T, Tengvall P. On the binding of complement to solid artificial surfaces in vitro. Biomaterials. 2002; 23: 981–991.
[27] Cheung AK, Parker CJ, Wilcox L, Janatova J. Activation of the alternative pathway of complement by cellulosic hemodialysis membranes. Kidney. 1989; 36: 257–267.

[28] Hakim RM. Clinical implications of hemodialysis membrane biocompatibility. Kidney Int. 1993; 44: 484–494.
[29] Holland FF, Gidden HE, Mason RG, Klein E. Thrombogenicity of heparin-bound DEAE cellulose hemodialysis membranes. ASAIO J. 1978; 1: 24–36
[30] Bowry SK, Rintelen TH. Synthetically modified cellulose (SMC): a cellulosic hemodialysis membrane with minimized complement activation. ASAIO J. 1998; 44: M579-M583
[31] Kishida A, Mishima K, Corretge E, Konishi H, Ikada Y. Interactions of polyethylenglycol-garfted cellulose membranes with proteins ad platelets. Biomaterials. 1992; 13: 113–118
[32] Sasaki M, Hosoya N, Saruhashi M. Vitamin E modified cellulose membrane. Artif. Organs. 2000; 24: 779–789
[33] Hoenich NA, Woffindin C, Cox PJ, Goldfinch ME, Roberts SJ. Clinical characterization of Dicea a new cellulose membrane for hemodialysis: Clin Nephrol. 1997; 48: 253–259
[34] Kandus A, Ponikvar R, Drinovec J, Kladnik S, Ivanovich P: Anaphylatoxins C3a and C5a adsorption on acrylonitrile membrane of hollow-fiber and plate dialyzer: an in vivo studycontact. Int J Artif Organs. 1990; 13: 176–180.
[35] Pascual M, Schifferli JA. Adsorption of complement factor D by polyacrylonitrile dialysis membranes. Kidney Int. 1993; 43: 903–911.
[36] Frank RD, Weber J, Dresbach H, Thelen H, Weiss C, Floege J. Role of contact system activation in hemodialyser-induced thrombogenicity. Kidney Int. 2001; 60: 1972–1981.
[37] Schultze G, Gohl H, Holmann S, Sinah P. Formation of thrombin-antithrombin complex using polyamid and cellulosic dializers. Int J Artif Organs. 1991; 14: 543.
[38] Verbelen D, Jochmans K, Herman AG, Van der Niepen P, Sennesael J, De Waele M. Evaluation of platelets and hemostasis during hemodialysis with six different membranes. Nephron. 1991; 59: 567–572.
[39] Cases A, Reverter JC, Escolar G, Sanz C, Lopez-Pedret J, Reverter L, Ordinas A. Platelet activation on hemodialysis: influence of dialysis membranes. Kidney Int. 1993; 43: S217–S220
[40] Verresen L, Fink E, Lemke HD, Vanrenterghem Y. Bradykinin is a mediator of anaphylactoid reactions during hemodialysis with AN69 membranes. Kidney Int. 1994; 45: 1497–1503.
[41] Tielemans C, Madhoun P, Lenaers M, Schandene L, Goldman M, Vanherweghem JL. Anaphylactoid reactions during hemodialysis on AN69 membranes in patients receiving ACE inhibitors. Kidney Int. 1990; 38: 982–984.
[42] Nagata K, Tsuji T, Todoroki N, Katagiri Y, Tanoue K, Yamazaki H, Hanai N, Irimura T. Activated platelets induce superoxide anion release by monocytes and neutrophils through P-selectin (CD62). J Immunol. 1993; 151: 3267–3273.
[43] Peters MJ, Dixon G, Kotowicz KT, Hatch DJ, Heyderman RS, Klein NJ. Circulating platelet-neutrophil complexes represent a subpopulation of activated neutrophils primed for adhesion, phagocytosis and intracellular killing. Br J Haematol. 1999; 106: 391–399.
[44] Bazzoni G, Dejana E, Del Maschio A. Platelet–neutrophil interactions. Possible relevance in the pathogenesis of thrombosis and inflammation. Haematologica. 1991; 76: 491–499.
[45] Li N, Hu H, Lindqvist M, Wikström-Jonsson E, Goodall AH, Hjemdahl P. Platelet–leukocyte cross talk in whole blood. Arterioscler Thromb Vasc Biol. 2000; 20: 2702–2708.
[46] Bonomini M, Stuard S, Carreno MP, Settefrati N, Santarelli P, Haeffner-Cavaillon N, Albertazzi A. Neutrophil reactive oxygen species production during hemodialysis: role of activated platelet adhesion to neutrophils through P-selectin. Nephron. 1997; 75: 402–411.
[47] Itoh S, Susuki C, Tsuji T. Platelet activation through interaction with hemodialysis membranes induces neutrophils to produce reactive oxygen species. J Biomed Mater Res A. 2006; 77: 294–303.
[48] Salvemini D, de Nucci G, Sneddon JM, Vane JR. Superoxide anions enhance platelet adhesion and aggregation. Br J Pharmacol. 1989; 97: 1145–1150.

[49] Tsunoda N, Kokubo K, Sakai K, Fukuda M, Miyazaki M, Hiyoshi T. Surface roughness of cellulose hollow fiber dialysis membranes and platelet adhesion. ASAIO J. 1999; 45: 418–423.
[50] Namekawa K, Fukuda M, Matsuda M, Yagi Y, Yamamoto K, Sakai K. Nanotechnological characterization of human serum albumin adsorption on wet synthetic polymer dialysis membrane surfaces. ASAIO J. 2009; 55: 236–242.
[51] De Bartolo L, Morelli S, Bader A, Drioli E. The influence of polymeric membrane surface free energy on cell metabolic functions. J Mater Sci: Materials in Medicine. 2001; 12: 959–963.
[52] De Bartolo L, Gugliuzza A, Morelli S, Cirillo B, Gordano A, Drioli E. Novel PEEK-WC membranes with low plasma protein affinity related to surface free energy parameters. J Mater Sci: Materials in Medicine. 2004; 15: 877–883.
[53] McGuigan AP, Sefton MV. The influence of biomaterials on endothelial cell thrombogenicity. Biomaterials. 2007; 28: 2547–2571.

5 Engineering of membrane bio-hybrid organs

5.1 Introduction

In this chapter we present the design criteria and the operative conditions enabling an efficient mass transport of nutrients and catabolites in a bioreactor, with particular attention paid to when transport occurs through polymeric membranes.

It is well accepted that the design of scaffolds and implants for clinical use, as well as of bioreactors for cell culture, is significantly affected by fluid dynamics; in general, a properly controlled environment in terms of mass transfer and reaction kinetics is able to reproduce specific functions and bioactive factors for long-term viability of cells.

Computer aided design (CAD) for advanced studies of mechanobiology and for mathematical modeling of culture conditions in a bioreactor increases its reproducibility and robustness, and allows a more realistic *in vitro* investigation and control of those parameters that – at the intersection of engineering, biology, and chemistry – reciprocally influence the behavior of cells and the performance biomaterials in complex tissue-engineered systems. Below we present and discuss experimental and computational approaches as indispensable tools for the development of bioreactors and functional scaffolds for tissue engineering. As a first step for a rational design of reactors, here we show the classical engineering approaches based on residence time distribution (RTD) analysis.

Recent progresses in computational fluid dynamics (CFD) for the prediction of oxygen, metabolites, and catabolite concentration profiles throughout high cell-density biodevices, including a wide collection of rate expressions for metabolic consumption of specific nutrients, and simulation cases related to lymphocytes, hepatocytes, and chondrocytes culture in bioreactors, are presented.

5.2 Fluid-dynamics of membrane bioreactors

The design of a bioreactor for cellular culture aims at creating an optimal artificial environment able to sustain the viability, differentiation, and functions of cells. In this respect, a well-engineered bioreactor is supposed to provide a supporting structure for adhesion of cells, an efficient supply of nutrients, and specific cell-signaling protein molecules (cytokines, growth factors etc.), an adequate removal of catabolites, and, in general, to assure controlled conditions with respect to eventual time-variant input.

Since the early description of a rudimental dialysis process made by of Graham [1], membranes play a significant role in biohybrid organs and bioreactors for clinical applications, due to their intrinsic characteristics of efficiency and operational

DOI 10.1515/9783110268010-006

simplicity, high permeability, and selectivity to molecular components, excellent stability under mild operational conditions, robust automation, and easy scale-up.

As detailed in Chapter 2, membranes are characterized by a high specific surface area (commercially available hollow fiber modules can reach a surface/volume ratio up to 10 000 m^2/m^3) making them an ideal support for the adhesion of anchorage-dependent cells; membranes show wide possibilities for chemical modification of the polymer matrix to promote specific cell–substrate interactions; membranes can be manufactured in multiform shapes (flat sheets, spiral wounds, tubular, hollow fibers, capillary, etc.) and tailored to cover a wide spectrum of MWCO; they provide both an efficient and selective transfer of metabolites to cells and a back-transport of catabolites, thus maintaining cellular viability and functions *in vitro* for weeks.

Similar to conventional chemical reactors, the first step in the design of a bioreactor is the analysis of flow patterns and mixing characteristics. In fact, due to the presence of membranes and eventually baffles, corners, preferential pathways, incomplete mixing, dead-zones, etc., the behavior of a real bioreactor can significantly differ from ideal ones.

In this respect, it is convenient to refer to ideal reactors representing the two opposite limits of flow field: the plug-flow reactor (PFR), characterized by the absence of mixing in the direction of flow (but complete radial mixing), and the completely mixed continuous stirred-tank reactor (CSTR); in most practical applications, appropriate analytical combinations of PFR and CSTR well approximate real situations.

A plug flow reactor (PFR) is schematized in Figure 5.1 (a) as a cylindrical tube that operates isothermally, at constant pressure, and under steady-state, with reaction mixture moving with nonbackmix flow, so that all fluid elements are perfectly ordered and none overtakes any adjacent fluid volumes. In a plug flow reactor, the fluid composition varies along the length of the reactor and – under the assumption of negligible changes of volume during the reaction – the design equation is

$$\frac{V_\mathrm{R}}{F_{i,0}} = \int_{x_{i,0}}^{x_{i,\mathrm{f}}} \frac{\mathrm{d}x_i}{(-\upsilon_i r_i)}, \tag{5.1}$$

where V_R is the volume of the reactor, r_i is the reaction rate, and $F_{i,0}$, x_i and υ_i are the initial molar flowrate, the fractional conversion, and the stoichiometric coefficient of the limiting reactant i (usually υ_i is set to 1), respectively. Subscripts 0 and f indicate the inlet and outlet conditions, respectively. In recycle reactors $x_{i,0} \neq 0$.

If changes in the volume flow rate due to chemical reaction are negligible (system at constant density),

$$\tau = \int_{c_{i,0}}^{c_{i,\mathrm{f}}} \frac{\mathrm{d}c_i}{(-\upsilon_i r_i)}, \tag{5.2}$$

where c is the concentration and τ is the *space time*, defined as the ratio between the volume of mixture in the reactor and the volumetric feed flowrate; the inverse of space time ($1/\tau$) is called *space velocity*. Equation (5.2)) is the same for a batch stirred tank reactor if τ is replaced by the real time.

Since a turbulent flow (i.e. Reynolds number < 2100) is required to approach a PFR behavior, in many practical applications including laboratory-scale reactors, hollow-fiber membrane bioreactors, and – in general – all devices facing the issue of cellular adhesion and sensitivity to mechanical shear stress, high flow rates are not achievable. Under these circumstances, laminar flow will occur and significant variations from ideality might be detected.

Continuous stirred tank reactors (CSTR) allow the direct measurement of steady-state reaction rates at constant temperature and pressure. This reactor is characterized by a total back-mixing, with fluid perfectly stirred and uniform in composition (Figure 5.1 (b)): therefore, the composition of effluent stream is the same as the composition of the fluid in the reactor. The design equation, expressed as

$$\frac{V_R}{F_{i,0}} = \frac{x_i}{(-\upsilon_i r_i)}, \tag{5.3}$$

correlates the reacting volume V_R to the reaction rate r_i, the molar feed rate $F_{i,0}$, the molar fraction x_i, and the stoichiometric coefficient υ_i of the limiting reactant *i*. In principle, the volume of the reacting system might differ from the reactor volume V_R; for cell culture applications, the reactant/product fluids are often in the liquid phase, and, when largely diluted, changes in reaction volume can be reasonably considered negligible. Under the assumption of constant volume, the design equation of a CSTR can be written as a function of concentration, and the reaction rate r is calculated as

$$(-\upsilon_i r_i) = \frac{c_{i,0} - c_i}{\tau}, \tag{5.4}$$

where c_i is the concentration of the limiting reactant; subscript 0 identifies the initial value.

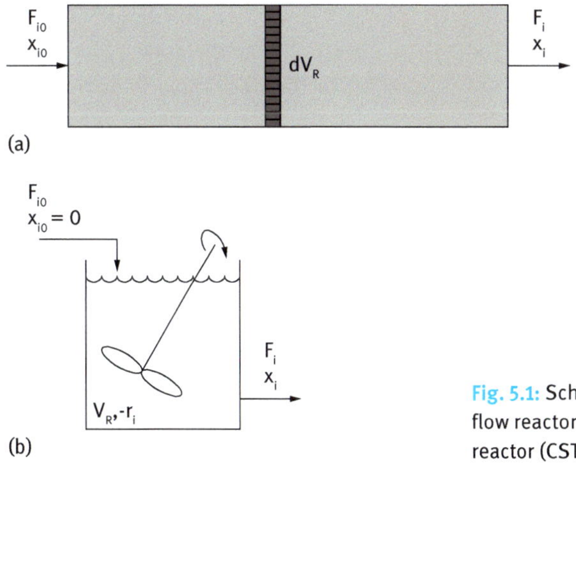

Fig. 5.1: Schematic representation of: (a) plug flow reactor (PFR); (b) continuous stirred tank reactor (CSTR).

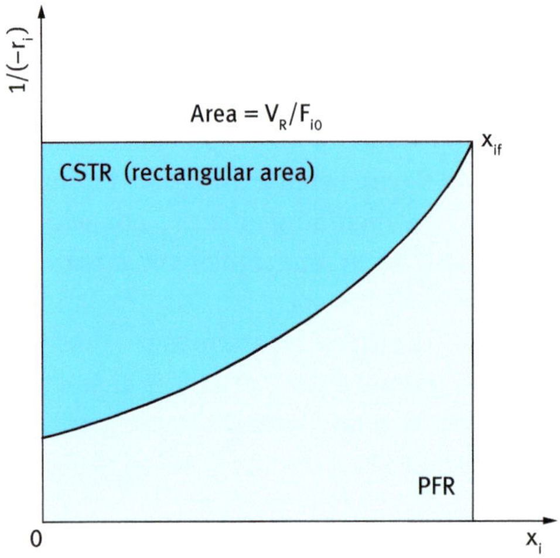

Fig. 5.2: Graphical resolution of design equations for a PFR (light blue) and for a CSTR (total rectangle area).

A comparison between PFR and CSTR suggests that these ideal reactors represent the maximum and minimum space time yield reactor configurations, respectively; therefore, for the same conversion, PFR always requires a lower reaction volume. From an analytical point of view, Figure 5.2 shows that the area below the curve $1/(-r_i)$ is always lower than the rectangular area for a CSTR.

In practical cases, the behavior of a real reactor might differ significantly from the extreme mixing characteristics of PFR and CSTR. Deviations from ideal conditions occur whenever the fluid moves through the reactor at different velocities, thus resulting in channeling, formation of preferential pathways, locally incomplete mixing and partial segregation. An additional reason for deviations might originate from local micromixing or diffusion in the direction of flow in tubular reactors.

Mixing conditions within a reactor are characterized in terms of residence time distribution (RTD), the probability distribution function describing the time that a fluid element spends inside the reactor. RTD analysis is an efficient diagnosis tool for inspection of troubleshooting in a reactor: it permits the achievement or preservation of a desired flow pattern, and the estimation of the performance of the real reactor.

Danckwerts [2] extensively investigated the concept of "fluid element", that is, a small fluid volume with defined physical properties. The RTD function, age distribution $E(t)$, measures the residence time of the various fluid elements in the reactor.

A comparison of the RTD function of a real reactor with the ideal ones allows a qualitative evaluation of the extent and reason of the pattern flow deviations. Quantitative analysis of the pattern flow analysis proceeds through the modeling of the real reactor by analytical combination of the RTD functions of ideal reactors that reasonably reproduces the experimental data.

Residence time distribution tests are experimentally carried out by stimulus-response technique: a nonreactive tracer is injected at the inlet of the reactor vessel (stimulus). The input signal to the reactor is a known function depending on the way of changing the concentration of the tracer in time: random, step, cyclic, pulse, discontinuous, wave signal, etc. The outlet stream is monitored as a function of time, and the response signal is recorded in terms of the concentration of the tracer leaving the reactor. The selected tracer should not modify the hydrodynamic conditions of the reactor (no adsorption on the walls or surfaces) with respect to usual operations (Figure 5.3).

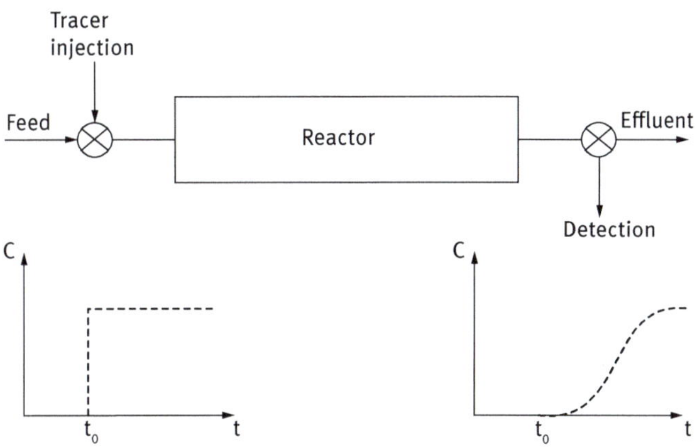

Fig. 5.3: Experimental setup for stimulus-response technique (step tracer input).

Colored and easy detectable materials, typically by spectrophotometric or fluorescence techniques, are the most common types of tracers; for example, trypan blue is used as macromolecular tracer in many biological applications. Figure 5.4 shows a typical laboratory setup for determining the residence time distribution of a membrane reactor. At time $t = t_0$, tracer (red-colored William's medium *E*) is pumped at constant concentration and flow rate through the membrane reactor (step input) initially filled with pure water; the outlet stream is processed in a spectrophotometer to measure and record the concentration in time.

From a practical point of view, laboratory practice is typically restricted to only two modes of tracer injection: pulse stimulus and step input signal.

The pulse injection method consists of the one-shot introduction of a small volume of concentrated tracer at the inlet of the reactor; the pulse is analytically described by the Dirac delta function

$$\delta(t - t_0) = \begin{cases} = 0, & t \neq t_0 \\ \neq 0, & t = t_0 . \end{cases} \tag{5.5}$$

Fig. 5.4: Laboratory setup for RTD analysis.

Although an instantaneous pulse cannot be produced, an input signal with injection time significantly smaller than the mean residence time of the reactor is considered appropriate.

On the other hand, a step function tracer input signal is very easy to perform and, consequently, preferentially used. With no tracer initially present in the inlet stream of the reactor, a step signal of concentration c_0 constant in time is introduced according to the following function:

$$c_0(t) = \begin{cases} = 0, & t < t_0 \\ = c_0, & t \geq t_0. \end{cases} \tag{5.6}$$

The concentration-time curve is recorded at the exit of the reactor until the concentration of the outlet stream becomes equal to c_0. The output signal, when evaluated as the ratio of the outlet concentration $c_{out}(t)$ on c_0, is called curve F (or cumulative function):

$$F(\theta) = \frac{c_{out}(t)}{c_0}. \tag{5.7}$$

In equation (5.7), θ is the reduced time (dimensionless), defined as

$$\theta = \frac{t}{\tau} = \frac{Qt}{V}, \tag{5.8}$$

where t is the actual time, τ the mean residence time, Q the volumetric flow rate, and V the volume of the reactor.

The residence time distribution is mathematically represented by the exit-age distribution $E(t)$ evaluated in the outlet stream of the reactor. According to the definition of $E(t)$,

$$\int_0^\infty E(t)\,\mathrm{d}t = 1. \tag{5.9}$$

The average residence time (or space time) τ is equal to the first moment of $E(t)$:

$$\tau = \int_0^\infty tE(t)\,\mathrm{d}t. \tag{5.10}$$

From step response results, $E(t)$ distribution can be obtained by differentiating the cumulative $F(t)$ response:

$$E(t) = \frac{\mathrm{d}F(t)}{\mathrm{d}t}. \tag{5.11}$$

The opportunity to differentiate $F(t)$ from the step response, thus obtaining $E(t)$, is challenging, due to the amplification of noises by numerical computing. However, an appropriate selection of curve-fitting algorithms significantly improves the differentiation procedure.

$E(t)$ function is applied to the prediction of the conversion in a real reactor. Under the assumption that the total conversion is the average conversion of all the fluid elements, the mean concentration of reactant i in the reactor outlet $\langle c_i \rangle$ is calculated as

$$\langle c_i \rangle = \int_0^\infty c_i(t)E(t)\,\mathrm{d}t. \tag{5.12}$$

RTD prediction of reactor conversion is only applicable to isothermal and single-phase systems.

The time evolution of a tracer-response technique for a flat sheet membrane reactor for pig liver cell culture, manufactured as a 65 cm^3 circular acrylic housing with 42 cm^2 membrane active area inserted at the bottom of chamber, is illustrated in Figure 5.5.

The cumulative RTD function evaluated during a stimulus response test under 0.6 ml/min feed flow rate of trypan blue and at recycle ratio of 1.44 shows significant discrepancies with respect to the ideal behavior of a perfectly mixed CSTR (Figure 5.6), mostly due to the presence of channeling through a preferential longitudinal pathway. However, it has been experimentally verified that a good mixing is obtained if the recycle ratio is increased to 5.52, and $F(\theta)$ curves overlaps the one for a ideal CSTR [3].

Hollow fiber membranes, usually manufactured in an asymmetric structure, having outer diameter in the range of 50–100 μm and installed in a bundle of several hundreds (or thousands) elements, are widely used for bioreactor applications. HF membrane modules provide a membrane area per unit volume of 2000–5000 $\mathrm{m}^2/\mathrm{m}^3$, ensure a stable permeation flux over time, can be easily manufactured by a spinning process and are relatively cheap.

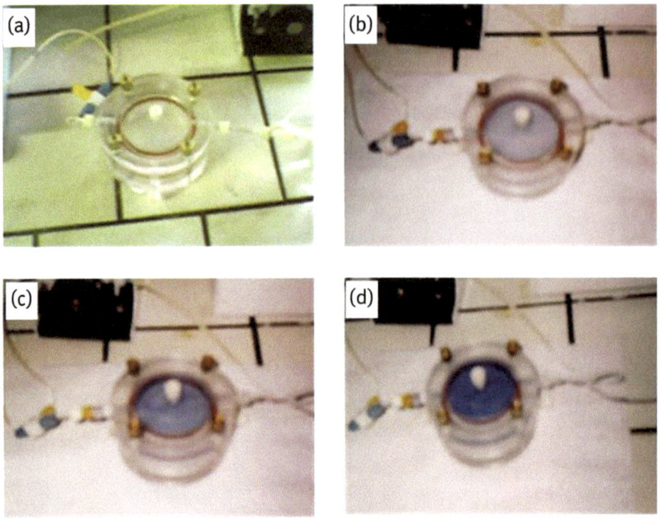

Fig. 5.5: RTD characterization of a flat membrane reactor (65 cm^3) for cell culture. Snapshots of reactor at different times under tracer (trypan blue) step input: (a) 0 min, (b) 5 min, (c) 15 min, and (d) 35 min.

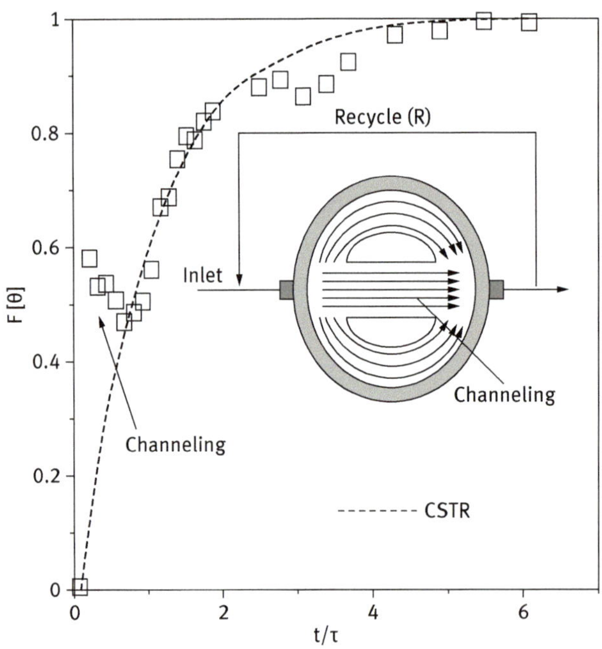

Fig. 5.6: Cumulative RTD function $F(\theta)$ as a function of the reduced time ($\theta = t/\tau$) of a 65 cm^3 flat-sheet bioreactor under a step stimulus (tracer feed flowrate: 0.6 ml/min) operating under a low recycle ratio ($R = 1.44$).

In a typical HF bioreactor, nutrients are fed through the lumen of hollow fibers and cells cultured in the shell where metabolic reactions take place. Accordingly, stimulus response techniques are implemented by feeding the tracer through the lumen side, and concentration is monitored at the outlet of the shell. In Figure 5.7, RTD analysis of a HF module for lymphocyte cell culture is reported.

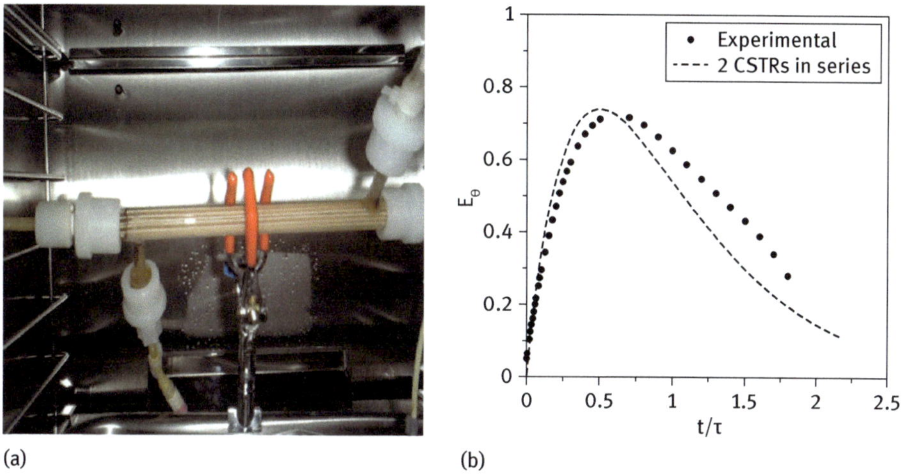

Fig. 5.7: (a) Hollow fiber reactor; (b) comparison between experimental RTD (inlet flowrate: 1 ml/min, shell volume: 200 ml) and residence Time distribution from 2-serial CSTRs model.

Hollow fibers can be properly assembled to obtain different reactor configurations. Snapshots of a tracer-step injection test carried out on a crossed-holow fiber membrane reactor for hepatocyte culture are shown in Figure 5.8 (a)–(d).

Specifically, two types of hollow fiber membranes (HFs) having different molecular weight cutoff, and the physico-chemical properties are perpendicularly bundled in the reactor: modified polyetheretherketone (PEEK-WC) and polyethersulfone (PES). Each type of fiber has a different function. PEEK-WC HFs are used to supply oxygenated medium that includes nutrients and metabolites to cells cultured in the extrafiber space, while microporous PES HFs are suitable for an efficient removal of catabolites and cell specific products from cell compartment. The reactor design aims at mimicking the *in vivo* arterious and venous blood vessels [4].

In this case, RTD is investigated by recording the system response at the exit of the PES fibers under a step input of tracer sent at the entrance of PEEK-WC fibers.

The experimental profile of $E_\theta(t)$, where $E_\theta(t) = \tau E(t)$, and its comparison with ideal CSTR residence time distribution is shown in Figure 5.9.

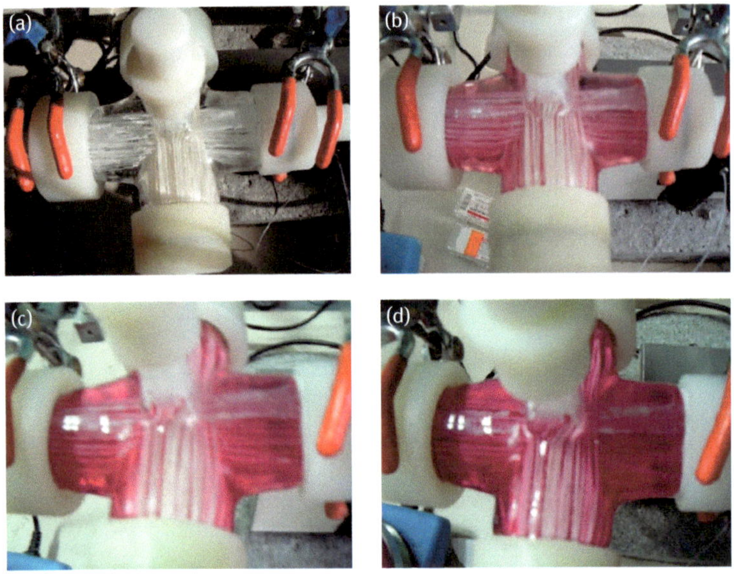

Fig. 5.8: Snapshots of a 40 cm^3 crossed-hollow fibers bioreactor under step-stimulus at: (a) 1 min; (b) 90 min; (c) 240 min; (d) 380 min. Williams' medium at flow rate of 1.5 ml/min is used as tracer.

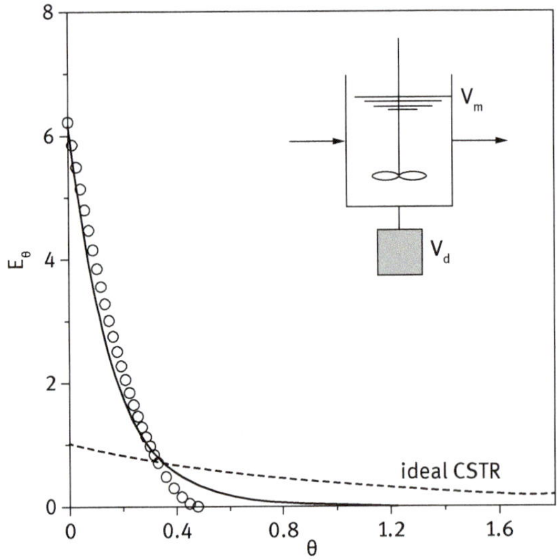

Fig. 5.9: Age distribution E_θ of the crossed-hollow fiber reactor illustrated in Figure 5.8; data are reported as a function of the reduced time θ. Symbols: experimental data; solid line: prediction given by CSTR with dead zone ($V/V_m = 3$) model; dotted line: RTD function for ideal CSTR.

In order to describe the deviation from ideality of the bioreactor (due to the existence of dead zones located in the edge of the device and of active zones where membrane hollow fibers intersect each others), a model is assumed consisting of CSTR with active volume V_m connected to a tank with volume V_d under stagnant flow. The mathematical relation characterizing a CSTR with dead zone is reported in Table 5.1.

In most cases, the real behavior of a bioreactor can be adequately described by simple models that combine PFR and CSTR with eventual fluid-dynamic variants (serial/parallel connections, dead volumes, internal recirculation, etc.); in most cases, these models (some of them reported in Table 5.1) are able to achieve results with good accuracy and low mathematical complexity [5].

RTD analysis of fluids is also used with the aim of identifing and correcting an improper flow inside a bioreactor caused by a poor fluid-dynamic efficiency of the device. Problems are often related to an asymmetric flow velocity, the presence of stagnant zones, time-lapse, channeling, bypass, etc.

Tab. 5.1: Brief overview of simple fluid-dynamic models useful to describe the real behavior of common bioreactors for cell culture.

Model	Reactor scheme	RTD function
Plug flow reactor (PFR)	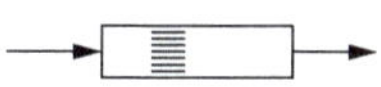	$E(t) = \delta(t - \tau)$
Plug flow with high longitudinal dispersion		$E = \frac{1}{2\sqrt{\pi\theta}\,(D/uL)} \exp\left[-\frac{(1-\theta)^2}{4\theta(D/uL)}\right]$ D: longitudinal dispersion coefficient u: flow velocity
Continuous stirred tank reactor (CSTR)	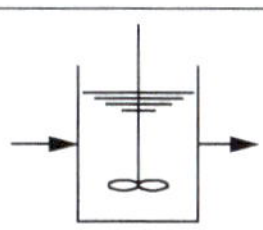	$E = \frac{1}{\tau} e^{-t/\tau}$
Series of N–CSTR	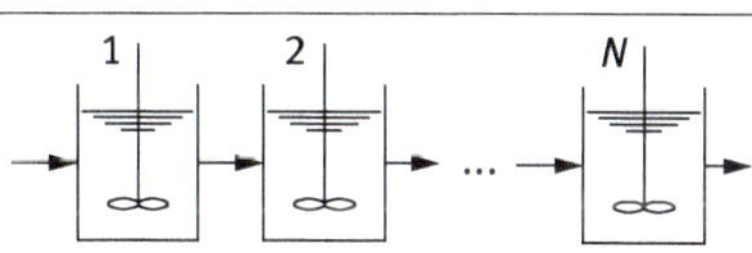	N tanks with equal volume: $E_\theta = \frac{N(N\theta)^{N-1}}{(N-1)!} e^{-N\theta}$
CSTR with dead zone	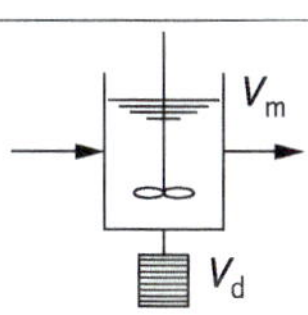	$V = V_m + V_d$ $E_\theta = \frac{V}{V_m} \exp\left(-\frac{V_d}{V_m}\theta\right)$
Serial connection of CSTR and PFR	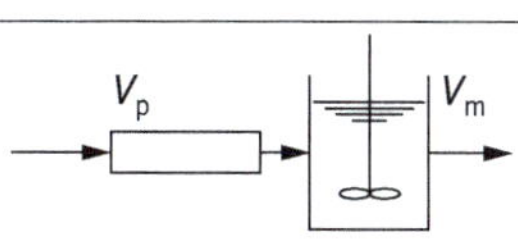	$V = V_m + V_d$ $E_\theta = \frac{V}{V_m} \exp\left[-\frac{V}{V_m}\left(\theta - \frac{V_p}{V}\right)\right]$

In particular, during culture experiments, an average residence time of the fluid inside the bioreactor should be chosen in order to obtain an appreciable metabolic conversion. Small-size lab scale reactors require low inlet flow rates to obtain an average residence time of a few hours; as a drawback, low flow rate typically results in an inadequate degree of mixing inside the bioreactor. A simple strategy to overcome this limit is to recycle part of the stream, leading the bioreactor back to the inlet where it mixed with fresh medium before entering the bioreactor. In such a case, residence time distribution analysis allows us to find the optimal recycle ratio for a complete mixing and uniform metabolite concentration in the bioreactor.

5.3 Reaction kinetics

The reaction rate of metabolic transformations (with particular focus on oxygen, glucose, and lactate) are generally modeled by the Michaelis–Menten equation

$$r = \frac{V_{\text{max}} S}{K_{\text{M}} + S}, \tag{5.13}$$

where S is the substrate (nutrient) concentration, V_{max} is the maximum reaction rate observed in the system, and constant K_{M} represents the substrate concentration corresponding to a reaction rate of $V_{\text{max}}/2$.

The basic reaction mechanism involves an enzyme E which binds the substrate S to form the complex ES; in the subsequent step, ES is converted to product P and free enzyme:

$$\text{E} + \text{S} \underset{k_{-1}}{\overset{k_1}{\rightleftarrows}} \text{ES} \overset{k_2}{\rightarrow} \text{P} + \text{E}, \tag{5.14}$$

with

$$K_{\text{M}} = \frac{k_{-1} + k_2}{k_1}. \tag{5.15}$$

At low substrate concentrations, the reaction rate is practically of first order, and enzyme concentration – in excess with substrate concentration – can be considered constant; at high substrate concentration the reaction rate approaches the constant value of V_{max}, and the kinetics is of zero order (enzyme saturation).

Cell culture experiments and tests are usually carried out at 37 °C. Reduction of temperature might cause a drastic decrement of cell viability or the degradation of mechanical properties of polymeric films supporting cell adhesion [6].

The terms in equation (5.13) can be obtained from experimental investigation. As an alternative, the literature provides a broad range of reference for Michaelis–Menten parameters; Table 5.2 reviews the reaction rates and kinetic parameters of oxygen, glucose, and lactate measured in bioreactors for mammalian cell cultures.

Tab. 5.2: Values of kinetic parameters for O_2 uptake.

Cell type	O_2 consumption rate	K_M	V_{max}	Reference
Chondrocytes (bovine)	$0.2–4.0 \times 10^{-18}$ mol/cell s	=	=	[7]
Chondrocytes (mini-pig)	10^{-10} mmol/cell h	0.01 mmol/l	=	[8]
Embryonic lung fibroblasts (human)	2.5×10^{-18} mol/cell s	=	=	[9]
Hepatocytes (primary, rat)	2.0×10^{-16} mol/cell s	=	=	[9]
Hybridoma	6.0×10^{-17} mol/cell s	=	=	[10]
Lymphocytes (human)	0.5–2 mM/h, 8×10^{6} cells/ml	=	=	[11]
Mesenchymal stem cell (human)	3.3×10^{-17} mol/cell s	=	=	[12]
Mesenchymal stem cells	$1.9–2.7 \times 10^{-17}$ mol/cell s	=	=	[13]
Murine hybridoma	0.078–0.086 µmol/h/10^{6} cells	=	=	[14]
Cartilage (human)	=	0.15 nm/mm^3	4×10^{-7} nmol/cell s	[15]
Chondrocytes (bovine)	=	0.006 mM	1.86×10^{-18} mol/cell s	[16]
Chondrocytes (rabbit)	=	6.3×10^{-2} mol/m^3	1.2×10^{-13} mol/cell h	[17]
Hepatocytes (porcine)	=	11.24 ± 0.4 mmHg	$0.35 \pm 0.05 \times 10^{-15}$ mol/cell s	[18]
Hepatocytes (rat)	=	0.44 mmHg	3×10^{-8} mol/cm^3 s	[19]
Hepatocytes (rat)	=	2.6 mmHg	0.43×10^{-15} mol/cell s	[20]
Keratinocytes (rat skin)	=	0.2 mg/l	9.7×10^{-17} mol/cell s	[21]
Mammalian cells (average values)	=	1.2 g/l	=	[22]
Tracheal epithelial cells (cow)	=	=	0.2×10^{-9} mmol/cell h	[23]

Tab. 5.3: Values of kinetic parameters for glucose uptake.

Cell type	Glucose consumption rate	K_M	V_{max}	Reference
Articular chondrocytes (bovine)	9.3×10^{-16} mol/cell s	=	=	[24]
Articular chondrocytes (rabbit)	0.7×10^{-17} mol/cell s	=	=	[25]
Hippocampal neurons (hamster)	100–300 μg/10^6 cells ml	=	=	[26]
Lymphocytes (human)	1-2 pg/h cell	=	=	[11]
Mesenchymal stem cells (human)	1.4×10^{-16} mol/cell s	=	=	[27]
Murine hybridoma	0.18–0.29 μmol/10^6 cells h	=	=	[14]
Mammalian cells (average values)	=	0.5416 ± 0.33 g/l		[22]

Tab. 5.4: Values of kinetic parameters for lactate uptake.

Cell type	Lactate consumption rate	K_M	V_{max}	Reference
Articular chondrocytes (rabbit)	1×10^{-17} mol/cell s	=	=	[25]
Lymphocytes (human)	0–20 mg/h l 8 × 10^6 cells/ml	=	=	[11]
Murine hybridoma	0.31–0.61 μmol/10^6 cells h	=	=	[14]
Mammalian cells (average values)	=	12.2 ± 6.0 g/l	=	[22]

During long-term cell culture experiments, the number of cells might vary in time due to proliferation or death caused by the poor assimilation of nutrients due to mass transfer limitations. In this case, the consumption rate is described on a cell basis, and Michaelis–Menten equation is modified in order to take into account the population cell dynamics:

$$\mu = \frac{\mu_{max} S}{K_M + S}, \tag{5.16}$$

where μ is the metabolic consumption rate of the nutrient normalized by cell number, and μ_{max} is the maximum rate obtained in correspondence of the highest value of cellular density. It is, in fact, assumed that cell number will reach a maximum value (X_{max}) being the proliferation limited by availability of surface area for adhesion or by the availability of nutrient.

If X indicates the number of cells per volume unit, cell growth can be adequately described by the following phenomenological equation:

$$\frac{dX}{dt} = \mu X \left(1 - \frac{X}{X_{max}}\right). \tag{5.17}$$

Contois kinetics modified by accounting for both nutrient and cell contact inhibition effects [28] and Moser equation [29] represent alternative reaction rate expressions, but are rarely used.

In general, appropriate reaction kinetics data are necessary to obtain realistic predictions from mathematical modeling.

5.4 Modeling HF membrane bioreactors

Modeling a bioreactor is preliminarily devoted to the analysis of hydraulic conditions, flow regime, velocity profiles, and share distribution throughout the biodevice. In particular, pressure drop and flow rate are critical operative parameters to prevent cell detachment from the supporting films and membranes, or to prevent the risk of scaffold deformation, to ensure a proper residence time of nutrients in the reactor, and to guarantee an appropriate delivery rate control of mass transfer resistances of metabolites (in particular, oxygen).

In general, may studies on cells in perfusion bioreactor converge towards an optimum flow rate of around 1 ml/min; a slightly higher flow rate increases nutrient delivery and stimulates the seeded cells by fluid shear forces [30]. A flow rate of 1 ml/min was proven to be sufficient for the survival of the initial cell mass of hepatocytes cultured, while an increase to 1.5 ml/min improved the oxygen delivery [31].

Moreover, evidence from RTD analysis of nonuniform flow patterns causes nonuniform and poor distribution of nutrients in the reactor volume, with significant impact on cell colonization.

The Navier–Stokes equation (5.18), proposed below in its symbolic form (to be adapted to most appropriate geometrical coordinates) and for an incompressible fluid, is at the basis of fluid flow analysis [32]:

$$\rho\left(\frac{\mathrm{d}v}{\mathrm{d}t}\right) + v\,\nabla v = -\nabla p + \mu\nabla^2 v + f, \tag{5.18}$$

where ρ is the density of the Newtonian fluid, v is the fluid velocity vector, p the pressure, μ is the dynamic (or absolute) viscosity, ∇^2 is the Laplacian operator and f the other field forces (if any), such as the gravity of centrifugal force.

Due to the extremely complex nature of the differential equations involved in the formulation of this equation, exact analytical solutions can be obtained only for very simple geometries and under reasonable simplifications.

Among the most common configurations, hollow fiber membrane bioreactors are largely used for mammalian cell culture,because they provide a large surface area per volume unit, a support to cellular adhesion, an efficient mass transfer of nutrients, metabolites,and specific components throughout the device. In axial-flow hollow fiber membrane bioreactor, a hollow-fiber bundle is placed in an appropriate housing. In the most common situation, cells are cultured in the extracapillary space (ECS), and nutrients fed through the lumen of the fibres and transported to ECS across the membrane. The transport across the membrane is both diffusive and convective. However, low membrane permeability, low molecular weight of transported components (i.e. oxygen, glucose, lactate, ammonia, urea, etc.), and low inlet pressure usually lead to the assumption of a bioreactor operating under diffusion control. Despite the conceptual simplicity of HF membrane bioreactor, the efficacy of such device is still limited for clinical applications due to the lack of information which might allow its optimal operation. In this respect, the development of appropriate mathematical models is necessary in order to create a well-controlled environment – with respect to mass transport phenomena and metabolic kinetics – which will maintain cell differentiation and functions for a long time.

When modeling hollow-fiber membrane reactors, commonly accepted assumptions are made in order to reduce the analytical description of the system to a reasonably small set of equations. The first geometrical assumption is to consider all fibers to be identical, cylindrical, and homogeneously distributed in the shell of the bioreactor. Within this ideally symmetric configuration, the next step is to assume that each fiber is surrounded by a uniform annulus of ECS, the so-called Krogh cylinder geometry (Figure 5.10).

The main limitation of the Krogh assumption is that the interstitial space between adjacent Krogh cylinders is neglected; in addition, probable nonuniform spacing among different fibers due to imperfect manufacture of the bioreactor is not considered. Nevertheless, advantages related to the possibility of restricting the modeling of mass transport and reaction to a single Krogh cylinder definitely outweighs the above-mentioned drawbacks.

A short literature survey related to the use of a Krogh cylinder in modeling membrane bioreactors for cell culture is reported in Table 5.5.

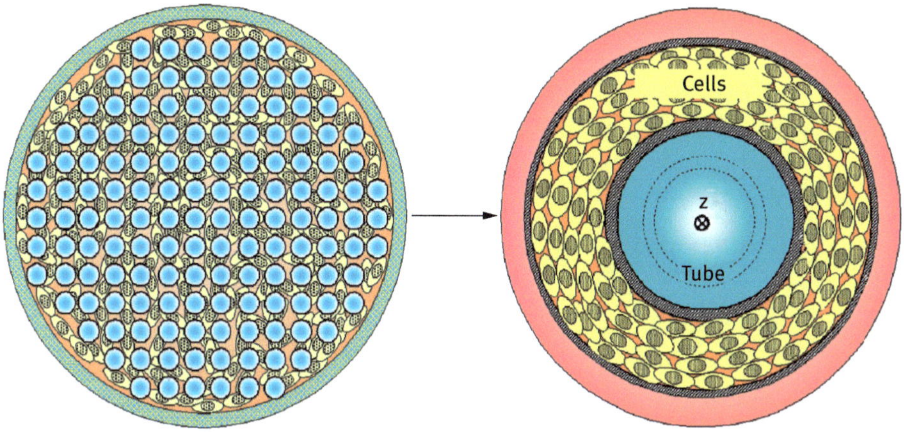

Fig. 5.10: Krogh cylinder concept.

Tab. 5.5: Krogh model in modeling axial-flow membrane bioreactors for mammalian cell culture.

Diffusion, zero order oxygen consumption	Isolated perfused rat hearts	[33]
Diffusion, zero order oxygen consumption	Human bone marrow	[34]
Myoglobin-facilitated oxygen diffusion, zero order oxygen consumption	Red blood cells	[35]
Myoglobin-facilitated oxygen diffusion, Michaelis–Menten kinetics	Skeletal muscle	[36]
Diffusion, Michaelis–Menten kinetics	Cortical and cancellous bone	[37]

Fluid flow in the lumen of a fiber is generally modeled by continuity and momentum equations for incompressible Newtonian fluids. This set of equations, analytically solved for boundary conditions that include symmetry at the lumen centerline and matching velocities and pressures at fiber wall with permeation data, give the profile of both axial and radial velocity along the hollow fiber membrane:

$$\begin{aligned} u_{\text{axial}} &= \frac{2Q_{\text{feed}}}{\pi R^2 N}\left(1 - \frac{Q_{\text{permeate}}}{Q_{\text{feed}}}\frac{z}{L}\right)\left(1 - \frac{r^2}{R^2}\right) \\ u_{\text{radial}} &= \frac{Q_{\text{permeate}}}{2\pi R L N}\left(\frac{2r}{R} - \frac{r^3}{R^3}\right), \end{aligned} \tag{5.19}$$

where Q is the flowrate, R is the fiber radius, N is the number of fibers, L is the fiber length, z and r the axial and radial coordinate, respectively. If the convective flux of species across the membrane is neglected ($Q_{\text{permeate}} = 0$), equations (5.19) reduce to the classical form:

$$u_{\text{axial}} = 2U\left(1 - \frac{r^2}{R^2}\right), \tag{5.20}$$

where U is the mean velocity. The axial component of the velocity profile for a fluid flow through hollow fibers is reported in Figure 5.11.

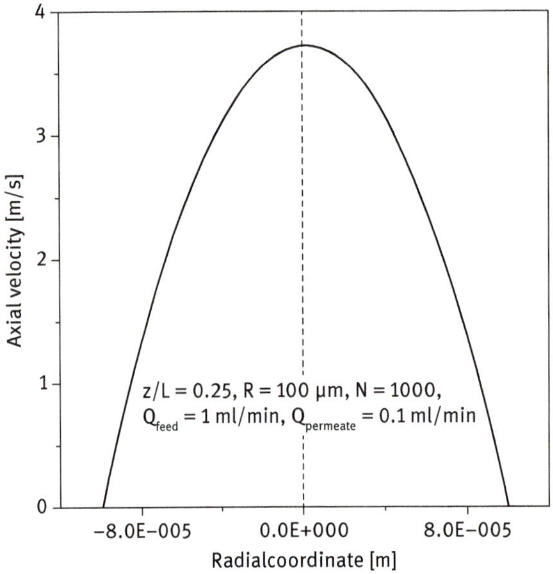

Fig. 5.11: Radial velocity profile of a well developed fluid flow through hollow fiber membranes.

Mass balance on the i-th component, written below for a system exhibiting a cylindrical symmetry and characterized by a constant density, is

$$\frac{\partial c_i}{\partial t} + \left(v_r \frac{\partial c_i}{\partial r} + v_\theta \frac{1}{r}\frac{\partial c_i}{\partial \theta} + v_z \frac{\partial c_i}{\partial z}\right) = D_i\left(\frac{1}{r}\frac{\partial}{\partial r}\left(r\frac{\partial c_i}{\partial r}\right) + \frac{1}{r^2}\frac{\partial^2 c_i}{\partial \theta^2} + \frac{\partial^2 c_i}{\partial z^2}\right) + \Psi_i, \tag{5.21}$$

where c is the concentration, v the velocity vector, D_i the diffusion coefficient, Ψ_i is the reaction term (negative for the consumption of a nutrient, most commonly dissolved oxygen or glucose), and r–z–θ the cylindrical coordinate system.

Equation (5.21) is independent on the θ coordinate in case of systems exhibiting radial symmetry; moreover, if steady-state conditions are also assumed, equation (5.21) shows the following simplified form:

$$v_r \frac{\partial c_i}{\partial r} + v_z \frac{\partial c_i}{\partial z} = D_i\left[\frac{1}{r}\frac{\partial}{\partial r}\left(r\frac{\partial c_i}{\partial r}\right) + \frac{\partial^2 c_i}{\partial z^2}\right] + \Psi_i. \tag{5.22}$$

The reaction term Ψ_i, in the common scenario, takes the form of the Michaelis–Menten kinetic rate as from equation (5.13).

For an axial flow hollow fiber membrane bioreactor, equation (5.22) (under common assumptions of radial Peclet number negligible with respect to axial Peclet number in the lumen, only radial transport in the membrane, no radial convection in the shell) is solved for a given set of boundary conditions:

– axial symmetry, lumen side

$$r = 0, \quad \forall z, \quad \frac{\partial c_{i,\text{lumen}}}{\partial r} = 0; \tag{5.23}$$

– feed concentration, lumen inlet

$$z = 0, \quad \forall r, \quad c_{i,\text{lumen}} = c_{i,\text{feed}}; \tag{5.24}$$

– lumen outlet

$$z = L, \quad \forall r, \quad \frac{\partial c_{i,\text{lumen}}}{\partial z} = 1; \tag{5.25}$$

– lumen/membrane interface

$$r = R_{\text{internal,fiber}}, \quad \forall z, \quad c_{i,\text{lumen}} = c_{i,\text{membrane}}; \tag{5.26}$$

– membrane/shell interface

$$r = R_{\text{external,fiber}}, \quad \forall z, \quad c_{i,\text{membrane}} = c_{i,\text{shell}}; \tag{5.27}$$

– symmetry at the Krogh radium, shell side

$$r = R_{\text{Krogh}}, \quad \forall z, \quad \frac{\partial c_{i,\text{shell}}}{\partial r} = 0. \tag{5.28}$$

Equation (5.22) and the corresponding boundary conditions (equations (5.23)–(5.28)) define a set of partial differential equations that describe, with an acceptable level of accuracy under the assumption made, the fluid dynamics of the bioreactor and the concentration profile of nutrients in the system.

The high complexity of the above equations often precludes the chance of obtaining an analytical solution. It is generally not possible to analytically solve the system using the nonlinear Michaelis–Menten reaction term; therefore, a nutrient concentration much higher than the value of K_M is often assumed, thus approximating the reaction term to a zero-order kinetics with $\Psi_i = V_{\max}$. Under these assumptions, an analytical solution to the problem was provided by [38].

Fortunately, today computational methods allow us to overcome these limitations, because they provide a powerful tool to solve such a complex mathematical problem in a short period of time. By using the finite element method (FEM), through appropriate mesh generation of varying degrees of refinement, the differential equations are discretized and numerically solved with high degree of accuracy [39]. As an example, Figure 5.12 shows the numerical solution of Navier–Stokes equations using the FEM software COMSOL Multiphysics to evaluate the glucose concentration profile inside an axial-flow hollow fiber membrane bioreactor for lymphocytes culture [40].

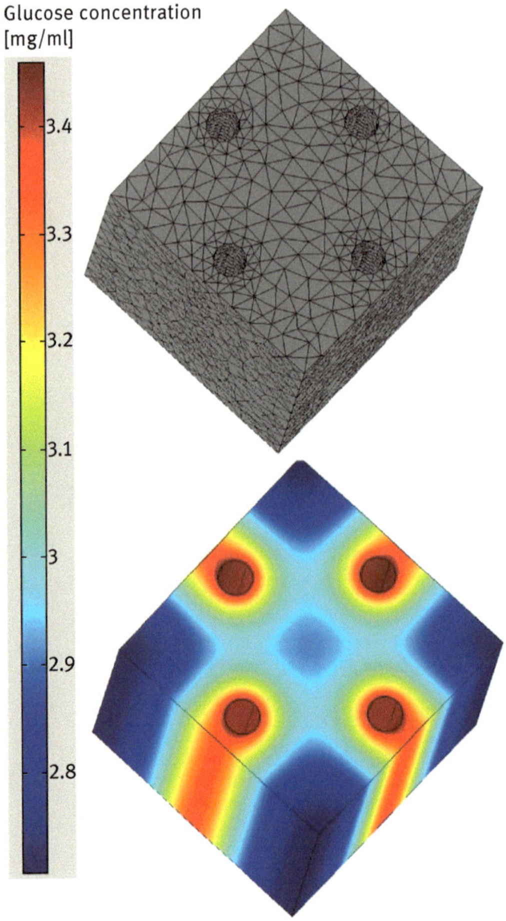

Fig. 5.12: Glucose concentration profile through a repeating unit of a hollow fiber membrane reactor for lymphocyte culture (diffusion coefficient: $2.1 \times 10^{-6}\ cm^2/s$, glucose uptake rate: $1 \times 10^{-4}\ \mu g/s\ mm^3$, glucose feed concentration: 3.5 mg/ml).

The determination of the optimal spacing between hollow fibers in a membrane bioreactor for cell culture is an important design issue, and becomes crucial for diffusion-limited transport of nutrients. This is particularly true for oxygen, considered the limiting reactant to mammalian cell cultures because of its low solubility. O_2 depletion due to cells uptake rate – increased at higher local cellular density – might result in a hypoxic environment in the core of extrafiber space. It is generally accepted that exposure to an oxygen partial pressure below 40 mmHg leads to necrosis.

Experimental observations showed that the interfiber spacing has to be lower than 250 µm for cell density of about 10^8 cells/cm^3 in order to avoid severe oxygen depletion [41]. At higher cell density (10^9 cells/cm^3), this threshold is reduced to about 80 µm [42]. Figure 5.13 shows the modelled oxygen concentration profile through hollow fiber bioreactor for hepatocytes cultured at cell density of 10^7 cells/cm^3 and fed with oxygenated medium (21 % O_2). The critical O_2 threshold, which is the minimum partial pressure of oxygen necessary for cells to perform all metabolic functions, was assumed

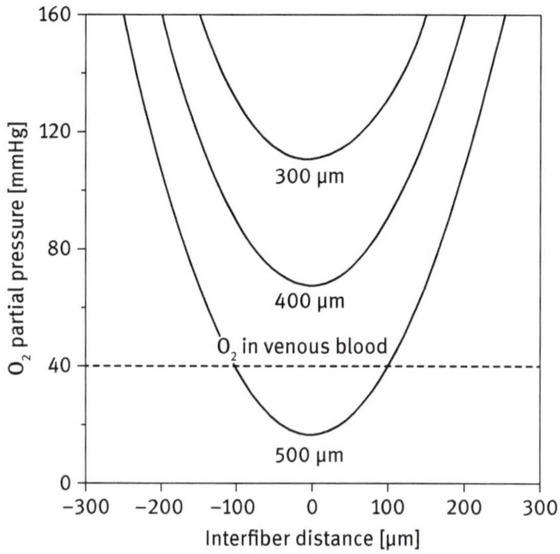

Fig. 5.13: O_2 concentration profiles in hepatocyte cell culture (diffusion coefficient: 1.8×10^{-5} cm^2/s, uptake rate: 17.3 μmol/h, cell density: 1×10^7 cells/cm^3).

to be 40 mmHg, which is the partial pressure of O_2 in the venous blood, corresponding to a dissolved O_2 concentration of 1.65×10^{-3} mg/ml. The critical limit is, in this case, shifted to about 450 μm. Figure 5.14 shows the simulated oxygen profiles inside a tracheal construct composed by an external region of respiratory epithelial cells and and internal compartment populated by chondrocyte cells. The computational FEM analysis was carried out as a function of the cell density (from 1×10^6 to 2×10^7 cells/cm^3). At a cell density of 5×10^6 cells/cm^3, O_2 concentration decreased progressively to a minimum value of 0.095 mM (72 mmHg); when the cell density is doubled, the minimum O_2 concentration reduces by 62 %, still above the critical threshold. For a cell density of 2×10^7 cells/cm^3, the highest investigated, the maximum penetration distance of oxygen is reduced to 350 μm [43].

5.5 Concluding remarks

Robust predictions of bioreactors for cell culture based on analytical solutions of complex fluid-dynamics equations or on numerical output from computational fluid dynamics (CFD) are an excellent tool for developing new methods and protocols for optimization and control of bioengineered scaffolds.

The validity of this approach is confirmed by the evidence that a well-controlled environment with respect to transport phenomena and metabolic kinetics is able to activate a specific cellular response and to maintain cell viability for a long time.

Quantitative measures for the degree of mixing and residence time distribution within the reactor allow an accurate kinetic modeling of the system, and permit us to achieve or preserve a desired flow pattern, to optimize the design of the bioreactor

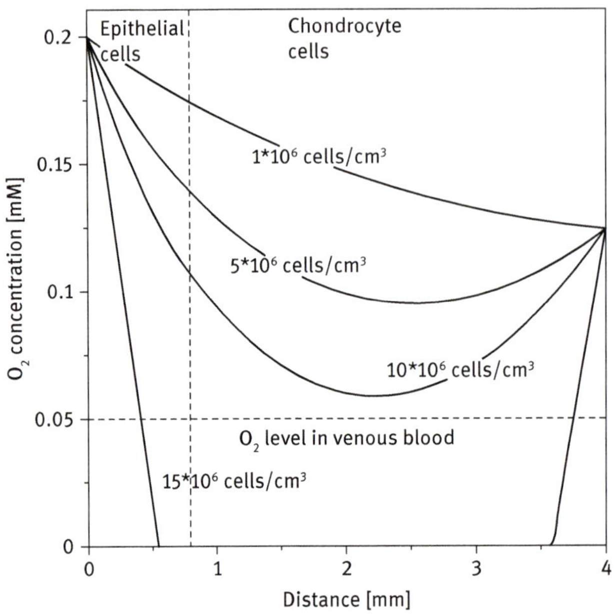

Fig. 5.14: Oxygen concentration profiles at different cell density in a tracheal construct.

shape, to identify appropriate location of the inlets and outlets, and to prevent incomplete radial mixing, stagnant regions and channeling.

Analysis of diffusive transport of metabolites to a cellular compartment through nano- or microporous membranes, as well as the evaluation of mass transfer limitations in conjunction with metabolic kinetics, allow us to predict the minimum nutrient concentration required to culture functional cells and the optimization of device geometry and operation parameters.

With the increasing complexity of fluid dynamics and the operation of future bioreactors which would require more advanced models, finite element methods offer the extraordinary opportunity to mathematically solve intricate sets of equations for which is impossible to obtain analytical solutions.

Driven by the increasing availability of computational resources, the simulation of fluid-dynamic and transport phenomena in tissue engineering and bioreactors represents today an essential tool for efficient multiscale predictions. However, the reliability of analytical and numerical approaches described in this chapter strongly depends on the availability of well-validated experimental data concerning the mechanical properties of supports for cell adhesion, reaction kinetic parameters on 3D scaffolds, and effective diffusivities; accurate experimental determination of these inputs represent the indispensable starting point before any further development of simulation tools.

5.6 References

[1] Graham T. The Bakerian lecture: Osmotic force. Philos Trans R Soc Lond 1. 1854; 44: 117–128.
[2] Danckwerts PV. The effect of incomplete mixing on homogeneous reactions. Chem Eng Sci. 1958; 8: 93–99.
[3] De Bartolo L, Salerno S, Curcio E, Piscioneri A, Rende M, Morelli S, Tasselli F, Bader A, Drioli E. Human hepatocyte functions in a crossed hollow fiber membrane bioreactor. Biomaterials. 2009; 30: 2531–2543.
[4] De Bartolo L, Salerno S, Giorno L, Morelli S, Barbieri G, Curcio E, Rende M, Drioli E. Membrane bioreactor using pig hepatocytes for in vitro evaluation of anti-inflammatory drugs. Catalysis Today. 2006; 118: 172–180.
[5] Levenspiel O. Chemical Reaction Engineering. 3rd edition. New York: John Wiley & Sons; 1999.
[6] Nishida K, Yamato M, Hayashida Y, Watanabe K, Yamamoto K, Adachi E, Nagai S., Kikuchi A, Maeda N, Watanabe H, Okano T, Tano Y. . Corneal reconstruction with tissue-engineered cell sheets composed of autologous oral mucosal epithelium. New Engl J Med. 2004; 351: 1187–1196.
[7] Malda J, Rouwkema J, Martens DE, Le Comte EP, Kooy FK, Tramper J, van Blitterswijk CA, Riesle J. Oxygen Gradients in Tissue-Engineered PEGT/PBT Cartilaginous Constructs: Measurement and Modeling. Biotechnol Bioeng. 2004; 86: 9–18.
[8] Nehring D, Adamietz P, Meenen NM, Portner R. Perfusion cultures and modelling of oxygen uptake with three-dimensional chondrocytes pellets. Biotechnol Tech. 1999; 13: 701–706.
[9] Guarino R, Dike L, Haq T, Rowley J, Pitner J, Timmins M. Method for determining oxygen consumption rates of static cultures from microplate measurements of pericellular dissolved oxygen concentration. Biotechnol Bioeng. 2004; 86: 775–787.
[10] Kim S, Yu D, Son J, Huebner H, Buchholz R. Calculations on O2 transfer in capsules with animal cell for the determination of maximum capsule size without O2 limitation. Biotechnol Lett. 1998; 20: 549–552.
[11] De Bartolo L, Piscioneri A, Cotroneo G, Salerno S, Tasselli F, Campana C, Morelli S, Rende M, Caroleo MC, Bossio M, Drioli E. Human lymphocyte PEEK-WC hollow fiber membrane bioreactor. J Biotechnology. 2007; 132: 65–74.
[12] Zhao F, Pathi P, Grayson W, Xing Q, Locke BR, Ma T. Effects of oxygen transport on 3-D human mesenchymal stem cell metabolic activity in perfusion and static cultures: experiments and mathematical model. Biotechnol Prog. 2004; 21: 1269–1280.
[13] Godara P, McFarland CD, Nordon RE. Design of bioreactors for mesenchymal stem cell tissue engineering. J Chem Technol Biotechnol. 2008; 83: 408–420.
[14] Sadettin SO, Palsson BO. Growth, Metabolic, and Antibody Production Kinetics of Hybridoma Cell Culture: 2. Effects of Serum Concentration, Dissolved Oxygen Concentration, and Medium pH in a Batch Reactor. Biotechnol Prog. 1991; 7: 481–494.
[15] Galbusera F, Ciuffi M, Raimondi MT, Pietrabissa R. Computational modelling of combined cell population dynamics and oxygen transport in engineered tissue subject to interstitial perfusion. Comput Methods Biomech Biomed Eng. 2007; 10(4): 279–287.
[16] Obradovic B, Carrier RL, Vunjak-Novakovic G, Freed LE. Gas exchange is essential for bioreactor cultivation of tissue engineered cartilage. Biotechnol Bioeng. 1999; 63: 197– 205 .
[17] Kino-Oka M, Maeda Y, Yamamoto T, Sugawara K, Taya M. A Kinetic Modeling of Chondrocyte Culture for Manufacture of Tissue-Engineered Cartilage. J Biosci Bioeng. 2005; 99(3): 197–207.
[18] Balis UJ, Behnia K, Dwarakanath B, Bhatia SN, Sullivan SJ, Yarmush ML, Toner M. Oxygen Consumption Characteristics of Porcine Hepatocytes. Metabolic Engineering. 1999; 1: 49–62.
[19] Glicklis R, Merchuk JC, Cohen S. Modeling Mass Transfer in Hepatocyte Spheroids via Cell Viability, Spheroid Size, and Hepatocellular Functions. Biotechnol Bioeng. 2004; 86(6): 672–680.

[20] Foy B, Rotem A, Toner M, Tompkins RG, Yarmush ML. A device to measure the oxygen uptake rates of attached cells: importance in bioartificial organ design. Cell Transplant. 1994; 3: 515–527.
[21] Acevedo CA, Weinstein-Oppenheimer C, Brown DI, Huebner H, Buchholz R, Young ME. A mathematical model for the design of fibrin microcapsules with skin cells. Bioprocess Biosyst Eng. 2009; 32: 341–351.
[22] Karra S, Sager B, Karim MN. Multi-Scale Modeling of Heterogeneities in Mammalian Cell Culture Processes. Ind Eng Chem Res. 2010; 49: 7990–8006.
[23] Kondo M, Tamaoki NDJ, Sakai A, Kameyama S, Kanoh S, Konno K. Increased oxidative metabolism in cow tracheal epithelial cells cultured at air-liquid interface. Am J Respir Cell Mol Biol. 1997; 16: 62–68.
[24] Windhaber RA, Wilkins RJ, Meredith D. Functional characterisation of glucose transport in bovine articular chondrocytes. Pflugers Arch Eur J Physiol. 2003; 446: 572–577.
[25] Tomita M, Sato EF, Nishikawa M, Yamano Y, Inoue M. Nitric oxide regulates mitochondrial respiration and functions of articular chondrocytes. Arthritis Rheum. 2001; 44: 96–104.
[26] De Bartolo L, Rende M, Morelli S, Giusi G, Salerno S, Piscioneri A, Gordano A, Di Vito A, Canonaco M, Drioli E. Influence of membrane surface properties on the growth of neuronal cells isolated from hippocampus. J Membrane Sci. 2008; 325: 139–149.
[27] Markusen JF, Mason C, Hull DA, Town MA, Tabor AB, Clements M, et al. Behavior of adult human mesenchymal stem cells entrapped in alginate-GRGDY beads. Tissue Eng. 2006; 12: 821–830.
[28] Liu D, Chua CK, Leong KF. A mathematical model for fluid shear sensitive 3D tissue construct development. Biomech Model Mechanobiol. 2013; 12: 19–31.
[29] Galban CJ, Locke BR. Analysis of cell growth kinetics and substrate diffusion in a polymer scaffold. Biotechnol Bioeng. 1999; 65: 121–132.
[30] Sikavitsas VI, Bancroft GN, Holtorf HL, Jansen JA, Mikos AG. Mineralized matrix deposition by marrow stromal osteoblasts in 3D perfusion culture increases with increasing fluid shear forces. Proc Natl Acad Sci USA. 2003; 100: 14683–14688.
[31] Kim SS, Sundback CA, Kaihara S, Benvenuto MS, Kim B-S, Mooney DJ, Vacanti JP. Dynamic seeding and in vitro culture of hepatocytes in a flow perfusion system. Tissue Eng. 2000; 6: 39–44.
[32] Battino R, Rettich TR, Tominaga T. The solubility of oxygen and ozone in liquids. J Phys Chem Ref Data. 1983; 12(2): 163–178.
[33] Grinberg O, Novozhilov B, Grinberg S, Friedman B, Swartz H. Axial Oxygen Diffusion in the Krogh Model Modifications to account for myocardial oxygen tension in isolated perfused rat hearts measured by EPR oximetry. Adv Experi Med Biol. 2005; 566: 127–134.
[34] Islam MA. A comparison between analytical and numerical solution of the Krogh's tissue cylinder model for human bone marrow. J Chem Eng. 2006; 24(1): 1–6.
[35] Endeward V, Gros G, Jurgens KD. Significance of myoglobin as an oxygen store and oxygen transporter in the intermittently perfused human heart: a model study. Cardiov Res. 2010: 87: 22–29.
[36] McGuire BJ, Secomb TW. A theoretical model for oxygen transport in skeletal muscle under conditions of high oxygen demand. J Appl Physiol. 2001; 91: 2255–2265.
[37] Zahm AM, Bucaro MA, Ayyaswamy PS, Srinivas V, Shapiro IM, Adams CS, Mukundakrishnan K. Numerical modeling of oxygen distributions in cortical and cancellous bone: oxygen availability governs osteonal and trabecular dimensions. Am J Physiol Cell Physiol. 2010; 299: C922–C929.
[38] Shipley RJ, Davidson AJ, Chan K, Chaudhuri JB, Waters SL, Ellis MJ. A strategy to determine operating parameters in tissue engineering hollow fiber bioreactors. Biotechnol Bioeng. 2011; 108(6): 1450–1461.

[39] Hutmacher DW, Singh H. Computational fluid dynamics for improved bioreactor design and 3D culture. Trends Biotechnol. 2008; 26(4): 166–172.
[40] Curcio E, Piscioneri A, Salerno S, Tasselli F, Morelli S, Drioli E, De Bartolo L. Human lymphocytes cultured in 3-D bioreactors: Influence of configuration on metabolite transport and reactions. Biomaterials 2012; 33: 8296–8303.
[41] Chresand TJ, Gillies RJ, Dale BE. Optimum fiber spacing in a hollow fiber bioreactor. Biotechnol Bioeng. 1988; 32(8): 983–992.
[42] Brotherton JD, Chau PC. Modelling of axial-flow hollow fiber cell culture bioreactors. Biotechnol Prog. 1996; 12: 575–590.
[43] Curcio E, Macchiarini P, De Bartolo L. Oxygen mass transfer in a human tissue-engineered trachea. Biomaterials. 2010; 31: 5131–5136.

6 Cell-membrane interactions

6.1 Mimicking *in vivo* environment

In vivo cells are surrounded by the extracellular matrix (ECM) that provides physical architecture and mechanical strength to the tissue. The native ECM is a highly hydrated, hierarchically organized, and dynamic structure that regulates adhesion, migration, proliferation, differentiation, morphogenesis, and gene expression [1]. In particular the extracellular microenvironment is a network hosting three different main effectors: (1) insoluble hydrated macromolecules (fibrillar proteins like collagens, noncollagenous glycoproteins such as elastin, laminin or fibronectin, and hydrophilic proteoglycans with large glycosaminoglycan (GAG) side chains) called physical signals; (2) soluble macromolecules; and (3) proteins on the surfaces of the neighboring cells that establish the cell-cell interaction [2, 3]. Mechanical properties of ECM are given by a complex structure of interwoven fibrous proteins of collagen and elastin, while other insoluble proteins such as fibronectin and laminin that are deposited on this backbone provide specific binding moieties to the cells for recognition and adhesion. Glycosaminoglycans such as hyaluronic acid and heparan sulphate fill the remaining space of this fibrous mesh serving as compression stress buffer and sequestering growth factors.

The interplay of different major signaling, environmental stress and physical cues from the ECM which surrounds the cells *in vivo* regulates the growth and differentiation of most cell types. In fact, the varied composition of ECM components not only provides the physical architecture and mechanical strength to the tissue, but also contains a reservoir of cell-signalling motifs (ligands) and growth factors that guide cellular anchorage and behavior. Therefore, dynamic changes in the composition and structure of the ECM affect both chemical and physical properties which result in alteration of cell adhesion.

It has been demonstrated that cells are able to sense and interpret the information coming from the ECM, responding and reorganizing in function of topography [4], mechanical properties (e.g. stiffness, viscosity, and elasticity) [5], molecules presented by the ECM [6], and concentration of soluble and tethered growth factors [7]. A multiple combination of physico-chemical and biological cues within a spatiotemporal context are received by cells through their interactions with neighboring cells, ECM, soluble/tethered factors [8]. The biological response is influenced by multiple cellular interactions with the individual and specific ECM molecules, and often with multiple sites within the same molecule, as well as by a highly dynamic and complex array of biophysical and biochemical properties of the ECM. The cells are able to receive the external signals through different cell surface receptors of the integrin family and integrate it by an intracellular signalling pathway which affects the cellular response in terms of gene expression, ultimately establishing the cell phenotype.

DOI 10.1515/9783110268010-007

Thus, the final decision of a cell to differentiate, proliferate, migrate, apoptose, or perform other specific functions is a coordinated response to the molecular interactions with these ECM effectors [2]. It is worth noting that the flow of information between cells and their ECM is highly bidirectional as, for example, observed in processes involving ECM degradation and remodeling. It is evident how native ECM exhibits macroscale to nanoscale patterns of chemistry and topography [3], and it is therefore somewhat unsurprising that cells respond to these various scales of chemically and/or topographically patterned features.

Cell survival and cell proliferation require interaction with the extracellular matrix. Epithelial cells in some tissues, such as skin, are continuously renewed from stem cells that rest on a basement membrane. Neighboring cells migrate into the space left empty by cells that have moved away to differentiate. In other epithelia that are not continuously renewed, interaction with the matrix appears to promote differentiation. ECM plays also a key role in the survival process of the cells indeed during involution, the basement membrane is dissolved by proteolysis, and the cells undergo apoptosis.

Nanostructured membranes are able to mimic the ECM, instructing cell adhesion and organization. Indeed they exhibit like ECM micro- to nanoscale of chemistry and topography providing physical, chemical, and mechanical signals to the cells that generate different responses (Figure 6.1) [9]. In particular, in the case of progenitor cells or stem cells a suitable membrane onto which cells can organize and develop is a key to the differentiation and maintenance of the differentiated phenotype at the morphological and functional levels. Some studies reported that chitosan membranes promoted the proliferation and differentiation of rat embryonic liver cells giving them the means to acquire and maintain specific functions [10]. Cells underwent a functional differentiation, showing their hepatocyte functions in terms of urea synthesis, albumin production, and diazepam biotransformation.

6.2 Receptors mediating cell interactions

The interaction of cells with the ECM is mediated by membrane receptors. The largest family of receptors, which mediates cell adhesion to fibronectin, laminins, and collagens, are integrins (Figure 6.2). Several other cellular receptors have also evolved to bind to various matrix components.

Integrins are membrane bound molecules (receptors) that can bind to extracellular matrix molecules ("adhesion proteins" and collagen) [11, 12]. They are the principal mechanism by which cells bind to and respond to the extracellular matrix. They are part of a large family of cell adhesion molecules which are involved in cell extracellular matrix and cell-cell interactions. Functional integrins consist of two transmembrane glycoprotein subunits that are noncovalently bound. Those subunits are called alpha and beta. The alpha subunits all have some homology to each other, as do the beta subunits. The receptors always contain one alpha chain and one beta chain and are

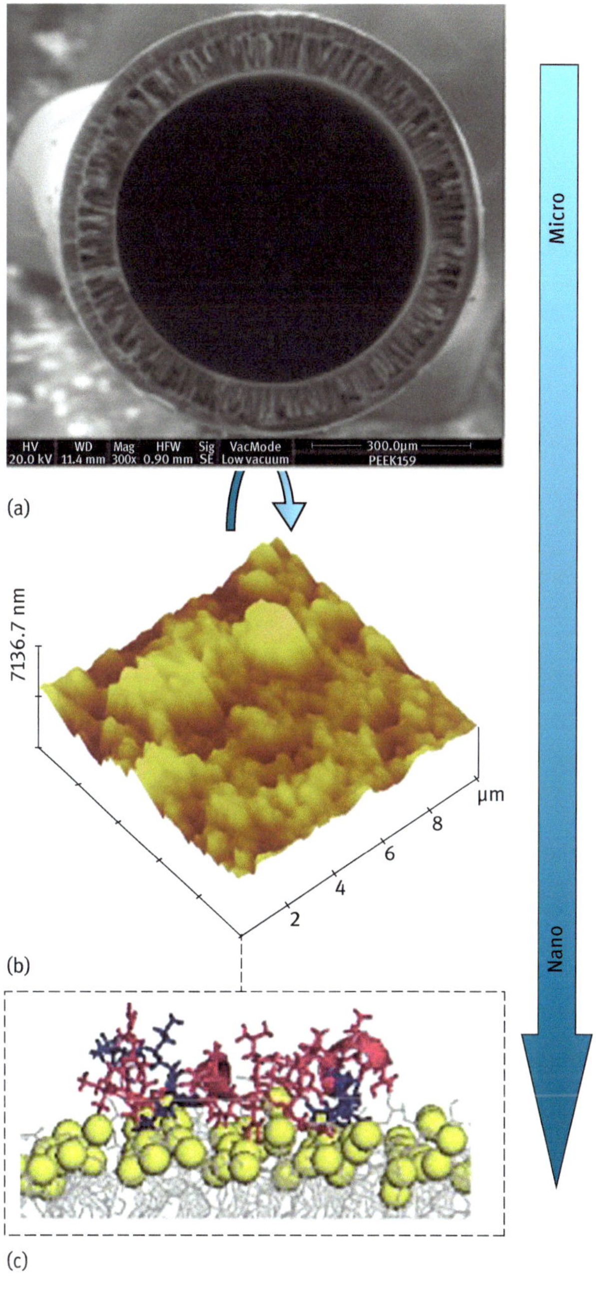

Fig. 6.1: Membrane as instructive extracellular matrix.

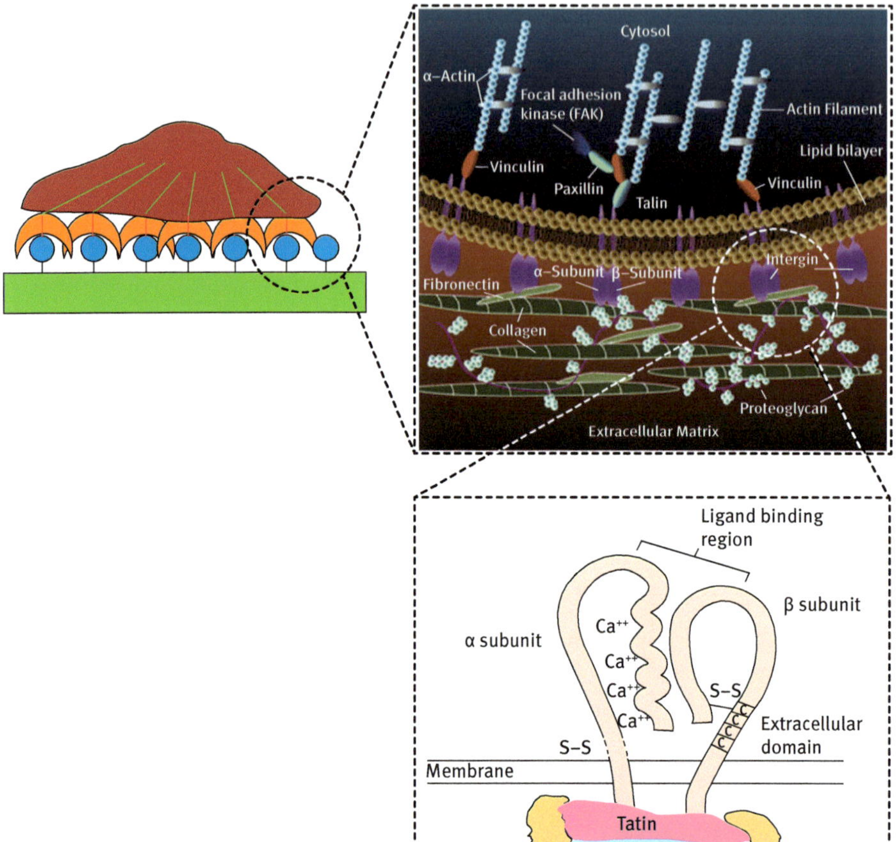

Fig. 6.2: Schematic representation of activated integrin and formation of ECM-integrin-cytoskeleton linkages in the focal adhesion.

thus called heterodimeric. Both subunits contribute to the binding of ligand. Until now 16 alpha and 8 beta subunits have been identified. From these subunits some 22 integrins are formed in nature, which implicates that not all possible combinations exist. The beta 4 subunit, for instance, can only form a heterodimer with the alpha 6 subunit. On the other hand, the beta 1 subunit can form heterodimers with 10 different alpha subunits [13]. Because not all the beta 1 alpha heterodimers have the same ligand specificities, it is believed that the alpha chain is at least partly involved in the ligand specificity. Integrins can bind to an array of ligands. Common ligands are fibronectin and laminin, which are both part of the extracellular matrix and basal lamina. Integrins recognize specific amino acid sequences in the ECM proteins such as the arginine-glycine-aspartic acid (RGD) amino acid sequence. This sequence is found in a number of ECM proteins as fibronectin, vitronectin, laminin, and type I collagen.

Integrins, like $\alpha_V\beta_3$ and $\alpha_5\beta_1$, specifically bind this sequence, while the former is a primary vitronectin receptor, and the latter is a primary receptor of fibronectin. Both of these ligands mentioned above are recognized by multiple integrins. For adhesion to ligands, both integrin subunits are needed, as is the presence of cations. The alpha chain has cation-binding sites. However, not all integrins bind the ECM proteins via RGD sequence: for example the integrin $\alpha_4\beta_1$ binds to a region of fibronectin that does not contain RGD [14]. Integrins are composed of long extracellular domains which adhere to their ligands, and short cytoplasmic domains which link the receptors to the cytoskeleton of the cell.

The structure of alpha subunits is very similar. All contain 7 homologous repeats of 30–40 amino acids in their extracellular domain, spaced by stretches of 20–30 amino acids. The three or four repeats that are most extracellular contain sequences with cation-binding properties. These sequences are supposed to be involved in the binding of ligands, because the interaction of integrins with their ligand is cation dependent. Integrins are responsible for cell attachment and organization of ECM and are involved in the transfer of signals from substratum to the cell interior through a highly organized aggregate of molecules, a focal adhesion that includes such cytoskeletal components (i.e., actin, talin, vinculin, FAK) [15]. This signaling is accompanied by the phosphorylation of different proteins on tyrosine residues, among them a 120 kDa protein called focal adhesion kinase FAK that is involved in the activation of genes responsible for protein synthesis and cell proliferation.

6.3 Cell adhesion

The native ECM exhibits from macro- to nanoscale patterns of chemistry and topography. For this reason the cells can respond to various chemically and/or topographically patterned features.

When cells are cultured *in vitro* they receive very different physical, chemical, and mechanical stimuli from the unfamiliar surrounding environment. Micro- and nanoscale mechanical properties of scaffolds are critical because the cells not only adhere to the surface but also "pull" on the substrate surface and on adjacent cells. The surface chemistry of the scaffolds affects the adhesion of cells through the ECM protein adsorption and stereospecific chemical interactions. Many tissue-engineering strategies consist of the modification of material surface structures aimed at mimicking the specific cues of the ECM in order to direct cell behavior, and trigger tissue regeneration [16–23]. When a material is in contact with a biological system, first water molecules interact with the material's surface, followed by protein adsorption. The surface-bound proteins provide the recognition sites that enable cell adhesion via specific cell receptors (e.g. integrins). The chemistry and topography of the surface affects protein adsorption in terms of quantity, conformation, orientation, and distribution over the surface. The adsorbed proteins may undergo conformational change in such a manner that prevent cell receptor recognition [24]. The cell adhesion is a prerequisite for the survival of different cell types. Cell adhesion and cytoskeleton adaptation to a material surface are crucial in the development of tissue constructs. Material surface properties can be improved by physico-chemical modification, biofunctionalization, and/or topographical surface structuring. Membranes that interact with cells should be chosen not only on the basis of their separation properties but it is necessary also to take into account their physico-chemical and morphological surface properties. It has been shown that the morphology of cells adherent to a substratum changes with its properties, and that the maintenance of cell morphology is important for cell functions.

6.4 Protein adsorption to the membrane surface

Protein adsorption to a membrane surface is influenced by bulk concentration of the protein solution, rate of diffusion, and affinity of the protein for the membrane. Solutions with higher concentrations of a protein lead to more protein on the surface. Another parameter is the diffusion rate of proteins, which depends on its size: smaller proteins are faster and tend to contact and adsorb first with respect to bigger proteins. An important factor is also the protein affinity for the membrane surface: proteins with high affinity likely adsorb and adhere strongly to the surface on the other hand proteins with low affinity form weak and small number of bond with the surface [25, 26]. Taking into consideration these aspects, the bonds between proteins and material surfaces are not completely static: they can break and reform randomly over and over in a process called the Vroman effect [27]. Therefore new proteins with high affinity to surface sites may take over the specific site from the lower affinity proteins. The higher affinity protein can replace previously preadsorbed lower affinity proteins (Figure 6.3).

Proteins contacting a membrane surface interact through intramolecular bonds that can be distinguished in: (1) hydrophobic interactions such as nonpolar domains of the protein which avoid polar groups of the surface; (2) ionic bonds between positive and negative charges; (3) charge transfer between molecules. The prevalence of these types of bonds is influenced by the properties of the proteins, as described in the Table 6.1.

The adsorption of proteins on a membrane surface depends on the properties of the membranes (Table 6.2). Before proteins adsorb on the membrane, water molecules from the solution interact with the material surface. If the material surface is hydrophobic, water molecules tend to interact each other more than with the surface, forming a layer which is ordered with decreased entropy. The break of this layer with proteins is favored by an increase of entropy, which is the reason for protein adsorption on hydrophobic surfaces [29, 30]. Therefore a great adsorption of proteins occurs with consequent change of conformation that can lead to the loss of activity. The interactions between the internal hydrophobic protein domains and the hydrophobic

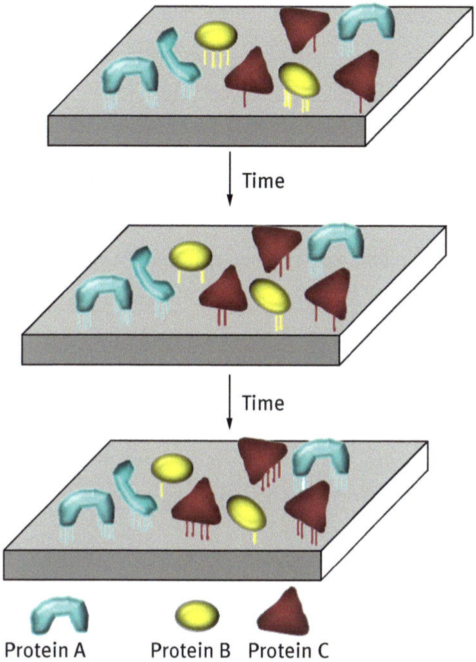

Fig. 6.3: Sequential adsorption of proteins as described by the Vroman effect. Protein A adsorbed on the surface in different conformation changes over time. The protein B with multiple bonds are replaced with time by a higher-affinity protein C (adapted from [28]).

Tab. 6.1: Protein properties affecting the protein adsorption on material surface.

Properties	Protein
Hydrophilicity/ hydrophobicity	Polar hydrophilic domains adsorb to polar surface/apolar hydrophobic domains adsorb to hydrophobic surfaces.
Size	Small proteins diffuse more quickly and reach the surface faster.
Charge	Charged proteins adsorb on the surface with opposite charge.
Structural stability/rigidity	Proteins that are less structurally stable exhibit greater unfolding upon adsorption on surface and form more contact points/bonds.

surface lead to denaturation. Indeed, proteins undergo a change of conformation to a larger extent on hydrophobic than on hydrophilic surfaces [31].

On the other hand, in the case of hydrophilic surfaces water molecules can form more hydrogen bonds with the surface, thus competition between water molecules and proteins results in a decrease of protein adsorption on such surfaces. Furthermore, the surfaces can have a distribution of charges that favor protein adsorption in the case of opposite charges.

A variety of functional species, such as amino, carbonyl, carboxyl, and aromatic groups, can be present on the surface of polymeric membranes. Depending on which species are exposed, proteins may have different affinities for various surfaces.

Tab. 6.2: Membrane properties affecting the protein adsorption.

Properties	Membrane surface
Hydrophilicity/hydrophobicity	Surface free energy parameters of membranes affect protein adsorption
Charge	Opposite charges between surface and protein promote protein adsorption
Chemical composition	Chemical nature of membranes affects the type of the bond between membranes and proteins
Topography	Roughness and features increase the surface area
Heterogeneity	Heterogeneity of the surface results in domains that interact differently with proteins

6.5 Topographical influences on cell adhesion and functions

The surface topography of a membrane can affect cellular adhesion, orientation, and ECM production.

Several methods, including electrospraying, electrospinning, layer-by-layer chemical deposition, and soft lithography, can be used for the realization of nanostructures [32]. Differently, polymer demixing, chemical etching, and colloidal lithography can create random patterns [33]. Cell behavior has been investigated in response to different surface features, such as ridges, grooves, pits, ribbons, islands, and spikes [34–38]. An important outcome of these studies is that the topography influences protein adsorption and cell behavior. Surfaces with nanoscale features, for example, provide relatively large surface areas compared to microscale features, favoring greater protein adsorption. It has been shown that the interaction between protein and surfaces is affected by the surface topography and shape of the proteins. In particular, proteins with different molecular weight and shape, like albumin (a small globular protein) and fibrinogen (a rod-like protein), have different adsorption profiles and conformations, depending upon the nanosphere size which was used as the model substrates of the topography [39]. The structure of albumin, a globular protein, is stabilized by high surface curvature, while fibrinogen, a rod-like protein, is distorted by wrapping around surface curvature, inducing secondary structure loss. It seems that fibrinogen, which has a rod-like shape, can adsorb in two possible orientations: side-on, with its long axis parallel to the particle radius, or end-on, with its long axis perpendicular to the radius. The side-on orientation is favored on smaller particles. A large change in fibrinogen conformation was observed when it adsorbs onto particles with radii less than 30 nm. Upon adsorption the molecule possibly wraps around the surface. The surface curvature of larger particles is too small to induce protein structural alteration, and an end-on orientation can be reached, driven by increased protein-protein interactions. Therefore it may be possible to design specific surfaces to allow fibrinogen or albumin, in order to allow platelet or cell adhesion and the integration of the implant into the host.

Protein adsorption and conformation, and consequently cell behavior, may also be influenced by an asymmetric distribution of nanostructures.

The topographical surface features affect cellular response not only through the protein adsorption profile, but also through the physical interactions with the nano- and microstructures of the surface. Different surface features can cause different physical stresses, which produce cytoskeletal tensions and which may activate various processes and gene expression.

It has been demonstrated that the roughness and pore size of polymeric membranes affect the viability and metabolic rates of isolated hepatocytes [40]. This study demonstrated that rougher membrane surfaces enhance cell adhesion, providing more anchorage points for the attachment of hepatocytes, which are anchorage-dependent cells. The microstructures over the surface induce the formation of cytoplasmatic membrane protrusions that allow the cell migration, the establishment of cell-cell contacts, and the aggregation in 3D structures. The morphological surface properties in terms of pore size, porosity, and roughness affect the morphofunctional behavior of hippocampal neurons [41]. It has been shown that neurons exhibit a different morphology and neurite outgrowth in response to varying the properties of the membrane surface. Membranes with roughness R_a ranging from 6 to 50 nm tend to strongly favor the formation of well-polarized neuronal structures which maintain an unaltered cytoskeletal features throughout their development processes. On the contrary, cells on rougher membrane surfaces (R_a = 200 nm) developed short neurites with the tendency to grow into the pores of the membrane and form aggregates.

Nano- and micropatterned membrane surfaces with channels, ridges, and bricks can influence the special orientation of cell cytoskeleton. Surface microstructures can serve as contact guidance which lead to morphological changes, cell migration, and cell alignment. For example, neurons have the capacity of responding to topographical features in their microenvironment, and they have been shown to adhere, migrate, and orient their axons in the same direction of surface features [42]. Cells "sense" topographical features. They explore the surrounding extracellular environment through neuritis filopodia and lamellipodia which emerged from the growth cone. As a consequence, if the surface exhibits a channel pattern, this is recognized by the cells in which cytoskeleton structures generate traction forces that depend on the assembling and orientation of microtubules and actin filaments within the cytoskeleton. The traction force exerted by a filopodia guide the direction of neuritis extension [43]. Nonpatterned and micropatterned biodegradable poly(L-lactic acid) membranes developed by phase separation micromolding have been investigated in the neuritis outgrowth and orientation of hippocampal cells. In particular the neuritis length, orientation, and specific functions of hippocampal cells have been explored on micropatterned membranes with ridges and channels of different width and with bricks up to 12 days of culture. As shown in Figure 6.4, neurites tend to emerge uniformly in all directions on nonpatterned membranes; on the other hand, patterned membranes with channels and ridges the height of the ridges limit the area for cell adhesion and the directions over which the microtubules and actin filaments accumulated and oriented the filopodia [42]. In the case of interconnected channel membranes, the elongation of neuronal processes and the formation of synaptic communication between cells followed with high fidelity the topography and brick geometry of interconnected channels of the membrane surface. Therefore patterned membranes induced high orientation of neurites in a restricted area of the grooves reducing the complexity of neuronal architecture with consequent creation of a highly ordered neuronal cell matrix.

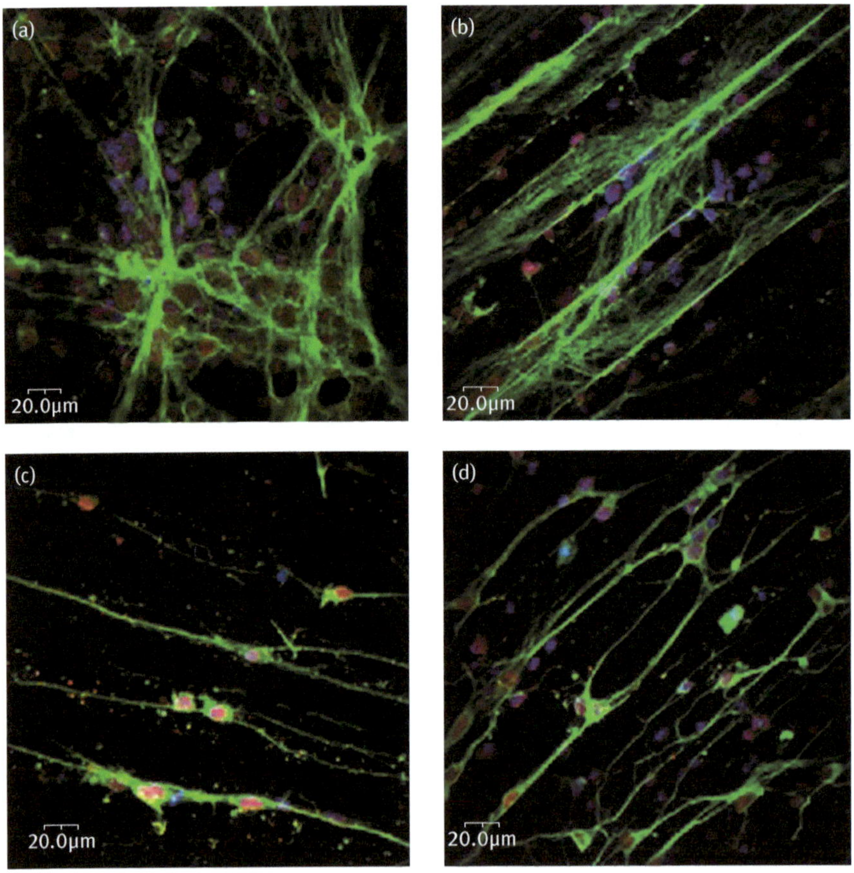

Fig. 6.4: Laser confocal micrographs of hippocampal neurons after 8 days of culture on: (a) nonpatterned membranes, (b) micropatterned membranes with 20 and 25 μm channel and ridge width, (c) micropatterned membranes with 20 and 17 μm channel and ridge width, and (d) micropatterned membranes with interconnected channels. Cells were stained for bIII-tubulin (green), GAP43 (red) and nuclei (blue). Reproduced in part from [42] with permission of Biomaterials.

Understanding the precise mechanisms of how cells respond to such patterns will enable creation of defined structures to guide and control the direction and degree of cell spreading, which is of paramount significance in tissue engineering applications.

Microgrooves have affected smooth muscle cell proliferation, performing an inhibitor role with a consequent reduction the incidence of intimal hyperplasia associated with endovascular stents [44]. Cardiomyocytes have also been extensively used to investigate the significant influence of topographical changes on cell alignment and elongation [45].

Recently it has been demonstrated that human bone marrow mesenchymal stem cells respond to different nanopattern designs with specific changes of their microtubule organization. In particular, groove patterns with width/spacing of 40/30 μm influenced stem cell alignment, elongation, and the neuronal-like cell differentiation [46, 47].

6.6 Influence of membrane physico-chemical properties on cell adhesion

The physico-chemical properties of membrane surfaces can affect protein adsorption and cell adhesion. Cell-substrate adhesion is a multistep process that involves, in sequence: interaction between cells and substrate; adsorption of ECM proteins onto the surface; recognition of ECM components by cell receptors; cytoskeletal rearrangements with formation of focal adhesion; cell spreading. In the initial step the interaction between cell and surface is governed by electrostatic, van der Waals and steric stabilisation forces (Figure 6.5).

The electrostatic forces can be either attractive or repulsive, depending on the charge associated with the substrate surface, considering that cells have a net negative charge [48]. The van der Waals forces are always attractive, while the steric stabilization forces are always repulsive. The steric stabilization forces originate from the osmotic imbalance in the gap between the cell and the surface, where the water

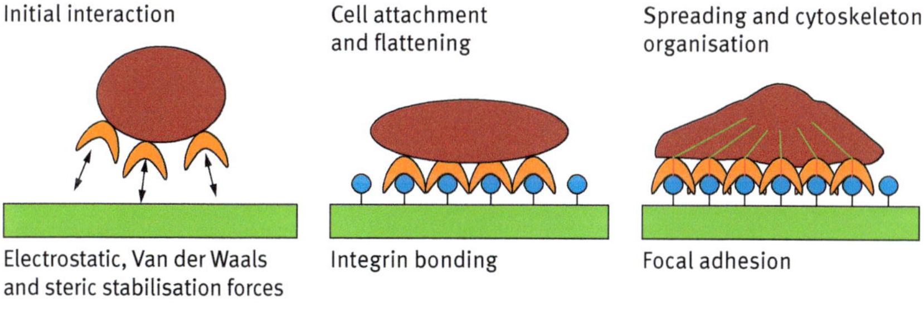

Fig. 6.5: Steps of *in vitro* cell adhesion.

molecules have been forced out. However, these are only unspecific interactions that are fairly weak. Subsequently, proteins present in the cell culture environment adsorb onto the surface. Besides the characteristics of the proteins, as discussed above, this process is also strongly dependent on the physico-chemical properties of the membranes. In particular, it has been shown that the cell adhesion is influenced by the surface free energy parameters of the membranes [49, 50]. Surface free energy is one of the parameters that characterizes the material surfaces [51]. There are various approaches to calculate the surface free energy from the contact angle measurements [52–56]. One of the approaches that gives the possibility of estimating the contribution of all components is the van Oss–Chaudhury–Good method [55, 56]. The authors divided the surface free energy (γ_S) into two components, one including the long-range interactions (London, Keesom, and Debye), the Lifshitz–van der Waals component (γ^{LW}), and the other that contains the short-range interactions (acid base), called the acid-base component (γ^{AB}). The latter component is considered to be equal where γ^+ (electron acceptor) and γ^- (electron donor) mean the acidic and basic constituents, respectively, which are associated with the acid-base interactions. On the basis of this approach, three reference liquids (e.g. ultrapure water, diiodomethane, and glycerol) can be used to determine the apolar γ^{LW}, the acid-base γ^{AB}, acid γ^+, base γ^-, components of surface free energy. Generally, the probe liquids have to be chosen taking into account the solubility of some polymers in common organic solvents. The Lifshitz–van der Waals component γ^{LW} of the membrane surface tension reflecting the dipole interactions can be calculated from the measured diiodomethane contact angles under the assumption that diiodomethane is an apolar test liquid:

$$\gamma_S^{LW} = \frac{\gamma_l^{LW}(1 + \cos\theta)^2}{4}. \tag{6.1}$$

After the γ^{LW} of membrane surface has been measured, it is possible to calculate the other components (γ^{AB}, γ^-, and γ^+) by using two polar liquids (glycerol and water):

$$\gamma_l(1 + \cos\theta) = 2\left(\sqrt{\gamma_S^{LW}\gamma_l^{LW}} + \sqrt{\gamma_S^+\gamma_l^-} + \sqrt{\gamma_S^-\gamma_l^+}\right) \tag{6.2}$$

and

$$\gamma_S^{AB} = 2\sqrt{\gamma_S^-\gamma_S^+}. \tag{6.3}$$

De Bartolo et al., investigated the cell adhesion and functionality on polymeric membranes with different physico-chemical properties: cellulose acetate, polysulfone, polycarbonate, polypropylene, and perfluoropolymer membranes. These membranes exhibited different acid-base components γ^{AB} of the surface free energy, due to different chemical structures and the amount of free end-groups of the polymer [49, 50]. The authors found a protein adsorption relatively large on hydrophobic membranes and relatively small on more hydrophilic membranes. Taking into account the van Oss–Chaudhury–Good equations, a surface can interact with water, exhibiting either

Lewis acid or base behavior, or both of them. Changes in surface structure that increase the concentration of ether groups enhance Lewis base parameter and hence its hydrophilicity, with little influence on the acid parameter. An equal increase of hydroxyl groups will increase both Lewis acid and base parameters. Therefore, a surface is hydrophobic on account of the LW character which it exhibits, with little or zero Lewis acid or base character, while it is hydrophilic accordingly the presence of Lewis base groups, Lewis acid groups, or both. The van Oss–Good–Chaudhury approach seems to give valuable chemical information on the investigated polymers when comparing only one parameter (γ^+ or γ^-) for different surfaces. However, it is also obvious that the base parameter γ^- is overdetermined.

Since the base parameter is an indicator of the hydrophilic character of the surface, the high amounts of adsorbed proteins occurred on membranes with γ^- ranging from 3 to 10 mJ/m^2, which corresponds to PP, PF, and PC membranes. The protein adsorption changed the surface free energy components of the membranes, increasing especially the base parameter. Cells adhered strongly on CA, PC, and PSf membrane surfaces with γ^- of 46 and 54 mJ/m^2, whereas on hydrophobic surfaces such as PP and PF membranes the greatest number of cells adhering to the membrane surface was measured on the surfaces with γ^- of 20 and 21 mJ/m^2, respectively. The results obtained in this study highlighted a relationship between the base parameter of the membrane surface free energy and cell adhesion and metabolic activity [50]. Regardless of the type of native polymeric membranes, it is possible to improve cell adhesion and specific functions by changing their surface free energy and related components.

6.7 Functionalization of membrane surfaces

A variety of techniques have been proposed in the last decades for chemical and/or biochemical modifications of the surface of materials in order to improve the interactions with cells and/or blood. The strategies undertaken by several groups to modify the surface of materials leaving unaltered the bulk and functional properties include: (1) modification of the material's surface via implantation of new atoms, or via oxidation of the outermost atomic layers; (2) the deposition and adsorption of noncovalently linked molecules or biomolecules onto the surface, producing a coating layer; and (3) covalently linked molecules or biomolecules forming a coating layer. Figure 6.6 reports some approaches to increase the biofunctionality of the membrane surface. Chemical modification of the polymeric membranes conferring charged endgroups (e.g. –OH–, –S–, –COO–, NH_3^+) to its surface may lead to protein adsorption and structural rearrangements via electrostatic interactions triggering the early stage attachment phase of cells. It has been shown that the grafting of N-containing functional groups on membranes allows an increase in the polarity of the surface and to have chemical groups typical also of proteins which could support cell adhesion and functions [57–59]. The grafting of N-functional groups can be achieved by a low-temperature plasma modification process which is widely used to modify the surface of biomaterials through etching, deposition and grafting without altering their bulk [60]. In particular, plasma treatments with NH_3 or N_2 feeds and plasma deposition processes with N-containing monomer feeds (e.g., allylamine) provide surfaces with N-containing functional groups, whose distribution and density can be tuned with the plasma parameters. NH_3 glow discharge plasma, for example, has been used to modify PEEK-WC-PU membrane surfaces with nitrogenated functionalities in order to enhance the maintenance of differentiated functions of human hepatocytes [57, 58]. In this study nitrogen functionalities grafted over the membrane trigger initial attachment of cells promoted by electrostatic interactions between amino groups positively charged at physiological pH, and negatively charged carboxylated groups of medium proteins and/or with proteoglycans of the pericellular membranes. Consequently, the liver-specific functions were maintained at high levels up to 19 days on these surface-modified membranes. The membranes plasma-grafted in NH_3 radiofrequency glow discharges have been employed to promote the expansion and differentiation of rat embryonic liver cells [59]. NH_3 plasma modification of PEEK-WC-PU membranes up-regulated and down-regulated the albumin and AFP gene expression, respectively, as a result of their enhanced differentiation in contrast to the other substrates used.

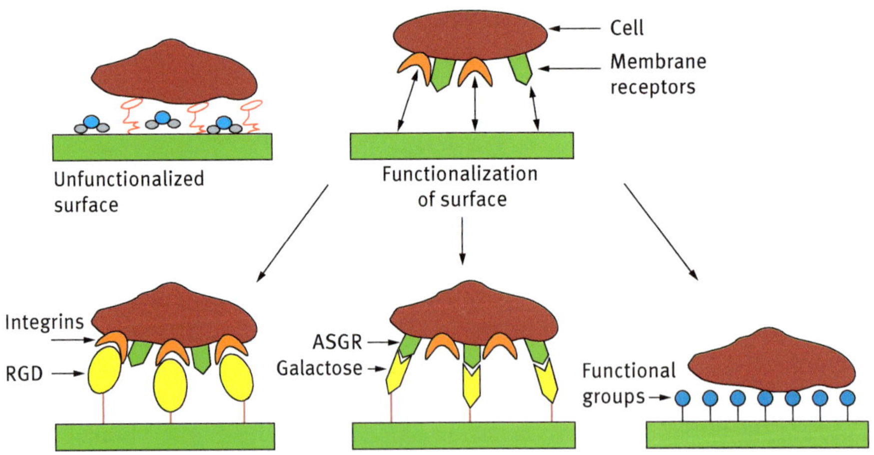

Fig. 6.6: Different approaches for improving the biofunctional specificity of the membrane surface.

Other approaches aimed at modifying the surface of materials involve the use of ECM molecules (collagen, fibronectin, laminin, poly-L-lysine, etc.) that interact with cell receptors. These proteins can be covalently linked to the surface or just adsorbed, producing a coating layer. The realization of biomaterials, including membranes that have selective cell adhesion and are able to promote a spatially organization of cells, can be achieved by surface functionalization of bioactive molecules like peptides. For example, RGD moieties have been immobilized onto the material surface in order to mimic the cell-binding domains of the ECM [61]. Indeed, this peptide represents the minimal adhesion domain of most of the ECM proteins (e.g. fibronectin, vitronectin, and collagen), and it has been shown to improve the adhesion of human hepatocytes on membranes [61]. A cyclic peptide of RGDfK (R: arginine, G: glycine, D: aspartic acid, f: D-phenylalanine, and K: lysine) was designed by Kessler et al. [62] in order to improve the cytocompatibility of the material. This cyclic peptide seems to be more stable and efficient in the conjugation with integrins with respect to the linear one. However, the cell response after surface modification of the material depends on structural parameters such as the density of the ligands, their spatial distribution, their steric hindrance, and their colocalization with ligands [63]. To this purpose the use of spacer arms (e.g. low molecular weight polyethylene glycol) to bind the biomolecule confers the required flexibility to cell binding [64].

Alternatively to biomolecules such as RGD or ECM molecules that interact with integrins present over the cell membranes, it is possible to immobilize molecules that interact specifically with a given cell population. For example, galactose motifs have been immobilized on membranes for enhancing the specific interaction with hepatocytes, due to the specific binding between the galactose moiety and the asyaloglycoprotein receptor present on hepatocyte cytoplasmatic membranes. Literature studies show that galactosylated polyethersulfone (PES) membranes support the longterm maintenance of liver-specific functions of human hepatocytes [65, 66]. The modification strategy reported in this study includes the deposition of a stable functional plasma deposited acrylic acid (pdAA) layer characterized by a known surface density of –COOH groups, followed by the immobilization of galactose derivatives in its acid form (galactonic acid) through a hydrophilic spacer arm molecule. The immobilization of galactose derivatives on PES membranes significantly improved the adhesion and metabolic functions of liver cells, in particular urea and albumin synthesis.

6.8 Influence of mechanical properties

The mechanical properties of the membranes, in addition to other properties, play a critical role in the process of cell adhesion and consequently in their behavior, including migration and spatial distribution [67–69]. Cells *in vivo* reside in a complex mechanical environment characterized by endogenous (generated by cells) and exogenous (applied to cells) forces. Cells on the ECM and in contact with neighboring cells generate endogenous forces through the cytoskeletal contractility [70]. Exogenous forces include gravity and tissue-specific interactions. Therefore cells sense and probe the stiffness of their environment, and they adhere and interact on the basis of the local mechanical stimuli which they receive [71]. They are able to convert mechanical stimuli into chemical signals through mechanotransduction process that involve stretch-mediated ion channels, primary cilia, integrins, G-protein receptors, cell-

cell adhesions, and the cytoskeleton. These transducers allow the conversion of mechanical input into complex intracellular signaling cascades that regulate cellular adhesion, spreading, migration, and proliferation [72]. Through cytoskeleton organization/tension cells transmit a series of intracellular signalling pathways, which at the same time activate or inhibit gene expression [73]. Fibroblasts and epithelial cells cultured on substrates ($E \approx 1$ kPa) promote focal adhesions forming dynamic and irregular punctate structures. On the contrary, an increase in the substrate stiffness ($E \approx 30$–100 kPa) favors the formation of arrays of elongated focal adhesions and an increase in tyrosine phosphorylation of focal adhesion kinase (FAK) and paxillin through intracellular signals. Stiff substrates increase both focal adhesion and cytoskeletal organization [74–76] with consequent changes in cell shape. For example, fibroblasts on stiff substrates exhibit an increase in the spread area, and actin stress fiber organization [75, 77] on compliant substrates shows a round shape with diffuse actin. Similar behavior has been reported for endothelial cells whose spreading increases with the substrate stiffness [78].

Since tissue stiffness ranges within several orders of magnitude, from adipose tissue (Young's modulus $E \approx$ several kPa) [79] to bone ($E \approx$ GPa) [80], the stiffness of the membranes should be taken into account for engineering a specific tissue or organ construct. In this context, the ability to develop biomaterials with elastic properties that recapitulate the same stiffness of the *in vivo* tissue is important for the realization of a new tissue and/or organs.

Additionally the stiffness of the material in contact with cells is not static but changes during the time from that which occurs *in vivo* during the physiological processes, especially in the case of biodegradable membranes that are subjected to degradation.

Membranes with different grades of tensile elasticity have been investigated in the neuronal survival and neuritis outgrowth. In particular membranes with Young's modulus ranging from 36 to 2228 MPa have been used in the fabrication of *in vitro* platforms for neuronal growth [81]. Chitosan membranes, owing to the rigid and brittle character (Young's modulus = 2288 MPa), turned out to be inappropriate candidates for nerve conduits because they could compress the regenerating nerve cells. On the other hand, membranes of polyurethane with Young's modulus of 36 MPa are too soft as nerve conduits, whereas membranes of polycarbonate and polycarbonate-polyurethane with Young's modulus of 208 MPa and 570 MPa, respectively, have been able to offer enough mechanical properties for nerve tissue. In another study membranes with elastic properties accelerated the fusion process of self-assembled spheroids consisting of fibroblasts and myoblasts leading to the formation of functionally active microtissues [82].

6.9 Mass-transport influence

One of the critical points that can affect the viability and functions of cells in contact with a biomaterial *in vitro* is the adequate supply of nutrients and the removal of wastes and catabolites from cell compartments. The mass transfer characteristics of the *in vivo* tissue environment can be described by the convective transport of nutrients in capillary vessels that are in close proximity of the cells (usually within 100 μm), and diffusive transport across the interstitial space to individual cells. Mimicking the *in vivo* environment, membranes should ensure adequate mass transfer. In general, membranes with pore sizes of lower than 1 μm are able to support free diffusion of molecules, but not cellular migration, whereas pores in the range of ≈ 10–100 μm readily allow host cells to migrate through the entire volume of the scaffold. However, the membrane pore size is defined on the basis of the molecules which are to be transferred. Transport complications arise because molecules present in the cell environment are comprized on a wide range of molecular sizes (from small electrolytes to large proteins) and physico-chemical properties (hydrophilic and hydrophobic molecules). The transport across the membrane occurs by diffusion and/or convection because of the transmembrane concentration or pressure gradients. The diffusion of molecules across the membrane is proportional to the transmembrane concentration gradient and diffusive permeability into the membrane. The molecule diffusion varies with membrane permeability and pore size and decreases as the membrane thickness increases. Each molecule has a different diffusion rate in liquid solution and the consumption or production rates of a given metabolite can be different, affecting the overall mass transfer. As previously described, the transport rate through the membrane depends on the MWCO, as well as the morphological and physico-chemical properties of both the membranes and the molecules of interest. Membranes with different permeability properties affected axonal outgrowth of hippocampal neurons. It has been shown that polyacrylonitrile HF membranes with hydraulic permeance of 0.215 l/h m^2 mbar ensured a sufficient mass transfer of nutrients and metabolites to the cells and the removal of catabolites from cell microenvironment, guaranteeing adequate levels of metabolic features. PAN HF membranes were able to recreate *in vitro* a 3D neural tissue-like structure, mimicking the hippocampal tissue.

6.10 References

[1] Tsang KY, Cheung MC, Chan D, Cheah KS. The developmental roles of the extracellular matrix: beyond structure to regulation. Cell and tissue research. 2010; 339(1), 93–110.

[2] Kleinman HK, Philip D, Hoffman MP. Role of the extracellular matrix in morphogenesis. Current Opinion in Biotechnology, 2003; 14: 526–532.

[3] Stevens MM, George JH. Exploring and engineering the cell surface interface. Science. 2005; 310 (5751): 1135–1138.

[4] Bauer AL, Jackson TL, Jiang Y. Topography of extracellular matrix mediates vascular morphogenesis and migration speeds in angiogenesis. LoS computational biology. 2009; 5: e1000445.

[5] Huebsch N, Arany PR, Mao AS, Shvartsman D, Ali OA, Bencherif SA, Rivera Feliciano J, Mooney DJ. Harnessing traction-mediated manipulation of the cell/matrix interface to control stem-cell fate. Nature materials. 2010; 9: 518–526.

[6] Rozario T, DeSimone DW. The extracellular matrix in development and morphogenesis: a dynamic view. Developmental biology. 2010; 341(1): 126–140.

[7] Cohen ED, Ihida-Stansbury K, Lu MM, Panettieri RA, Jones PL, Morrisey EE. Wnt signaling regulates smooth muscle precursor development in the mouse lung via a tenascin C/PDGFR pathway. The Journal of clinical investigation. 2009; 119(9): 2538–2549.

[8] Fisher OZ, Khademhosseini A, Langer R, Peppas NA. Bioinspired materials for controlling stem cell fate. Accounts of Chemical Research. 2010; 43(3): 419–428.

[9] De Bartolo L, Bader A. Biomaterials for Stem Cell Therapy: State of Art and Vision for the Future. Boca Raton FL: CRC Press; 2013.

[10] Piscioneri A, Campana C, Salerno S, Morelli S, Bader A, Giordano F, Drioli E, De Bartolo L. Biodegradable and synthetic membranes for the expansion and functional differentiation of rat embryonic liver cells. Acta Biomaterialia. 2011; 7: 171–179.

[11] Hynes RO. Integrins: bidirectional, allosteric signaling machines. Cell. 2002; 110: 673–68.

[12] Plow EF, Haas TA, et al. Ligand binding to integrins. J Biol Chem. 2000; 275(29), 21785–21788.

[13] Takada Y, Ye X, Simon S. The integrins. Genome Biol. 2007; 8(5): 215.

[14] Ruoslahti E, Pierschbacher MD. New perspectives in cell adhesion: RGD and integrins. Science 1987; 238(4826): 491–497.

[15] Geiger B, Bershadsky A, Pankov R, Yamada KM. Transmembrane extracellular matrix-cytoskeleton crosstalk. Nat Rev Mol Cell Biol. 2001; 2: 793–805.

[16] Hamerli P, Weigel T, Groth T, Paul D. Surface properties of and cell adhesion onto allylamine-plasma-coated polyethylenterephtalat membranes. Biomaterials. 2003; 24: 3989–3999.

[17] Hersel U, Dahmen C. RGD modified polymers: biomaterials for stimulated cell adhesion and beyond. Biomaterials. 2003; 24: 4385–4415.

[18] De Bartolo L, Morelli S, Piscioneri A, Lopez LC, Favia P, d'Agostino R, Drioli E. Novel Membranes and surface modification able to activate specific cellular responses. Biomolecular Engineering. 2007; 24: 23–26.

[19] Salerno S, Piscioneri A, Laera S, Morelli S, Favia P, Bader A, Drioli E, De Bartolo L. Improved functions of human hepatocytes on NH3 Plasma–grafted PEEK-WC-PU membranes. Biomaterials. 2009; 30: 4348–4356.

[20] Ma Z, Mao Z, Gao C. Surface modification and property analysis of biomedical polymers used for tissue engineering. Colloids Surf B. 2007; 60: 137–157.

[21] Ben Rejeb S, Tatoulian M, Khonsari FA, Durand FA, Martel A, Lawrence JF, et al. Functionalization of nitrocellulose membranes using ammonia plasma for the covalent attachment of antibodies for use in membrane-based immunoassays. Anal Chim Acta. 1998; 376: 133–138.

[22] Puleo DA, Kissling RA, Sheu MS. A technique to immobilize bioactive proteins, including bone morphogenetic protein-4 (BMP-4), on titanium alloy. Biomaterials 2002; 23: 2079–87.

[23] Griesser HJ, Chatelier RC, Gengenbach TR, Johnson G, Steele JG. Growth of human cells on plasma polymers: putative role of amine and amide groups. J Biomater Sci Polym Ed. 1994; 5: 531–554.

[24] Roach P, Farrar D, Perry CC. Interpretation of protein adsorption: surface-induced conformational changes. J Am Chem Soc. 2005; 127(22): 8168–8173.

[25] Dee KC, Puleo DA, Bizios R. Protein-surface interactions. In: Dee KC, Puleo DA, Bizios R . An Introduction to Tissue-Biomaterial Interactions. New York, USA: John Wiley & Sons, Inc.; 2002.

[26] Horbett TA. The role of adsorbed proteins in tissue respone to biomaterials. In: Ratner BD . Biomaterials Science: An Introduction to Materials in Medicine. Boston, New York, London: Elsevier Academic Press; 2004. pp. 237–244.
[27] Vroman L, Adams AL, Fischer GC, Munoz PC. Interaction of high molecular weight kininogen, factor XII, and fibrinogen in plasma at interfaces. Blood. 1980; 55(1): 156–159.
[28] Puleo DA, Bizios R. Biological interactions on material surfaces. Dordrecht, Heidelberg, London, New York: Springer; 2009.
[29] Vogler EA. Structure and reactivity of water at biomaterial surfaces. Adv Colloid Interface Sci. 1998; 74: 69–117.
[30] Jell G, Minelli C, Stevens MM. Biomaterial-Related Approaches: Surface Structuring. In: Meyer U, Meyer T, Handschel J, Wiesmann HP (eds.). Fundamentals of Tissue Engineering and Regenerative Medicine. Berlin, Heidelberg: Springer-Verlag; 2009.
[31] Tengvall P. How surfaces interact with the biological environment. In: Ellingsen JE, Lyngstadaas SP (eds.). Bioimplant interface improving biomaterials and tissue reactions. Boca Raton: CRC Press; 2003. pp. 285–303.
[32] Qin D, Xia Y, Whitesides GM. Soft lithography for micro- and nanoscale patterning. Nat Protoc. 2010; 5: 491–502.
[33] Engel E, Michiardi A, Navarro M, Lacroix D, Planell JA. Nanotechnology in regenerative medicine: the materials side. Trends Biotechnol. 2008; 26: 39–47.
[34] Bruinink A, Wintermantel E. Grooves affect primary bone marrow but not osteoblastic MC3T3-E1 cell cultures. Biomaterials. 2001; 22: 2465–2473.
[35] Curtis A, Wilkinson C. Topographical control of cells. Biomaterials. 1997; 18: 1573–1583,
[36] den Braber ET, de Ruijter JE, Smits HTJ, Ginsel LA, von Recum AF, Jansen JA. Quantitative analysis of cell proliferation and orientation on substrata with uniform parallel surface microgrooves. Biomaterials. 1996; 17: 1093–1099.
[37] Desai TA. Micro- and nanoscale structures for tissue engineering constructs. Med Eng Phys. 2000; 22: 595–606
[38] Flemming RG, Murphy CJ, Abrams GA, Goodman SL, Nealey PF. Effects of synthetic micro- and nano-structured surfaces on cell behavior. Biomaterials. 1999; 20: 573–588.
[39] Roach P, Farrar D, Perry CC. Surface Tailoring for Controlled Protein Adsorption: Effect of Topography at the Nanometer Scale and Chemistry. J. Am Chem Soc. 2006; 128: 3939–3945.
[40] De Bartolo L, Catapano G, Della Volpe C, Drioli E. The effect of surface roughness of microporous membranes on the kinetics of oxygen consumption and ammonia elimination by adherent hepatocytes. J. Biomaterials Science-Polymer Edition. 1999; 10(6): 641–655.
[41] De Bartolo L, Rende M, Morelli S, Giusi G, Salerno S, Piscioneri A, Gordano A, Di Vito A, Canonaco M, Drioli E. Influence of membrane surface properties on the growth of neuronal cells isolated from hippocampus. J Mem Sci. 2008; 325: 139–149.
[42] Morelli S, Salerno S, Piscioneri A, Papenburg BJ, Di Vito A, Giusi G, Canonaco M, Stamatialis D, Drioli E, De Bartolo L. Influence of micro-patterned PLLA membranes on outgrowth and orientation of hippocampal neurites. Biomaterials. 2010; 31: 7000–7011.
[43] Ngo TTB, Waggoner PJ, Romero AA, Nelson KD, Eberhart RC, Smith GM. Poly (L-lactide) microfilaments enhance peripheral nerve regeneration across extended nerve lesions. J Neurosci Res. 2003; 72: 227e38.
[44] Taneja V, Vertegel A, Langan III EM, Laberge M. Influence of topography of an endovascular stent material on smooth muscle cell response. Ann Vasc Surg. 2011; 25: 675.
[45] Orlova Y, Magome N, Liu L, Chen Y, Agladze K. Electrospun nanofibers as a tool for architecture control in engineered cardiac tissue. Biomaterials. 2011; 32: 5615.

[46] D'Angelo F, Armentano I, Mattioli S, Crispoltoni L, Tiribuzi R, Cerulli GG, et al. Micropatterned hydrogenated amorphous carbon guides mesenchymal stem cells towards neuronal differentiation. Eur Cell Mater. 2010; 20: 231–244.
[47] Biggs MJ, Richards RG, Dalby MJ. Nanotopographical modification: a regulator of cellular function through focal adhesions. Nanomedicine. 2010; 6: 619–633.
[48] Elul, RJ. Fixed charge in the cell membrane. Physiol. 1967; 189: 351–365.
[49] De Bartolo L, Morelli S, Rende M, Gordano A, Drioli E. New modified polyetheretherketone membrane for liver cell culture in biohybrid systems: adhesion and specific functions of isolated hepatocytes. Biomaterials. 2004; 25: 3621–3629.
[50] De Bartolo L, Morelli S, Bader A, Drioli E. Evaluation of cell behaviour related to physico-chemical properties of polymeric membranes to be used in bioartificial organs. Biomaterials. 2002; 23(12): 2485–2497.
[51] van Oss CJ. Interfacial forces in aqueous media. New York: Marcel Dekker; 1994.
[52] Wu S. Polymer Interface and Adhesion. New York: Marcel Dekker; 1982.
[53] Owens DK, Wendt RC. Estimation of the surface free energy of polymers. J Appl Polym Sci. 1969; 13: 1741.
[54] Kaelble DH. Dispersion-Polar Surface Tension Properties of Organic Solids. J Adhesion. 1970; 2: 66.
[55] Good RJ, van Oss CJ. In: Schrader ME, Loeb GL (eds.). Modern approaches to wettability. New York: Plenum Press; 1992.
[56] van Oss CJ, Chaudhury MK, Good RJ. Monopolar surfaces. Adv Colloid Interface Sci. 1987; 18: 35.
[57] Salerno S, Piscioneri A, Laera S, Morelli S, Favia P, Bader A, Drioli E, De Bartolo L. Improved functions of human hepatocytes on NH3 Plasma–grafted PEEK-WC-PU membranes. Biomaterials. 2009; 30: 4348–4356.
[58] Laera S, Lopez LC, De Bartolo L, Morelli S, Salerno S, Piscioneri A, Nardulli M, Gristina R, d'Agostino R, Favia P. H2-NH3 Plasma-grafting of PEEK-WC-PU membrane sto improve their cytocompatibility with hepatocytes. Plasma Processes and Polymers. 2009; 6: S81–S84.
[59] Pavlica S, Piscioneri A, Peinemann F, Keller M, Milosevic J, Staeudte A, Heilmann A, Schulz-Siegmund M, Laera S, Favia P, De Bartolo L, Bader A. Rat embryonic liver cell expansion and differentiation on NH3 plasma-grafted PEEK-WC-PU membranes. Biomaterials. 2009; 30: 6514–6521.
[60] Sardella E, Favia P, Gristina R, Nardulli M, d'Agostino R. Plasma-aided micro and nanopatterning processes for biomedical applications. Plasma Process Polym. 2006; 3: 456–69.
[61] De Bartolo L, Morelli S, Lopez L, Giorno L, Campana C, Salerno S, Rende M, Favia P, Detomaso L, Gristina R, d'Agostino R, Drioli E. Biotransformation and liver specific functions of human hepatocytes in culture on RGD-immobilised plasma-processed membranes. Biomaterials. 2005; 26(21): 4432–4441.
[62] Kantlehner M, Schaffner P, Finsinger D, Meyer J, Jonczyk A, Diefenbach B, Nies B, Holzemann G, Goodman SL, Kessler H. Surface coating with cyclic RGD peptides stimulates osteoblast adhesion and proliferation as well as bone formation. Chem Bio Chem. 2000; 1: 107–114.
[63] Shin H, Jo S, Mikos AG. Biomimetic materials for tissue engineering. Biomaterials. 2003; 24: 4353–4364.
[64] Shin H, Jo S, Mikos AG. Modulation of marrow stromal osteoblast adhesion on biomimetic oligo[poly(ethylene glycol) fumarate] hydrogels modified with Arg-Gly-Asp peptides and a poly(ethyleneglycol) spacer. J Biomed Mater Res. 2002; 61: 169–179.
[65] Memoli B, De Bartolo L, Favia P, Morelli S, Lopez LC, Procino A, Barbieri G, Curcio E, Giorno L, Esposito P, Cozzolino M, Brancaccio D, Andreucci VE, d'Agostino R, Drioli E. Fetuin-A gene ex-

pression, synthesis and release in primary human hepatocytes cultured in a galactosylated membrane bioreactor. Biomaterials. 2007; 28: 4836–4844.
[66] De Bartolo L, Morelli S, Rende M, Salerno S, Giorno L, Lopez LC, Favia P, d'Agostino R, Drioli E. Galactose derivative immobilized glow discharge processed PES membranes maintain the metabolic activity of human and pig liver cells. J Nanoscience and Nanotechnology. 2006; 6: 2344–2353.
[67] Discher D E, Janmey P, Wang YL. Tissue cells feel and respond to the stiffness of their substrate. Science. 2005; 310: 1139–1143.
[68] Engler AJ, Rehfeldt F, Sen S, Discher DE, Microtissue elasticity: measuerements by atomic force microscopy and its influence on cell differentiation. Methods in cell biology. 2007; 83: 521–545.
[69] Saez A, Ghibaudo M, Buguin A, Silberzan P, Ladoux B. Rigidity-driven growth and migration of epithelial cells on microstructured anisotropic substrates. Proc Natl Acad Sci USA. 2007; 104: 8281–8286.
[70] Burridge K, Chrzanowska-Wodnicka M. Focal adhesions, contractility, and signaling. Annu Rev Cell Dev Biol. 1996; 12: 463–518.
[71] Discher DE, Mooney DJ, et al. Growth factors, matrices, and forces combine and control stem cells. Science. 2009; 324: 1673–1677.
[72] Jaalouk DE, Lammerding J. Mechanotransduction gone awry. Nat Rev Mol Cell Biol. 2009; 10: 63–73.
[73] Discher DE, Janmey P, Wang YL. Tissue cells feel and respond to the stiffness of their substrate. Science. 2005; 310(5751): 1139–1143.
[74] Pelham Jr RJ, Wang Y. Cell locomotion and focal adhesions are regulated by substrate flexibility. Proc Natl Acad Sci USA. 1997; A94: 13661–13665.
[75] Yeung T, Georges PC, et al. Effects of substrate stiffness on cell morphology, cytoskeletal structure, and adhesion. Cell Motil Cytoskeleton. 2005; 60: 24–34.
[76] Genes NG, Rowley JA, et al. Effect of substrate mechanics on chondrocyte adhesion to modified alginate surfaces. Arch Biochem Biophys. 2004; 422: 161–167.
[77] Georges PC, Janmey PA. Cell type-specific response to growth on soft materials. J Appl Physiol. 2005; 98: 1547–1553.
[78] Reinhart-King CA, Dembo M, et al. The dynamics and mechanics of endothelial cell spreading. Biophys J. 2005; 89: 676–689.
[79] Samani A, Plewes D. A method to measure the hyperelastic parameters of ex vivo breast tissue samples. Phys Med Biol. 2004; 49: 4395–4405.
[80] Rho JY, Ashman RB, et al. Young's modulus of trabecular and cortical bone material: ultrasonic and microtensile measurements. J Biomech. 1993; 26: 111–119.
[81] Morelli S, Piscioneri A, Messina A, Salerno S, Al-Fageeh MB, Drioli E, De Bartolo L. Neuronal growth and differentiation on biodegradable membranes. J Tissue Eng Regen Med. 2015; 9: 106–117.
[82] Messina A, Morelli S, Forgacs G, Barbieri G, Drioli E, De Bartolo L. Self-assemblying of tissue spheroids on polymeric membranes. J Tissue Eng Regen Med. 2015. doi: 10.1002/term.2105.

7 Membrane bioartificial organs

7.1 Membrane bioartificial organs

Health care requests for patients with tissue loss or end-stage organ failure is increasing due to the aging of the worldwide population and the development of pathologies related to organ functions. Transplantation represents the gold standard for lifesaving but is limited by the scarce availability of donor organs, and as a consequence, there has been a growing demand for biological substitutes that are able to restore or to temporarily maintain or improve organ functions. The development of a bioartificial organ involves the creation of a biomimetic microenvironment which is able to support the growth and the maintenance of biological tissue or cells in a natural or synthetic framework.

In this context selective membranes of suitable molecular weight cutoff are able to compartmentalize cells in a microstructured environment as selective barriers to prevent immune system components from getting into contact with the implant, while allowing nutrients and metabolites to permeate freely to and from cells [1]. During recent years several studies demonstrated that synthetic membranes due to their characteristics of biostability, biocompatibility and selectivity can be used for the development of bioartificial systems using isolated cells that works as a bioartificial organ. Membranes are used, for example, in bioartificial liver, pancreas, and kidneys, or to encapsulate cells to be implanted in a specific site of the human body.

Production of sufficient tissue or organ *in vitro* for using in the treatment of patients with organ failure requires adequate oxygenation and appropriate transport of nutrients to, and waste product from, the tissue. To overcome the limitations of static culture methods, various dynamic systems such as bioreactors have been developed. Here the bioreactors allow the culture of cells under tissue-specific mechanical forces such as pressure, shear stress, and interstitial flow. Furthermore, in these systems a constant turnover of tissue culture medium enhances the gas and nutrient exchange, which together with the complete fluid dynamics control ensures the long-term maintenance of cell viability and function [2]. Membrane bioreactors are designed to reproduce the functions of organs which are supposed to replace, rather than its anatomical structure. For this reason they represent a chance for restoring, maintaining or improving tissue functions, or for designing *in vitro* physiological models (e.g. liver, pancreas, neurons, etc.).

DOI 10.1515/9783110268010-008

In these devices biological components, cells, are integrated into synthetic membranes that act as a support for the adhesion and compartmentalization of cells. Different types of devices have been proposed using flat sheet or hollow fiber membranes, capsules or coatings for the immobilization of cells [3]. Various types of bioreactors have been investigated for tissue culture, including the spinner flask bioreactor, the rotating vessel bioreactor, the rotating wall perfused vessel bioreactor, and the perfused column bioreactor, as reviewed by Freed and Vunjak-Novakovic [4]. Also, perfused chambers have been used to encourage culture medium to reach the center of the bioconstruct through microchannels [4].

In recent years the *in vitro* use of membrane biohybrid systems including bioreactor systems contributed in attaining important information about the effect of various drugs, such as diclofenac, rofecoxib, or paracetamol – the effects of which are not completely known – on the specific functions of human hepatocytes [5–7].

Membranes of suitable molecular weight cutoff are used in bioartificial organs (e.g., pancreas, liver) as selective barriers to prevent immune system components from getting into contact with the implant, while allowing nutrients and metabolites to permeate freely to and from cells. The use of membranes in bioartificial substitutes and tissue engineering date back to the year 1933, when Vincenzo Bisceglie in Bari, Italy encased mouse tumor cells in a nitrocellulose membrane and inserted them into the abdominal cavity of a guinea pig, to show that cells were not killed by an immune reaction in the pig [8]. Subsequently many researchers focused on the development of immunoprotective membranes to prolong the life of a transplant. Currently two important areas of interest are represented by bioartificial pancreas for the treatment of insulin-dependent diabetes and liver assistance device for the temporary treatment of acute liver failure. Membrane capsules containing dopamine-secreting cells also are being explored for treating Parkinson's disease, a progressive brain disorder characterized by a deficiency of the neurotransmitter dopamine [9]. Immunoprotective membrane cell transplants are being investigated to treat other nervous system disorders. Polymer membranes are also being explored to block cell adhesion or scar tissue formation, for example after surgery, and thus improve wound healing. In addition, membranes are being investigated for prevention of restenosis (coronary artery narrowing) after angioplasty [8].

In membrane bioartificial organs, cells are compartmentalized by means of selective membranes that permit the transport of nutrients and metabolites to cells and the transport of catabolites and specific metabolic products to the blood. The membrane must prevent contact between xenogeneic cells and the patient's blood to avoid an immunological response and the rejection of the xenograft. Membranes act as a means for cell oxygenation, and in the case of anchorage-dependent cells as substrata for cell attachment and culture. As a result, the type of membrane to be used in a bioartificial organ must be chosen on the basis of its permeability characteristics as well as on its physico-chemical properties related to the separation process.

7.2 Bioartificial liver

7.2.1 Liver structure and functions

The liver, one of the largest organs located in the abdominal cavity performs numerous functions vital for the maintenance of body homeostasis. The supply of blood to the liver is from the hepatic artery that carries blood from the aorta, and from the portal vein, which carries blood containing digested nutrients from the gastrointestinal tract, the spleen, and pancreas. These blood vessels subdivide into capillaries, which then lead to lobules, which are the functional units of the liver. The liver is drained by the hepatic veins and the bile which is produced is drained via the biliar canals (Figure 7.1).

The liver is divided into lobules, which are polyhedrons centered on a central vein. At the lobule vertices, a bile duct, a branch of the hepatic artery, and portal vein form "the portal triad". The bulk of the liver is primarily composed of parenchymal cells such as hepatocytes, making up 70 % of liver mass, and nonparenchymal cells such as hepatocyte precursor cells (oval cells or Ito cells), stellate cells, kuppfer cells, epithelial cells, sinusoidal epithelial cells, biliary epithelial cells, and fibroblasts [10]. Hepatocytes are organized in cords between sinusoidal capillaries within the lobule. Liver cells are spatially organized to optimize communication and transport. Cells communicate directly through cellular and gap junctions, and via soluble chemical signals or via signals associated to the macromolecules which form the surrounding ECM, enabling cells to differentiate, grow, function, and undergo apoptosis [11]. Factors affecting the cellular environment control the size, shape, and population of the colony. Structurally, hepatic cells are attached to a basal membrane composed of laminin and type IV collagen. They are connected to other cells through homotypic or heterotypic cellular junctions, alerted by the cell adhesion molecules (CAMS) binding to receptors, and surrounded by an extracellular matrix (ECM) that includes fibronectin and type I and III collagen [12]. Furthermore, hepatocytes exhibit a striking polarity which is expressed at multiple levels: in the overall cell shape, distribution of the cytoskeleton and organelles, and in the division of the plasma membrane into three functionally and compositionally distinct domains: basolateral, canalicular, and lateral. At the basal surface the transport of small molecules across membranes and the exchange of metabolites with blood take place, whereas the secretion of bile acids and detoxification products occurs on the apical domain. Thus, the metabolic activities of the hepatocytes change spatially along the length of the sinusoid and are apparently regulated by gradients of oxygen, hormones and ECM composition, which create a phenomenon of "liver zoning".

The development of constructs with metabolic functions equivalent to those of the liver poses technical challenges due to the complexity of liver cell physical-chemical requirements. The liver performs numerous functions, including synthesis of serum proteins such as albumin and that occurs exclusively in the hepatocytes, regulation of nutrients, glucose and fatty acid metabolism, detoxification, production of bile,

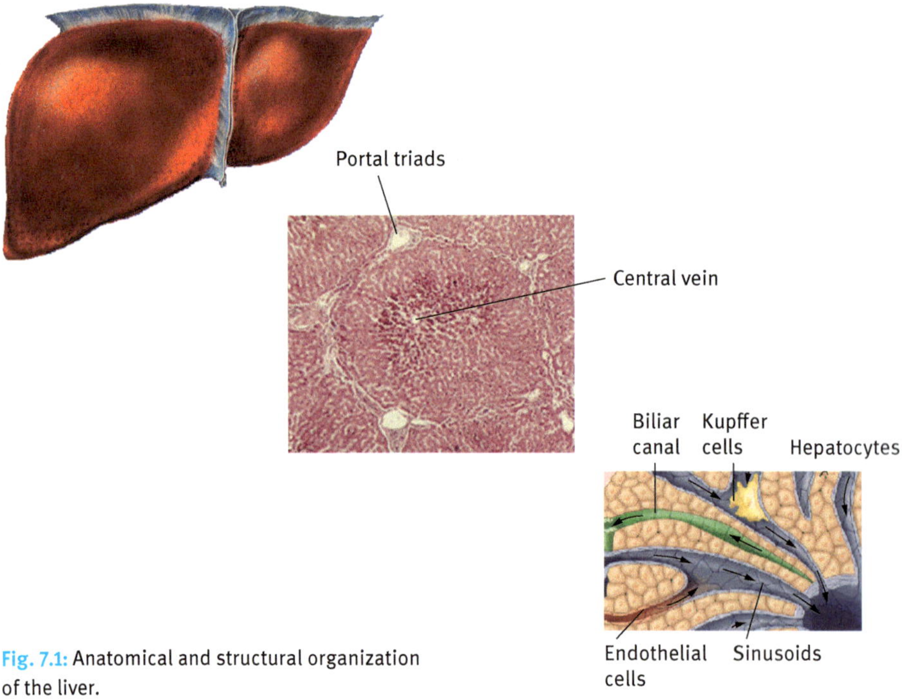

Fig. 7.1: Anatomical and structural organization of the liver.

urea synthesis, and the biotransformation of xenobiotic and endogenous substances through the phase I and phase II biotransformation reactions. The hepatocytes primarily exhibit the characteristic hepatic functions and are the preferred candidates for a bioartificial liver device.

7.2.2 Liver disease

The loss of liver function results in the disruption of essential body functions which may lead to death. Patients with liver failure develop hepatic encephalopathy and have a dramatic rise in intracranial pressure. The liver is normally able to regenerate after acute injury and recover its function under appropriate physiological stimuli [10]. However, liver failure occurs when the normal regenerative process is compromised and the residual functional capacity of the damaged liver is unable to sustain life.

Liver disease is now the fifth most common cause of death after heart disease, stroke, chest disease, and cancer. Several causes can lead liver injury: cirrhosis, primary liver cancer, chronic viral hepatitis B and C, and nonalcoholic fatty liver disease are among the most frequent causes of liver disease (Figure 7.2).

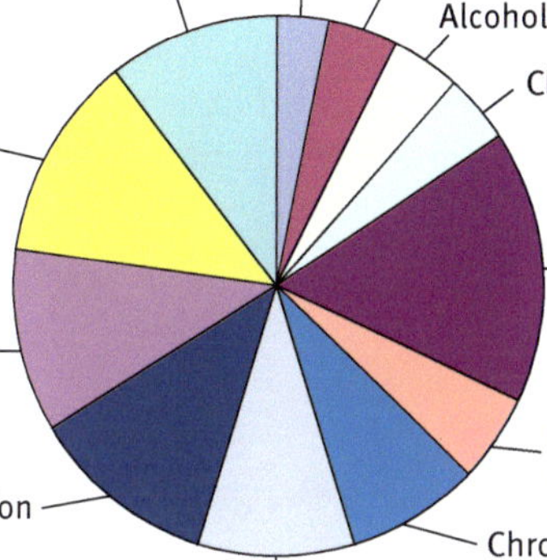

Fig. 7.2: Main causes of liver diseases.

Liver transplantation is the only established treatment for liver failure; every year, the number of patients needing a hepatic transplant increases. The European Liver Transplant Registry (ELTR) has cumulated data concerning 130 441 liver transplants from 168 centres in 33 countries from May 1968 to June 2015[1]. Cirrhosis is the most frequent indication for transplantation in Europe, followed by cholestatic disease, primary liver tumors, and acute hepatic failure. Every year 30 000 people die of end-stage liver disease in United States, and the current treatment of liver disease not only generates a list of more recipients than donors, but also produces a cost of $ 9 billion in the United States[2] only. Currently there are 118 333 patients on the waiting list for a donor organ, and of those there are 14 369 candidates awaiting a liver transplant[3]. The World Health Organization estimates that about 170 million people are infected with hepatitis C virus (HCV), and more than 32 million are in South East Asia.

Insufficient number of donor organs for orthotopic liver transplantation worldwide has urgently increased the requirement for new therapies for acute and chronic liver disease. For this purpose a liver support device is needed as a bridge to transplantation or as a support for the patient to recover the liver functions. The classical extracorporeal detoxification systems that are based purely on physical separation of toxins from the patient's blood (e.g. dialysis, hemofiltration, plasmapheresis) are inadequate, considering the multiple and complex functions that the liver performs. The introduction of active functioning hepatocytes able to perform all synthetic and

1 European Liver Transplantation, www.eltr.org

2 United Network for Organ Sharing, unos.org

3 Organ procurement and transplantation network, http://optn.transplant.hrsa.gov (data as of March 15, 2017)

detoxification functions as in the native liver allowed the creation of a bioartificial liver (BAL), which is a promising approach for the treatment of hepatic failure. Generally, a BAL system consists of functional liver cells supported by an artificial cell culture material. In particular, it incorporates hepatocytes into a bioreactor where the cells are immobilized, cultured, and induced to perform the hepatic functions by processing the blood or plasma of patients.

7.2.3 BAL design issues

The development of a BAL system involves many design considerations. The cells used in BAL devices must perform all synthetic and detoxifying functions as *in vivo*. The membranes must provide: (1) an adhesion support to the cells; (2) an adequate mass transfer of oxygen, nutrients, and toxic substances from the blood or plasma of patients to the cell compartments, and of proteins and other specific compounds and catabolites produced by cells from the cell compartment to the blood or plasma; (3) immunoprotection of cells; (4) biocompatibility. BAL devices are classified on the basis of cell source, type of culture system for the hepatocytes, and type of bioreactor (Tables 7.1–7.3).

Cell source for BAL

In the field of liver therapeutic tissue engineering the choice of the cell type and cell source is a key point, since the cellular component plays a critical role in the performance of a BAL device (Table 7.1). The cells inside the BAL must retain their differentiated functions. Due to a lack of human organ availability, one of the main sources of hepatocytes for bioartificial systems is xenogeneic material. Primary porcine hepatocytes, for example, which can be obtained in large quantities, have been widely used as the cell source for hybrid artificial livers. Porcine hepatocytes exhibit biotransformations functions, synthesis of urea, albumin, and other proteins, and are activated by growth factors that also activate human cells [13]. Although these cells are easily obtained in large quantities and demonstrate the same qualities and therapeutic effects of human hepatocytes, this type of cell source carries the risk of xenogeneic infections, and since they are not of human origin they are not genetically matched, resulting in immune rejection by the patient [14]. An alternative approach is to use an immortalized hepatoblast carcinoma line which has the necessary functional and survival characteristics. The advantages of using established cell lines include the ability to culture large quantities of cells for an extended period of time and the ability to control the degree of hepatocyte function displayed [15]. However, because these cells are cancerous in nature, it is important to maintain safe handling practices when considering the possibility of using these cells clinically. Primary hepatocytes are harvested via perfusion and are the precursor cells to mature hepatocytes. Furthermore, primary human hepatocytes are very difficult to culture, and human cells obtained via perfusion do not survive beyond several divisions. In addition, not only are human hepatocytes difficult to maintain in culture, they also may loose their normal functions or undergo dedifferentiation [16]. Current attention is being focused on finding a reliable source of human hepatocytes. Stem cells have been suggested as interesting cell sources. These cells, found in sources such as bone marrow, are the most flexible cells in terms of being undetermined in their pathway and expressing a remarkable ability to differentiate into a desired cell type. Factors that contribute to a particular cell signaling pathway are currently being investigated. These include growth factor, cellular signaling within the interstitial fluid and extracellular matrix, and cues from other cells of either the same or different cell types. Hepatic progenitor cells (oval or Ito cells) found within the liver have already begun to differentiate, but still have several options before becoming destined to a specific cell line. These cells will not necessarily become mature hepatocytes but could differentiate into other functional cells of the liver [17]. The generation of hepatic endoderm from pluripotent stem cells is now being identified as one resource to meet the big demand for functional human hepatocytes. Human embryonic stem cells are derived from the inner cell mass of preimplantation embryos and possess the ability to self-renew and differentiate to all cell types. These attributes in theory give them the potential to provide an unlimited supply of replacement cells for regenerative medicine [18]. Although the human embryonic stem cells could potentially provide an infinite source of hepatocytes, stem cells and somatic cells generally have limited function without the specialized tissue microenvironment. Thus, the recreation of such a niche is of primary importance for the development of scalable and high fidelity resources for drug testing or BAL reconstruction.

Tab. 7.1: Cell source to be used in a bioartificial liver.

Source of hepatocytes	Advantage	Disadvantage
Primary human hepatocytes	Optimal function	Scarce availability
Porcine hepatocytes	Large source	Immunologic problems, PERV transmission
C3A cells	Express some liver functions	Functionality and safety uncertain
Immortalization of hepatocyte lines	Capacity of division growth	Tumorigenic potentiality
Stem cells	Self duplication, multi-directional differentiation	Complicated micro-environment
Coculture cells	Improve hepatocyte morphology and functions	High technical standard

Culture techniques

A successful BAL device requires a favorable culture model for hepatocytes to maintain their viability and differentiated functions. Hepatocytes are involved in many important liver functions: blood detoxification; bile secretion; protein, steroid, or fat metabolism; vitamin, iron, or sugar storage. This multifunctionality implies a great number of biological parameters, which are difficult to reproduce *in vitro* to maintain all the functionalities of the hepatocytes. Moreover, primary cultured hepatocytes rapidly lose liver-specific functions when maintained under standard *in vitro* culture conditions. To overcome such limitations, several different culture models for maintenance of hepatocytes *in vitro* have been developed. As mentioned before, a stationary suspension culture of isolated hepatocytes is ineffective, since in this circumstance hepatocytes lose differentiated function within hours. An alternative technique, resulting in improved cell viability and functional activity, is attachment culture. An example is given by the use of an overlying layer of collagen gel developed by Dunn et al.; this culture model is described as a "collagen sandwich culture". In this model, cells are spread on a monolayer and then covered with another collagen layer [19]. It has become increasingly clear that three-dimensional rather than monolayer growth is particularly important for maintaining differentiated hepatocyte function in culture. A means of establishing three-dimensional hepatocyte growth is the creation of multicellular spheroid aggregates. Early methods utilizing stationary culture techniques had the disadvantage of requiring several days for spheroid formation [20], rendering this approach impractical for clinical use. More easily and rapidly hepatocyte spheroids – usually with diameters of 50–500 μm – can be formed from suspended isolated cells by a spinner or rotational flask [21], or by incubation with a small number of collagen-coated dextran, polystyrene, or glass microcarriers as a nidus for cell aggregation [22]. These induced hepatocyte aggregates maintain their specific functions and their peculiar ultrastructure resembling that of a normal liver.

Another method of hepatocytes immobilization, cell encapsulation within alginate-polylysine, polyacrylate, or cellulose acetate, also promotes the development of three-dimensional hepatocyte growth. The potential for three-dimensional structure formation can be further enhanced by enriching the solution with natural extracellular matrix proteins prior to encapsulation, or can include supplementation of factors in the culture medium and the cocultivation of hepatocytes with other type of cells, the so-called "feeder cells" [23]. Thus cells are protected from mechanical damage, and additionally, immunoisolation from xenogeneic cells should be possible. *In vivo* studies have demonstrated that encapsulated hepatocytes transplanted in Gunn rats restored liver function without immunosuppression [24]. Among the techniques for maintaining hepatocellular function the use of artificial spherical bodies called microcarriers, ensures a larger surface for cell adhesion. For instance cellulose multiporous microcarriers (MCs) are capable of immobilizing isolated cells in their micropores, and thus they are a suitable extracellular matrix for maintenance of cellular function [25]. An example is provided by Wu et al. who fabricated by using chitosan as a raw material, a suitable size of porous microcarrier in which rat hepatocytes were cultured. The microcarrier was modified with lactose and maltose. The cells were able to maintain a morphologic structure similar to that *in vivo* and showed higher metabolic activities on lactose-modified microcarriers [26]. Coculture with nonparenchymal cells significantly enhances hepatocyte viability and function, probably due to the mixed effects of autocrine and paracrine stimulation by growth factors derived from the nonparenchymal cells and the reproduction of extracellular matrix. In the liver, hepatocytes and endothelial cells are closely associated but separated by the extracellular matrix of the space of Disse. To establish the heterotypic cell–cell interaction essential for the maintenance of proper liver function, Jindal et al. created an *in vitro* coculture system able to provide native cues, consisting of hepatocytes, collagen, and an overlaying endothelial monolayer. The authors demonstrated that this culture configuration induces the early recovery of hepatocytes following cell isolation as evidenced by the increased albumin and fibrinogen protein secretion as well as gene expression [27].

Another study regarding the coculture system was carried out in order to maintain the differentiated status by means of hepatic stellate cells (HSCs), their soluble and particulate factors, and lipid extracts. The central question of the study was to underline the fact that HSCs maintained hepatocytes functions and structure by multiple signals, preferentially cell bound, pointing in particular to nonsoluble membranous ligands contacts. This showed once again the pivotal role played by the surrounding microenvironment for the maintenance of differentiated hepatocytes [28].

Bioreactor

Considering the several functions that the liver performs, a hybrid liver-support device is one of the most complex bioreactors. Therefore, its design must be optimized in order to ensure: the rapid detoxification of neural and hepatic toxins; the return of liver-specific hepatotrophic factors, as well as liver-specific coagulation factors, back into the patient's blood; and the maintenance of liver-cell detoxification and synthetic functions until liver tissue regeneration or organ transplantation. On the basis of these considerations, an efficient culture device must be designed considering the following important design criteria: (1) to use a sufficient number of differentiated hepatocytes which can maintain the long-term functions; (2) to reduce mass transfer resistances and eliminate substrate limitations so that the device can function at maximum efficiency; (3) to minimize the dead volume in the device, thereby reducing plasma dilution effects in the patient. The ideal bioreactor design would maximize mass transfer to the hepatocytes, thereby enabling nutrients, including oxygen, and toxins from the patient's blood or plasma to reach the hepatocytes. The treated blood or plasma, including metabolites and synthetic products, would then be returned to the patient's circulation. To achieve this task a large surface area is important for cell adhesion. An ideal device should integrate efficient mass transport, scalability, and maintenance of hepatocyte functions. One of the most promising bioreactors is the membrane bioreactor. Membranes with suitable molecular weight cutoff (MWCO) have been proven to provide an effective immunoisolation barrier to immunocompetent species present in the patient's blood. In principle, xenogeneic or allogenic implants might be used without need of immunosuppression therapy. Membranes also permit the transport of nutrients and metabolites to cells and the transport of catabolites and specific metabolic products to the blood. In the case of anchorage-dependent cells, they offer high surface area available for cell attachment and culture. In these bioreactors, mass transfer is determined by the membrane MWCO or pore diameter and occurs by diffusion and/or convection in response to existing transmembrane concentration or pressure gradients. Some bioreactors use membranes with MWCO ranging from 70 to 100 kDa which are suitable for the transport of serum albumin but exclude proteins with high MW such as immunoglobulins and cells. The advantage of using these membranes is to provide immunoprotection. Other bioreactors use microporous membranes with a large pore diameter (0.2 μm) which allow the free passage of plasma proteins, toxins, and clotting factors between the blood or plasma and cells, but they exclude the passage of cells. The advantage of using a membrane with a large pore diameter is to increase the fluid convection in order to improve mass transfer conditions. Polymeric membranes with different morphology and chemico-physical properties have been used in BAL devices [29]. The majority of extracorporeal BAL has used cellulose and polysulfone derivatives (Tables 7.2 and 7.3). Morphological properties (e.g. pore size, pore size distribution, and roughness) and physico-chemical characteristics (e.g. surface charge, wettability, surface free energy) affect all the adhesion and metabolic functions of hepatocytes [30, 31]. Presently, most of the commercial membranes used for liver cell culture are developed for hemodialysis, which are optimized to be inert with blood proteins and cells. As a result membranes express poor properties regarding cell interaction and functions. Thus, the development of membranes able to favor the adhesion and the expression of liver specific functions is quite important for the design of a tissue-engineered liver bioreactor.

Another important issue in bioartificial liver design is the maintenance of sufficient oxygen supply to the hepatocytes. Since hepatocytes exhibit high metabolic rate with significant oxygen uptake, in order to oxygenate the circulating blood or plasma some devices incorporate an oxygenator in the bioreactor, while others use an inline oxygenator in the extracorporeal perfusion circuit. The modern bioreactor design should simulate the complex physiologic environments *in vivo*, such as physicochemical parameters (e.g. pH, temperature, concentration of nutrients and metabolites), mass transfer rates, and biomechanical conditions. Various bioreactors have been designed to meet these demands. They range from rather simple hollow-fibre systems or dialyzers to more complex bioreactors which try to mimic the architecture of the liver. However, most of these systems have their limitations, either in that they offer a suboptimal environment for cell growth, or their function is poorly reproducible.

Tab. 7.2: Main characteristics of membrane BAL systems in clinical evaluation.

Bioartificial system	Bioreactor configuration	Membrane material	Bioreactor cell capacity	Cell Type	Culture Technique	Cell position
Kiil dyalizer bio-artificial liver [39]	Plate	Cellulose (MWCO 20 kDa)	1×10^{10}	Primary rabbit hepatocytes	Suspension	Dialysate compartment
Membrane bio-reactor [40]	Cartridge	PVC	4×10^{7}	Porcine hepatocytes	Suspension	Shell
ELAD Amphioxus cell technology [41]	Hollow fiber	Cellulose acetate (MWCO = 70 kDa)	2×10^{11}	Human cell line (C3A)	Aggregates	Shell
LSS Charité, Humboldt Univ. Germany [42]	Hollow fiber	Polyamide (MWCO = 100 kDa) Polyethersulfone (MWCO = 80 kDa) Silastic Polypropylene (0.2 μm pore size)	2.5×10^{9}	Pig primary hepatocytes – endothelial cells	Aggregates	Shell
Hepat Assist Circe Biomedical [43]	Hollow fiber	Polysulfone (0.2 μm pore size)	5×10^{9}	Cryopreserved porcine hepatocytes	Microcarrier attached irregular aggregates	Shell
AMC-BAL Univ. Amsterdam [44]	Spiral wound	Nonwoven polyester matrix, polypropylene (0.2 μm pore size)	1×10^{10}	Pig primary liver cells	Small aggregates	Shell/on the non-woven polyester matrix
BLSS Excorp Medical Inc. [45]	Hollow fiber	Cellulose acetate (MWCO = 100 kDa)	70–120 g	Porcine primary hepatocytes	Collagen gel entrapped	Shell
HBAL [46]	Hollow fiber	Polysulfone (MWCO = 100 kDa)	1×10^{10}	Porcine hepatocytes	Aggregates	Shell
TECA-HALSS Corp. [47]	Hollow fiber	Not reported	$1–2 \times 10^{10}$	Porcine hepatocytes	Suspension	Shell
RFB [48]		Polyester screen	200–230 g	Porcine hepatocytes		

Tab. 7.3: Membrane BAL systems for *in vitro* and preclinical tests.

Bioartificial system	Bioreactor configuration	Membrane material	Bioreactor cell capacity	Cell Type	Matrices	Culture Technique	Cell position
Liver x2000 [49]	Hollow fiber	Polysulfone (MWCO = 100 kDa)	1×10^8	Porcine hepatocytes	Collagen type I	Entrapment	Lumen
BAL [50]	Hollow fiber	Polysulfone (0.2 μm pore size) – agarose microcapsule	9×10^7	Rat primary hepatocytes, Hep G2	Encapsulation in agarose	Multicellular spheroids	Extrafiber space
BAL [51]	Hollow fiber	Polyolefin fiber (0.4 μm pore size)	10×10^{10}	Porcine hepatocytes Hep G2	Matrigel	Entrapment	Extrafiber space
FMB-BAL [52]	Flat	Polyterafluoroethylene, polycarbonate (0.2 μm pore size)	1×10^{10}	Pig hepatocytes	Collagen	Sandwich	Between flat sheet membranes
BLSS [54]	Hollow fiber	Polyethylene (Plasma Flo) (0.3 μm pore size)	5.4×10^9	Porcine hepatocytes	Collagen	Entrapment	Extrafiber space
Oxy-HFB [55]	Crosswise hollow fiber	Polyethylene HF (0.2 μm pore size) Polypropylene HF	$1–5 \times 10^9$	Pig liver cells	–	–	Extrafiber space
LLS HALLS [56]	Hollow fiber multicapillary	Polyethylene coated with EVAL polyurethane foam and capillary	0.5–100 g	Porcine hepatocytes	–	Organoids spheroids	Extrafiber space
Mini-bioreactor [58]	Flat	Polyterafluoroethylene dense	6×10^6	Porcine hepatocytes	Collagen	monolayer	Over the surface
Multibore fiber bioreactor [59]	Multibore capillary	Modified polyethersulfone (0.2 μm pore size)	7.5×10^6	Human hepatocytes	–	Small aggregates	Lumen

Tab. 7.3: (continued)

Bioartificial system	Bioreactor configuration	Membrane material	Bioreactor cell capacity	Cell Type	Matrices	Culture Technique	Cell position
PVDF-hollow fiber [60]	Hollow fiber	Polyvinylidene difluoride (0.5 μm pore size)	5×10^7	Rat hepatocytes	Galactose-tehered pluronic copolymer	Aggregates	Extracapillary space
Flat membrane bioreactor [61]	Flat	Galactosylated polyether-sulfone (0.2 μm pore size)	4.7×10^7	Human hepatocytes	Galactose immobilisation	Small aggregates	Over the surface
RWMS [62]	Flat	Fluorocarbon dense	$7.5–9 \times 10^5$	Rat hepatocytes	–	Spheroids	Over the surface
LSS [63]	Hollow fiber	Cellulose acetate (MWCO = 100 kDa)	2×10^6	RTH33 cell line	PMA-30 modification	Monolayer	Extracapillary space
Slide reactor [64]	Hollow fiber	Polyethersulfone (0.2 μm pore size)	8×10^4	Human hepatoma cells	–	Aggregates	Between hollow fibers
PDMS micro-bioreactor [65]	Flat	Polydimethylsiloxane and polyester (0.4 μm pore size)	5×10^5	Rat hepatocytes	Collagen	Monolayer	Over the surface
Flat-plate*	Microchannel bioreactor	Polyurethane dense	2×10^6	Rat hepatocytes	–	Monolayer	Over the surface
Crossed HF bioreactor [67]	Crossed hol-low fibers	Polyethersulfone (0.2 μm pore size) and modified polyetherether-ketone (MWCO = 190 kDa)	2×10^7	Human hepatocytes	–	Small aggregates	Among fiber network

* American Diabetes Association, www.diabetes.org

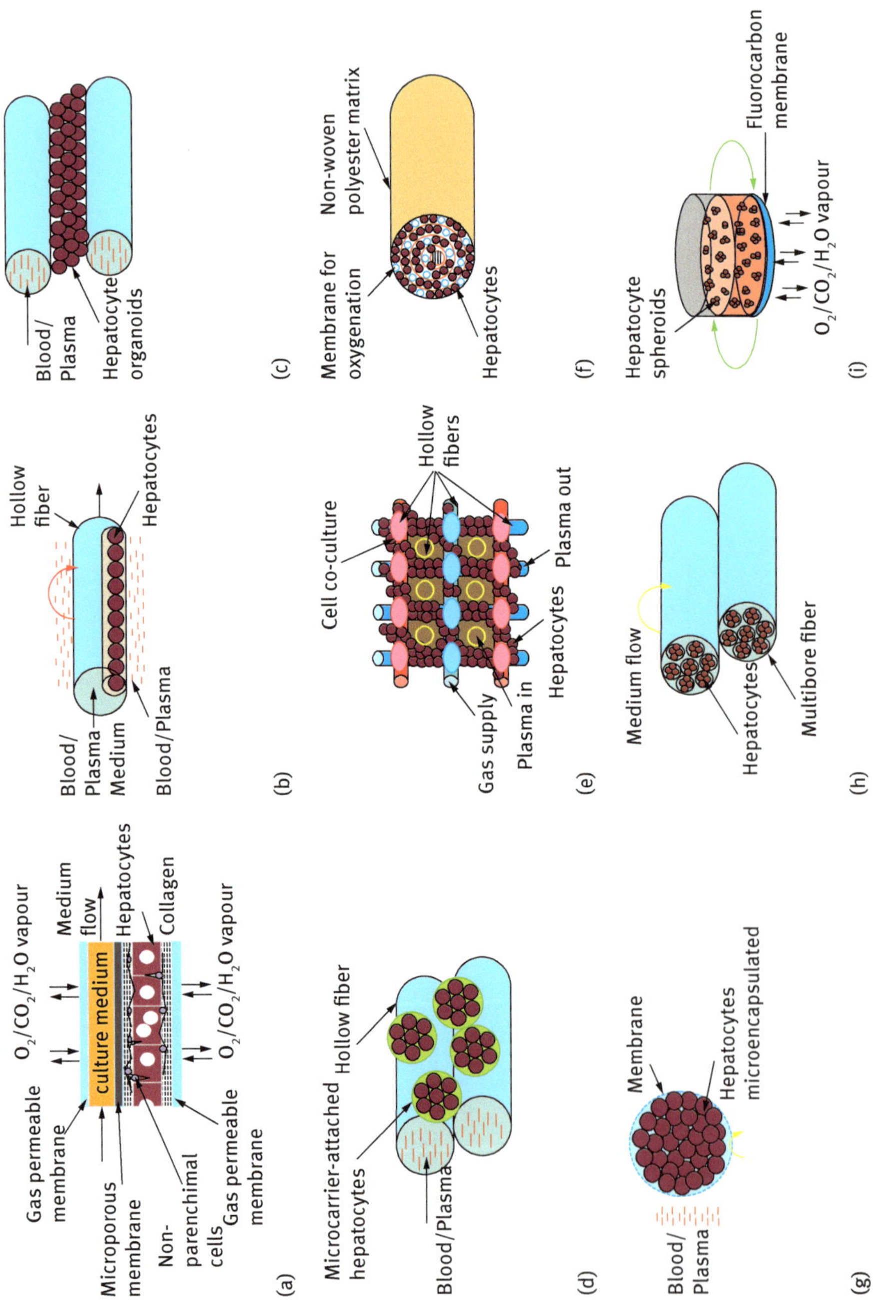

Fig. 7.3: Configuration of membrane bioreactors using hepatocytes cultured: (a) between flat-sheet membranes; (b) entrapped in a three-dimensional contracted gel matrix inside hollow fiber membranes; (c) outside hollow fiber membranes organized in organoids; (d) outside hollow fiber membranes attached to microcarriers; (e) in a network formed by four capillary membranes with different functions; (f) in a spirally wound nonwoven polyester matrix inside hollow fibers; (g) in microcapsules; (h) in the intraluminal compartment of a multibore fiber bioreactor; (i) in a rotating-wall gas-permeable membrane system.

Hepatocytes have been cultured in membrane bioreactors in different configurations (Figure 7.3): (a) between flat-sheet membranes in a sandwich configuration; (b) in the lumen of hollow fibre membranes entrapped in a collagen layer; (c) in the shell of hollow fibre membranes in monolayer, (d) aggregate or spheroid structure, and (e) attached to microcarriers; (f) in a network of hollow fiber membranes with different functions; (g) in a spirally wound device in which hollow fibers are used to provide oxygen to the cells; (h) in multibore capillaries; (i) microencapsulated, and (j) in an oxygen permeable membrane rotating system under microgravity conditions [32]. In all of these bioreactor configurations, hepatocytes are cultured in contact with the membranes therefore membranes with adequate surface and structural properties have to be used into the bioreactor.

Mass transport

For the design of an efficient membrane bioreactor, transport phenomena determining the overall transmembrane flux of metabolites, catabolites, cell regulatory, and immune-related soluble factors need to be considered. Transport complications arise because solutes are comprised in a wide range of molecular sizes (from small electrolytes to large proteins) and physico-chemical properties (hydrophilic, hydrophobic molecules). In order to decrease the complexity of transport analysis, a large number of the papers dedicated to hollow fiber bioreactors assume that the transport of low MW nutrients and metabolites occurs by diffusion [33, 34], whereas convective contributions to mass transfer are generally neglected since pressure shear stress decreases cell viability. Conversely, convection phenomena are supposed to regulate the mass transfer of macromolecular species and proteins [35], and transport properties of cytocompatible membranes are analyzed by standard sieving experiments.

However, in the liver, oxygen and nutrient supply to the cells occurs by means of sinusoidal structures located on two sides of each cell; sinusoids act as transport channels for convective flow, and they enable efficient diffusive nutrient transport through the cellular space and extracellular matrix.

An efficient transport of metabolites and nutrients is required for *in vitro* maintenance of the hepatocyte viability and functions.

In a BAL system, three compartments must be distinguished: the blood/plasma compartment, the membrane compartment, and the cell compartment. Mass transfer in the blood compartment is primarily due to the convection mechanism, because the osmotic forces during the steady-state operation of BAL are negligible. The rate at which the molecules are transported is dependent on the fluid dynamics of the blood compartment and on the physico-chemical properties of the molecules. Considering that the blood is a fluid with a non-Newtonian behavior [36], it is difficult to model the mass transfer. In many devices a cartridge separates the cellular component of the blood before entering into the BAL, and therefore only plasma flows through the BAL. In this case the plasma is a Newtonian fluid, and the mass transfer coefficient increases with increasing plasma flow, and the membrane acts as static selective barrier promoting secondary flows and mass transfer. The optimization of plasma flow is dependent on the physiological plasma parameters and bioreactor geometry.

Mass transport through the membrane depends on the morphological and physico-chemical properties of the membrane and molecules. The transport across the membrane occurs by diffusion and/or convection because of the transmembrane concentration or pressure gradients. The diffusion of molecules across the membrane is proportional to the transmembrane concentration gradient and diffusive permeability into the membrane. Molecular diffusion varies with membrane permeability and pore size and decreases as the membrane thickness increases. Diffusion coefficients in the membrane decrease with molecular weight and Stokes radius of solutes [37]. The membrane transport of molecules is also dependent on their physico-chemical properties (e.g. hydrophilicity/hydrophobicity, charge) related to the membrane properties: for example polymeric membrane charge can affect the diffusion of a charged molecule.

In convective transport, the hydraulic permeance and the sieving coefficient (molecule rejection by the membrane) affect the transport rate of the molecules [38]. Therefore, the mean pore size, pore size distribution, porosity, and thickness of the membrane are important in the determination of the hydraulic permeance and of the membrane separation properties. The accumulation of molecules at the membrane interface and the irreversible and reversible adsorption of molecules in the pores may also affect mass transfer across the membrane.

Cells can be compartmentalized in the lumen or shell of hollow fiber membranes or between flat sheet membranes or microencapsulated. In the cell compartment molecule transport, metabolite consumption and production occur simultaneously. In the cell compartment the solutes are transported by diffusion with a rate dependent on cell density and on the presence of extracellular matrices (ECM) (collagen, fibronectin, etc.) as well diffusion distance between membrane and cells. The presence of ECM components may limit the solute diffusion and increase the distance between cells and membrane, thereby increasing the concentration gradient. Therefore, the transport in the cell compartment can be described in terms of diffusion-reaction kinetics. Concentration gradients of nutrients and metabolites are determined by the metabolic activity of cells. The metabolite concentration decreases with increasing metabolic activity of cells. As a result the concentration of large molecules that have a low diffusion coefficient decreases more steeply across the cell compartment than the concentration of small molecules. The cell concentration and their distribution in the cell compartment play an important role in the transport of oxygen, nutrients and metabolites: cells metabolically active and high density assemblies may cause steep concentration gradients which may hinder cell viability. The kinetics of metabolic reactions may also affect the mass transfer. Most metabolic reactions follow Michaelis–Menten kinetics. As the nutrient or metabolite is transported into the cell compartment, the nutrient is also consumed, with a consequent reduction of its concentration. As a result the combination of the cell consumption with the molecule mobility and cell distribution loading determine the concentration profile of a metabolite in the cell compartment. The cell compartment should provide the optimal culture microenvironment for hepatocytes while limiting diffusion distances. Moreover, information about the rate equations for the given liver specific metabolic reactions is necessary for the correct design of BAL.

7.2.4 Membrane bioreactors for BAL systems in clinical applications

Currently, several BAL devices are in various stages of clinical evaluation and are listed in Table 7.3. Many of these devices use hollow fiber membranes as supports for the cultured hepatocytes and as immunoselective barriers between the plasma of patients and the hepatocytes. In 1987 Matsumura et al. reported an early clinical trial of a bioartificial liver [39]. The device was developed on the principle of hemodialysis against a suspension of functioning hepatocytes. The liver suspension was placed in a dialysate compartment on one side of a cellulosic selective membrane. The blood flows through a compartment on the opposite side of the membrane. Afterwards, one of the first large clinical studies was performed by Margulis et al., in which 20 ml capsules filled with pig hepatocytes in suspension were used [40]. These reports are the only two reports based on suspension culture. Since hepatocytes are anchorage-dependent cells, they rapidly lose their viability and functions in suspension. The BAL devices developed later are based on adhesion culture.

The most common bioreactors developed as a bioartificial liver use hollow fiber membranes. This bioreactor meets the main requirements for cell culture: a wide area for cell adhesion, oxygen and nutrient transfer, removal of catabolites, and protection from shear stress. In particular hollow fiber membranes may serve to initiate tissue formation, to guide the microarchitecture of the developing tissue by providing histo-appropriate mechanical forces, and to confer mechanical strength.

Since the 1990s several BALs have been proposed. Sussman and colleagues developed an extracorporeal liver-assist device (ELAD) in which human hepatocyte cell line C3A, which is derived from hepatoblastoma cell line (Hep G2), are located outside the hollow fiber, and blood flows through the lumen of hollow fibers. A portion of the patient's plasma is ultrafiltrated through a cellulose acetate membrane (70 kDa) and is in direct contact with the C3A cells [41]. This device is commercialized by the Amphioxus Cell Technology. A hollow fiber device that uses cryopreserved porcine hepatocytes attached to collagen-coated dextran microcarriers, called Hepatic Assist, was developed by Demetriou and coworkers. In this system hepatocytes are loaded into the extracapillary space, and patient plasma flows through the capillary lumen of membranes with a pore size of 0.2 μm. This size is sufficiently small to block the passage of whole cells [42]. Plasma first passes through an activated charcoal column and flows through the lumen of the hollow fibres. A more complex system was proposed by Gerlach et al. The liver support system (LSS) or modular extracorporeal liver system (MELS) consists of a bioreactor with four interwoven independent capillary membrane systems which serve different functions. The cells are cultured on the outer surface and among the capillaries. Each fiber type exhibits a different function: silastic membranes for oxygen supply and removal of carbon dioxide, polyamide fiber for the plasma inflow, polyethersulfone fiber for the plasma outflow, and hydrophilic polypropylene membranes for the sinusoidal endothelial coculture [43]. With this capillary array, decentralized metabolite and gas exchange with small gradients are possible. Due to the independent plasma inflow and plasma outflow compartments, decentralized perfusion of cells between these capillaries is achieved. Additional functions could be integrated into the module.

The bioartificial liver support system (BLSS) is a hollow fiber device that uses porcine hepatocytes embedded in a collagen matrix. This system uses cellulose acetate hollow fibers with a 100 kDa MWCO containing more than 70 g of primary porcine hepatocytes embedded in a collagen matrix. The patient's blood is perfused through the capillary lumen [44]. In the circuit nutrient stream directly perfuses the hepatocytes providing specific nutrients. The Academic Medical Center bioartificial liver (AMC-BAL) developed by Flendrig and coworkers uses a three-dimensional, spirally wound, nonwoven polyester matrix for hepatocyte attachment with integrated hollow fibers for oxygen delivery to the cells [45]. In contrast with previously described systems, the AMC-BAL uses direct contact between the patient's plasma and the matrix attached hepatocyte to improve bidirectional mass transfer.

A radial flow bioreactor (RFB) was developed by Morsiani et al. at the University of Ferrara. In this system porcine hepatocytes are entrapped in woven-nonwoven polyester fabrics. The patient's plasma passes from the center to the periphery of the module [46].

Another BAL system that is currently in clinical testing is the TECA-hybrid artificial liver support system (TECA-HALSS) developed in Bejing. This bioreactor from TECA Corp. consists of hollow fiber membranes which compartmentalize a suspension of 10 to 20×10^9 porcine hepatocytes [47].

A hybrid bioartificial liver (HBAL) consisting of polysulfone hollow fiber membranes with MWCO of 100 kDa has been developed by Ding et al. [48]. In this device 1×10^{10} porcine hepatocytes cultured into the extrafiber compartment have been used. The HBAL system was used in the phase I trial for the treatment of 12 patients.

7.2.5 Membrane BAL systems in preclinical and *in vitro* evaluation

Several BAL systems have been evaluated preclinically, *in vitro* experiments, and in large animal models of liver failure.

In the Liverx2000 system of Hu and coworkers, hepatocytes are suspended in a collagen gel and injected into the lumen of hollow fibres with a MWCO of 100 kDa and the extracapillary compartment is perfused with a recirculating medium. Medium flowing through the luminal space provides hepatocyte nutrients [49] (Table 7.2). A bioartificial liver support system consisting of a hollow fiber cartridge using encapsulated multicellular spheroids of rat hepatocytes was developed by Shiraha et al. The spheroids, formed in a positively charged polystyrene dish were encapsulated into microdroplets of agarose that contained about 9×10^7 rat hepatocytes. The medium was circulated in a closed circuit where the cartridge was inserted [50].

Several alternative device configurations have advanced to the stage of large animal, preclinical evaluation. Naka et al. developed a system using primary porcine hepatocytes similar to the BLSS. The difference is in the use of microporous polysulphone hollow fiber membranes in the hepatocyte bioreactor and in the perfusion of plasma through the bioreactor. The system has shown some efficacy in support of ischemic pig liver failure model [51].

The flat membrane bioreactor (FMB) developed by De Bartolo and coworkers consists of primary porcine hepatocytes cultured between flat membranes. This is a reproducible model with total hepatectomy in pigs, suitable to test the safety and efficacy of liver support systems. Isolated hepatocytes were cultured within an extracellular matrix between oxygen-permeable flat-sheet membranes. In particular both sides of the outside shell consist of PTFE membranes permeable to oxygen, carbon dioxide, and water vapor allowing direct oxygenation of cells adhered to the surface and of the medium overlying the cells. Porcine hepatocytes are maintained in a three-dimensional coculture with nonparenchymal cells. A microporous polycarbonate membrane separates the medium from the cell compartment. The FMB maintained stable cell specific functions and was proven as a safe and efficient device [52].

Nagaki and coworkers developed a hybrid liver support system which consists of plasma perfusion through porous hollow fiber modules inoculated with 10 billion porcine hepatocytes entrapped in a basement membrane matrix, Engelbreth–Holm–Swarm gel. This system was applied to pigs with ischemic liver failure. It was demonstrated that the use of a BAL support device in combination with a hollow fiber module and hepatocytes entrapped in the gel has potential advantages for clinical use in patients with hepatic failure [53].

Another bioreactor developed on a large scale was the oxygenating hollow fiber bioreactor (OXY-HFB) BAL system developed by Jasmund et al. In this device primary liver cells are seeded on the surface of the fibers in the extrafiber space. Plasma from the patient is perfused through an extrafiber space and brought into direct hepatocellular contact [55].

Driven by the necessity for improving the maintenance of liver-specific functions of hepatocytes into the bioreactor, Mizumoto et al. developed two types of hybrid artificial liver support system (HALSS) that use hepatocyte in spheroid and organoid structure [56]:

- a PUF-HALSS comprising an artificial liver module made of polyurethane foam (PUF), in which hepatocytes form spheroids in its pores, and maintain liver-specific functions for at least ten days *in vitro*;
- an LLS-HALSS that uses a liver lobule-like structure (LLS) module containing hollow fibers with a microregular arrangement in which hepatocytes cultured in the extra-fiber space of the module form the organoids that maintain liver-specific functions for at least two months *in vitro*.

In preclinical experiments, a PUF-HALSS was applied to a pig with liver failure. To evaluate the effect of liver regeneration, PUF- and LLS-HALSS were applied to a rat having reversible hepatic failure. These HALSS may be useful to treat liver failure patients until liver transplantation or until regeneration of the native liver [56, 57].

A bioartificial liver support system, in which encapsulated multicellular spheroids of rat hepatocytes were utilized as a bioreactor in a hollow fiber cartridge, was designed by Shiraha et al. [50].

A new small-scale bioreactor with the hepatic sandwich model and a gas-permeable polytetrafluorethylene (PTFE) membrane at the bottom allowing a definable oxygen exchange was constructed by Schmitmeier and coworkers and compared with the conventional well plates. This is an attractive *in vitro* model system for pharmacological studies [58].

A multibore fiber bioreactor was developed by De Bartolo et al. as an *in vitro* liver tissue model to study disease, drugs and therapeutic molecules alternatively to animal experimentation [59]. This bioreactor, thanks to the membrane configuration, combines the advantage of seven compartments represented by seven capillaries arranged in one single fiber with high stability and mechanical resistance. Human hepatocytes were cultured in the intraluminal compartment of the multibore fiber bioreactor. The multibore fiber bioreactor ensures a sufficient oxygenation process, nutrient feeding, end-product removalm and distribution of fluid molecules inside the cell environment and can be used as *in vitro* liver tissue model for the evaluation of hepatic metabolic transformation of compounds and therapeutic molecules.

Several membrane bioreactors using galactose-modified membranes have been developed to improve adhesion and specific functions of liver cells. An important example is the galactosylated polyvinylidene difluoride hollow fiber bioreactor developed by Lu et al., in which isolated rat hepatocytes were seeded and cultured in the extralumenal space of the hollow fiber cartridge for 18 days in a continuously circulated system [60].

A flat bioreactor using galactosylated-polyethersulfone membrane promoted the long-term maintenance of differentiated functions of human hepatocytes outside of the body [61]. This galactosylated membrane bioreactor works perfectly as an *in vitro* model system reproducing hepatic acute-phase response that occurs during the inflammation process; in fact, it has the capability to simulate an IL-6-induced acute phase response with a reduced synthesis and release of albumin and an increased production of C reactive protein. Moreover this system gave evidence that IL-6 down-regulated the gene expression and synthesis of fetuin-A by primary human hepatocytes.

Based on the use of a gas permeable membrane, a rotating-wall gas-permeable membrane system was developed and used for the formation and culture of hepatocyte spheroids. Microgravity conditions were obtained in rotating-wall systems where hepatocyte aggregates were formed by cells protected from gravitational forces and acceleration. Thanks to the high O_2 permeability of the rotating-wall membrane system the viability and functions of cells improved with respect to a polystyrene rotating wall system [62].

Ye et al. developed a bifunctional cellulose acetate (CA) hollow fiber membrane bioreactor modified with 2-methacryloyloxyethyl phosphorylcholine (MPC) copolymers (PMB30) (MPC-co-n-butyl methacrylate) (BMA) and PMA30 (MPC-co-methacrylic) for preparing a novel liver-assist hollow-fiber membrane bioreactor [63].

At the time of the development of small scale bioreactors, microdevices have been proposed for testing of materials and/or drug toxicity. The slide reactor, developed by Sauer et al., is a simple hollow fiber-based bioreactor suitable for light microscopy and time-lapse video observation [64]. The slide reactor offers a cell compartment separated from a medium inflow and outflow compartment. Due to its simple design and the use of materials available in most laboratories, a slide reactor is a simple valuable tool to assess the cell-to-cell and cell-to-membrane interaction and enables the comparison of different types or arrangements of hollow fibers, e.g. for use in bioreactor-based extracorporeal liver assist devices, or for analysis of the influence of medium supplements on cell viability and tissue integrity.

Ostrovidov and coworkers developed two types of membrane microbioreactors to improve the maintenance of primary rat hepatocytes: one with a commercially available polyester membrane, and the other with a polydimethylsiloxane (PDMS) membrane. These microbioreactors closely mimic the *in vivo* liver architecture and proved to be very promising tools for future applications in drug screening or liver tissue engineering [65].

A flat plate microchannel bioreactor with an internal membrane oxygenator to improve the oxygen supply to the cells was developed by Roy et al. The hepatocytes, attached to a glass substrate, are in direct contact with the perfusing medium. A polyurethane gas permeable membrane separates the liquid compartment from the oxygenating gas compartment. This design allows oxygen delivery to the hepatocytes to be decoupled from the medium flow, thereby allowing oxygen delivery and flow to be studied independently [66].

A crossed hollow fiber membrane bioreactor was developed to support the long-term maintenance and differentiation of human hepatocytes. The bioreactor consists of two types of hollow fiber membranes, having different MWCO and physico-chemical properties, cross-assembled in alternating manner: modified polyetheretherketone (PEEK-WC) and polyethersulfone (PES), used for medium inflow and outflow, respectively. The combination of these two fiber sets produces an extracapillary network for the adhesion of cells and an efficient mass exchange through the cross-flow of culture medium. The optimized perfusion conditions of the bioreactor allowed the maintenance of liver functions with regard to urea synthesis, albumin secretion, and diazepam biotransformation up to 18 days of culture [67].

7.3 Bioartificial pancreas

7.3.1 Anatomy of the pancreas

The pancreas is an elongated organ located behind the stomach across the back of the abdomen. The pancreas is made up of two types of glands: exocrine and endocrine. The exocrine gland secretes digestive enzymes into a network of ducts that join the main pancreatic duct, which runs the length of the pancreas. The endocrine gland, which consists of the islets of Langerhans, secretes hormones into the bloodstream (Figure 7.4).

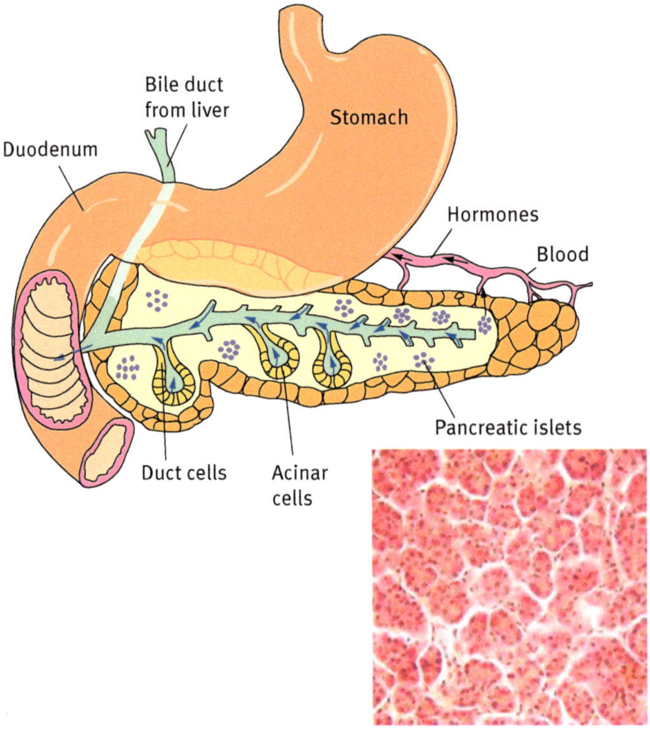

Fig. 7.4: Anatomy and structural organization of the pancreas.

The pancreas has digestive and hormonal functions. The enzymes secreted by the exocrine gland in the pancreas help break down carbohydrates, fats, proteins, and acids in the duodenum. These enzymes travel down the pancreatic duct into the bile duct in an inactive form. When they enter the duodenum, they are activated. The exocrine tissue also secretes bicarbonate to neutralize stomach acid in the duodenum. The hormones secreted by the endocrine gland in the pancreas are insulin and glucagon, which regulate the level of glucose in the blood, and somatostatin, which in turn prevents the release of the other two hormones.

The exocrine pancreas consists of **acini,** which resemble bunches of grapes. Each acinus consists of a single layer of 40–50 pyramidal epithelial cells surrounding a lumen. The epithelial cells produce a secretion (pancreatic juice) that contains enzymes, ions, and water. The cells become wider during active secretion. The base of the acinar cells is strongly basophilic, owing to the presence of endoplasmic reticulum, where there is a high concentration of RNA. This part of the cell therefore stains darker with hematoxylin and eosin. The apex of the cells is abundant with secretory granules containing the zymogen precursors of the pancreatic enzymes. The number of secretory granules increases after fasting, and decreases after a meal.

In the endocrine pancreas, the **islets of Langerhans** are embedded in the exocrine tissue. Each islet is composed of 2–3 thousand epithelial cells. The epithelial cells are arranged in a compact structure that is pervaded by a capillary network. A thin layer of reticular fibres separates the islets from the surrounding exocrine tissue. There are *four* different cell types within the islets of Langerhans, each of which produce different hormones; they include: **α-cells,** which produce glucagon, typically located at the periphery of the islet (they are not present in all islets); **β-cells** that secrete insulin; this is predominant cell type, located in the center of the islet and contributing to 70 % of all cells; **δ-cells** secreting somatostatin (there are low numbers in all islets); **F-cells,** which produce pancreatic polypeptide and are few in number (they may also be present in the exocrine tissue) [68].

7.3.2 Pancreatic diseases

There are over 400 million diabetics in the world today. Diabetes is the primary cause of death for 71 382 Americans per year. Diabetes contributes to the death of 231 404 Americans annually. According to the American Diabetes Association, the total cost of diagnosed diabetes amounted to $ 245 billion in 2012.[4] Diabetes mellitus is a disease in which high levels of sugar occur in the blood and urine. The cause of the elevated sugar levels is the insufficient secretion of the hormone insulin by the pancreas. In the absence of this hormone, the body's cells are not able to absorb sugar from the

4 American Diabetes Association, www.diabetes.org

blood stream in the normal fashion, and the excess sugar is excreted in the urine. While therapeutic solutions such as injectable insulin allow diabetics to live longer, diabetes remains the third major killer, after heart disease and cancer. Diabetes is a very disabling disease, because the usual therapies are not able to keep blood sugar levels constant and to prevent swings between high and low blood sugar levels, which cause damages to other organs, such as the kidneys, eyes, and blood vessels.

Diabetes is classified into two main types. Type 1, or insulin dependent diabetes, is usually associated to a complete lack of insulin due to autoimmune destruction of the insulin-producing β-cells [69]. The events that cause Type 1 diabetes are unknown, probably viral or environmental triggers act upon a genetically susceptible population. Type 2, or non-insulin-dependent diabetes, arises from peripheral resistance to insulin and a relative insufficiency of insulin, resulting in an initial attempt by the β-cells to compensate with the release of higher amount of insulin with respect to normal values. As Type 2 diabetes progresses, β-cells become desensitized to persistently high glucose concentrations, and normal responses to glucose signaling are lost [70].

The treatment of diabetes mellitus was limited to dietary manipulation prior to the discovery of insulin by Banting and Best (1921). The discovery of insulin changed an often rapidly fatal disease to a chronic condition requiring life-long treatment. Current treatment for diabetes, both Type 1 and Type 2, includes exogenous insulin therapy and endocrine replacement by transplantation. Both of these clinical approaches have considerable inherent drawbacks. Exogenous insulin treatment implies a poor control of blood glucose levels that leads to severe secondary complications, such as retinopathy, neuropathy, nephropathy, and cardiovascular disease [71]. An alternative therapy is a pancreas transplantation. Since 1966, more than 30 000 pancreas transplants have been performed worldwide. Transplantation, however, requires major surgery and dependence on lifelong immunosuppression to prevent rejection. Because of the limited availability of human pancreases and the need for immunosuppression, relatively few pancreas transplants are done in comparison to the entire diabetic population. Improvements in surgical technique or immunotherapy are unlikely to make whole organ pancreas transplantations available to the majority of patients with diabetes. Islet transplantation promises to be a cure at least as effective as pancreas transplantation, while being much less invasive. The efficiency of islet recovery from the whole organ pancreas and the susceptibility of allogeneic islet to immune attack are the two major barriers to successful islet transplantation. There are approximately 1 million islets in an adult human pancreas; thus islet transplantation usually requires islets isolated from two or more donor pancreases. Because islet isolation requires manipulation of human tissue, the process must be carried out in good manufacturing process (GMP) facility, which increases the cost of the procedure.

7.3.3 Cell source

An alternative to organ transplantation being tested in animal and pre-linical models is the bioartifical pancreas. The most important issues that must be considered in the design of membrane devices concern the cell source and configuration.

In vitro treatments can induce insulin-producing cells from a wide range of cell sources (Table 7.4).

Tab. 7.4: Cell sources in a bioartificial pancreas.

Cell source	Drawbacks
Adult cells derived from bone marrow, umbilical cord blood, liver, intestine, after introduction of pancreatic lineage-related genes such as Pdx1.	Cells are not so mature as β-cells in most cases, safety concerns if they are gene-engineered.
Pancreas-derived cells	Slow growth
Embryonic stem cell	insulin-producing cells appear in certain neuronal cells.
Induced pluripotent stem cell	The final differentiation is not clear.

Several reports have indicated that insulin-producing cells similar to the β-cell can be induced in various type of adult cells derived from bone marrow, umbilical cord blood, the liver, or intestine, after introduction of pancreatic lineage-related genes such as Pdx1 [72].

Another cell source is pancreas-derived cells, since exocrine pancreatic tissue is a waste product of islet isolation for transplantation. Pancreatic duct epithelial cells obtained from adult nonobese diabetic mice can continuously produce islet-like cell clusters after long-term culture [73]. Similarly, *in vitro* generation of β-like cells from cultured duct-like cells has been reported by several groups. Acinar cells are shown not to contribute to endocrine cell growth *in vivo*, but they are able to generate β-like cells *in vitro* through the formation of cell clusters in suspension culture [74]. Furthermore, recent reports showed that exocrine cells can be directly converted to β-cells *in vivo* by introducing three transcription factors: Pdx1, Ngn3, and MafA 31 [75]. In addition to duct and acinar cells, cells with a less differentiated fibroblast-like appearance can be cultured from pancreatic tissue.

Pancreatic duct-like cells, and even acinar cells, seem to be a potential source for β-like cells. MSC-like cells from the pancreas, with high proliferative ability, may be another candidate.

However, an attractive cell source for any kind of cell therapy is embryonic stem cells (ESC), since they proliferate almost indefinitely and have pluripotency [76]. Many groups have studied β-cell differentiation from ESC: insulin-producing cell clusters were generated through nestin-positive cells by step-wise procedures somewhat similar to the ones for neural cell differentiation. However, these cells can hardly be regarded as β-like cells, considering that insulin-positive cells were shown to result mostly through insulin uptake from the culture medium. It is known that several types of insulin-producing cells other than β-cells appear in certain neuronal cells, yolk sacs, and liver during embryonic development [77]. Therefore, insulin staining and mRNA expression in these early studies are now considered to be attributed mostly to differentiation toward neuron-like cells or extra-embryonic endoderm.

Induced pluripotent stem (iPS) cells generated from somatic cells to mimic ESC can be considered to be an alternative cell source [78]. Human iPS cells avoid the ethical difficulties concerning the use of human embryos and tissue rejection after transplantation if they are generated from the patient's own cells. ESC and iPS cells are a potential and indefinite source for cell therapies. However, further studies toward the final differentiation of β-like cells, and acceptable safety protocols for clinical usage are needed.

7.3.4 Membrane bioartificial pancreas

A membrane bioartificial pancreas is based on the use of isolated islets of Langerhans or single β-cells, which are capable of sensing the plasmatic glucose concentration and produce insulin amounts related to the actual glycemia, entrapped by means of membranes. In 1970 W.L. Chick and colleagues transplanted isolated islets protected by a hollow-fiber ultrafiltration membrane (an acrylonitrile-vinyl chloride copolymer) into dogs made diabetic by surgically removing the pancreas. The device consists of a chamber, containing hollow tubes made of a semipermeable acrylic copolymer, connected to standard vascular grafts. Islets are placed inside the chamber, through ports in the housing into the cavity, but are outside of the blood stream. A nominal MWCO of 80 kDa permits free diffusion of nutrients and insulin across the membrane, but inhibits the transport of immunoglobulins and immunocytes from the blood stream into the chamber (Table 7.5) [79].

Since 1970 research efforts have been devoted to the development of a hybrid bioartificial membrane pancreas [80–88]. Different bioartificial pancreas were designed in four main configurations: hollow fibers, capsule, coatings, and sheet. The many different types of prosthetic devices proposed so far can be grouped in three main categories: extravascular devices, intravascular devices, and microencapsulated islets of Langerhans. In the first case, the tissue is compartmentalized between membranes, if in a flat sheet configuration, or in the lumen of hollow fibre membranes, and then implanted in an extravascular site (see Figure 7.5). Other researchers [84] proposed a hollow fiber device as an extravascular bioartificial pancreas. This device consists of hollow fibers 0.5 mm in diameter, 20 mm in length, and containing 80 000 islets/ml (Figure 7.5 (g)). The extravascular systems generally suffer from an

Tab. 7.5: Characteristics of a membrane bioartificial pancreas.

Reference	Bioreactor configuration	Membrane material	Membrane configuration	MWCO/ Pore size	Culture technique
Chick et al., 1970 [79]	Hollow fiber	Acrylonitrile-vinyl chloride copolymer	Tubular	50 kDa	Immobilisation in lumen space
Colton C.K., 1995 [80]	Planar islet sheets	Cellulose ester; alginate film	Flat-sheet (thickness = 25–150 μm)	80 kDa	Microencapsulated islets between flat sheet membrane
Chang 1995 [81]	Hollow fiber	Cuprophan	Hollow fiber diameter = 0.5 mm	50 kDa	Immobilisation in lumen space
Calafiore et al., 1997 [82]	Microcapsules	Poly-l-ornithine	Circular	50–70 kDa	Encapsulation
De Vos et al., 1997 [83]	Microcapsules	Alginate-polylysine	Circular	50 kDa	Encapsulation
Petersen et al., 2001 [84]	Hollow fiber	Polysulfone	Capillary internal diameter = 600–800 μm, wall thickness = 100 μm	50 kDa	Macroencapsulation in intra-capillary space
Sakai et al., 2001 [85]	Microcapsule	Amynopropyl-silicate	Circular	60 kDa	Coated alginate beads
Silva et al., 2009 [86]	Hollow fiber	Polysulfone	Hollow fiber internal diameter = 1 mm, thickness 280 μm	100 kDa	Sepharose microspheres
Kang et al., 2010 [87]	Hollow-ribbon nitinol scaffold	Amphiphilic conetwork	Fiber thickness 5–10 μm	Pore size 11 nm	Immobilization
Lee et al., 2009 [88]	TheraCyte™	Polytethrafluoro-ethylene		Pore size 0.4 μm	Aggregates

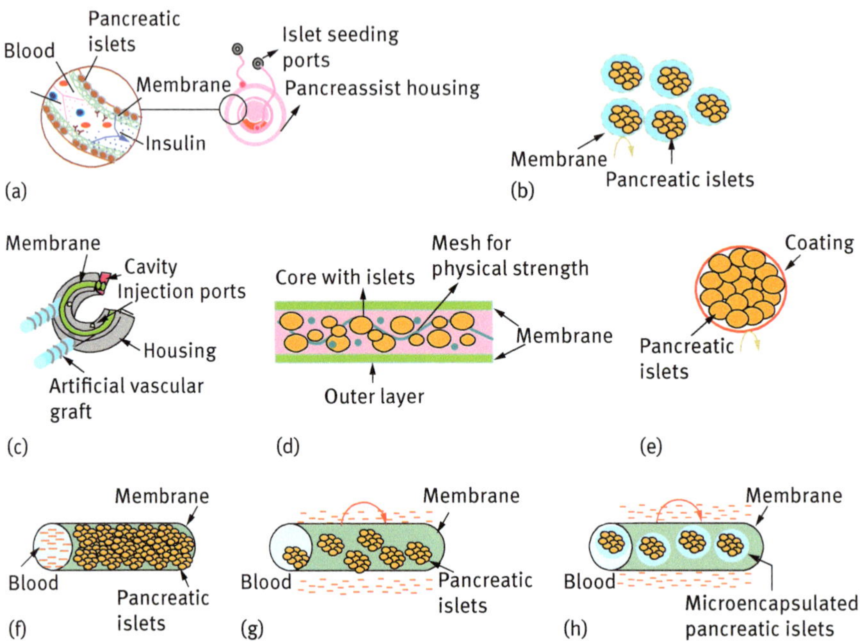

Fig. 7.5: Schematic representation of bioartificial pancreas configurations: (a) device developed by Chick et al., where pancreatic cells are loaded outside the tubular membrane; (b) microencapsulated pancreatic cells; (c) cells loaded outside of tubular membrane; (d) pancreatic cells cultured in islet sheets between flat sheet membranes; (e) cells coated with membrane; (f) pancreatic cells loaded outside of hollow fibres in extracapillary compartment; (g) pancreatic cells loaded in the lumen of hollow fiber; (h) microencapsulated cells loaded in the lumen of hollow fiber.

intrinsically slow insulin response following changes of blood glucose concentration, limited by the purely diffusive mass transport and by the fibroblastic response of the host.

Microencapsulated systems are realized by forming a membrane layer around the the islets of Langerhans. Hydrogels are very attractive for microencapsulation, because they provide higher permeability for low molecular weight nutrients and metabolites. The most commonly applied materials for microencapsulation are alginate, chitosan, agarose, cellulose, poly(hydroxyethylmetacrylate-methyl methacrylate) (HEMA-MMA), copolymers of acrylonitrile, and polyethylene glycol (PEG). Microcapsules are formed by using a membrane in the form of alginate gel around the islets of Langerhans, obtaining thus microcapsules with diameters of 300–400 m (Figure 7.5 (b)) [89]. Normoglycemia has been reported after intraperitoneal transplantation of alginate-polylysine microencapsulated allogenic and xenogeneic islets in diabetic animal models and recently also in humans. However, graft survival is always limited to several weeks. Graft failure is interpreted as a nonspecific immune response i.e. foreign body reaction against the microcapsules resulting in a progressive overgrowth of the capsule and subsequent necrosis of the islets.

Researchers have focused on highly purified alginate and other biochemicals more biocompatible, and on novel membranes able to prevent permeation of low molecular weight humoral molecules released by xenogeneic islets. A major limitation to the encapsulation device is that they are incapable of efficiently encapsulating a large number of islets in a reasonable time. This may result in hypoxic stress and loss of functionality of islets in large scale experiments [90]. An alternative microencapsulation method, having the advantage to rapidly encapsulate a large number of islets into microcapsules, utilizes multichannel air jacket microfluidic devices [91]. A reduction in capsule size is beneficial for islets and also exponentially decreases the total transplant volume. A smaller diameter of the capsules results in a better diffusion of nutrients to the islets, and Omer et al. demonstrated that capsules with a diameter of 600 ± 100 μm showed improved stability *in vivo* over larger capsules with diameters of 1000 ± 100 μm [92]. Furthermore islets are clinically transplanted into the liver through the portal veins, so that microcapsules with large diameters are expected to plug vessels. The diameter of the encapsulated islets must be much smaller than that currently attained to allow transplantation. Therefore, several efforts have been made to develop coating techniques which permit to coat islets with a very thin membrane layer. Islets can be covered with a thin polyion complex membrane using a layer-by-layer method, or with alginate/PLL/alginate multilayer coating, or with PEG. The advantage of using coating technique or conformal coating is the reduction of the microcapsules diameter [93].

Other systems are represented by macrocapsules, which can be distinguished as extravascular and intravascular, based on their transplant location. Intravascular systems contain islets which are seeded enclosed within a larger tube and implanted into the vessels of the host. Hollow fibers are perfused by blood flow. This device has been successful in inducing normoglycemia in various diabetic animal models including rats, dogs, and monkeys [94], although it requires intense systemic anticoagulation due to direct contact of the material with blood. Extravascular devices, in contrast, have the advantage that biocompatibility issues do not pose a serious risk to patients. They have been designed in both flat sheet membrane and hollow fiber configuration. A selective membrane around the sheet allows diffusion of nutrients and secreted hormones, but rejects macrophages. They are usually coated by hydrogels to achieve a smooth outer surface for an enhanced biocompatibility. Initial studies with extravascular macrocapsules were based on encapsulation of multiple islets in one or several large capsules. Islets aggregated in large clumps were not successful, due to necrosis at the center of the clumps [95]. Later, this problem was addressed by isolated immobilization of the islets in a matrix before encapsulation.

Silva et al. developed a vascular hollow fiber BAP by using polysulfone membranes with MWCO of 100 kDa and islets of Langerhans. This system was established as appropriate for *in vitro* testing [86]. Intravascular membrane devices are designed so that the membrane separates the graft directly from blood stream of the host. In Figure 7.5 (f), cells are cultured outside hollow fiber membranes arranged in housing according to a shell-and-tube configuration. These devices suffer from blood clotting at the interface between blood and the synthetic membrane or point of access, but they are extremely attractive in terms of design flexibility and use. Additionally, the implant site can be chosen on the basis of reducing the response time of the prosthesis following an increase of blood glucose concentration.

A device currently under preclinical development, licensed from Circe Biomedical, is the PancreAssist™ System.[5] This device consists of a single tubular membrane surrounded by insulin-producing porcine islets, which are in turn enclosed within a disk-shaped housing (Figure 7.5 (a)). The porous tubular membrane permits the transport of nutrients and glucose to cells and the transport of insulin from cells to the blood. The membrane also prevents contact between immunological species present in the patient's blood and islets. This device should be implanted near the kidney and surgically connected directly to circulatory's system using a vascular graft. Preclinical studies used 11 devices implanted more than 250 animals. The key advantages of this technology is the use of a biocompatible and immunoprotective membrane, primary cells, direct blood contact for oxygen supply, rapid insulin release, physiological blood glucose feedback, and no anticoagulation or immunosuppression.

The islet sheet is a thin planar bioartificial pancreas, licensed from Islet Sheet Medical Company, containing live, functional islets in an artificial polymer matrix (Figure 7.5 (d)).[6] Each sheet is several cm in diameter, contains 2–3 million cells, and islets microencapsulated within a mesh to increase the physical strength between two layers of alginate membrane. A 0.2 μm cellulose ester filter membrane is saturated with crosslinking solution. The 4–6 sheets of cells contain enough islet tissue to cure diabetes in an adult. The sheet is so thin (the overall thickness is 250 μm) so that diffusion is sufficient to supply nutrients up to the center of the sheet. Coating on the exterior of the sheet prevents contact between the cells inside and the immune effector cells of the host, and inhibits diffusion of antibodies and complement. The alginate membranes show high permeability to different solutes and excluded immunocompetent species. No immune suppression drugs are needed. The sheet may be removed or replaced at any time. Experiments on islet sheets began at the University of Chicago in September 1998, and moved to the University of Cincinnati and the University of Alberta in 2000. Large animal studies have given encouraging results. The implantation of islet sheets in omentum from a pancreatectomized dog permitted to normalize the level of glucose into blood, and at 60 days the blood sugar reached the lowest measured values [96]. Lee et al. [88] investigated human fetal pancreatic

---

5 www.trademarkia.com/pancreassist-75261053.html

6 www.hanumanmedicalfoundation.org/type-1-diabetes-research/project-team/islet-sheet-medical.html

islet-like cell clusters enclosed in devices made of polytetrafluoroethylene with a pore size of 0.4 μm, transplanted at a subcutaneous site. Ten weeks after transplantation in nonobese diabetic mice differentiated β-cell progenitors were found in the device, and glucose level was normalized, thus confirming the efficacy of this approach.

7.4 Bioartificial kidneys

7.4.1 Why bioartificial kidneys?

The kidney was the first solid organ whose function was substituted by an artificial device. Current therapy for ischemic or toxic acute kidney injury (AKI) or acute tubular necrosis (ATN) predominantly consists of hemodialysis (HD) or peritoneal dialysis (PD), which represent the only successful long-term ex vivo organ substitution therapy. Patients with AKI still have a high mortality rate, greater than 50 %, due to their propensity to develop systemic inflammatory response syndrome. In particular CVD (cardiovascular disease) is the leading cause of morbidity and mortality in ESRD (end-stage renal disease): approx. 50 % of ESRD patients die from CVD, and cardiovascular mortality is 15–30 times higher than in the age-adjusted general population. In ESRD, in addition to traditional Framingham risk factors, a considerable number of nonclassical factors are known to play a role in CVD progression, such as inflammation [97], vascular calcification [98], and LVH (left ventricular hypertrophy).

Although hemodialysis or peritoneal dialysis has dramatically changed the prognosis of renal failure, it cannot be considered a complete replacement therapy, because it provides only a filtration function and does not replace the homeostatic, regulatory, metabolic, and endocrine functions of the kidney. Therefore dialysis should be considered as a partial substitution rather than a renal replacement therapy.

The addition of metabolic activity, (such as ammoniagenesis and glutathione reclamation) endocrine activity, (such as activation of vitamin D3, whose low level seems to correlate with high mortality rates in hospitalized patients), immunoregulatory support and cytokine homeostasis, might provide additional physiologic replacement activities to the current history of the disease. The development of an implantable bioartificial kidney incorporating both biologic and synthetic components could result in substantial benefits for patients by increasing life expectancy, mobility, and quality of life, with less risk of infection and reduced costs.

A bioartificial kidney, mimicking the human kidney, requires two main units, the glomerulus and the tubule, to replace the excretory and metabolic functions of the kidney (Figure 7.6) [99, 100]. Currently hemodialysis, and peritoneal dialysis replace the excretory functions of the kidney, but they do not provide the lost metabolic function.

In order to develop a more complete and functional device, research in the past decade has focused on the engineering of a bioartificial kidney (BAK).

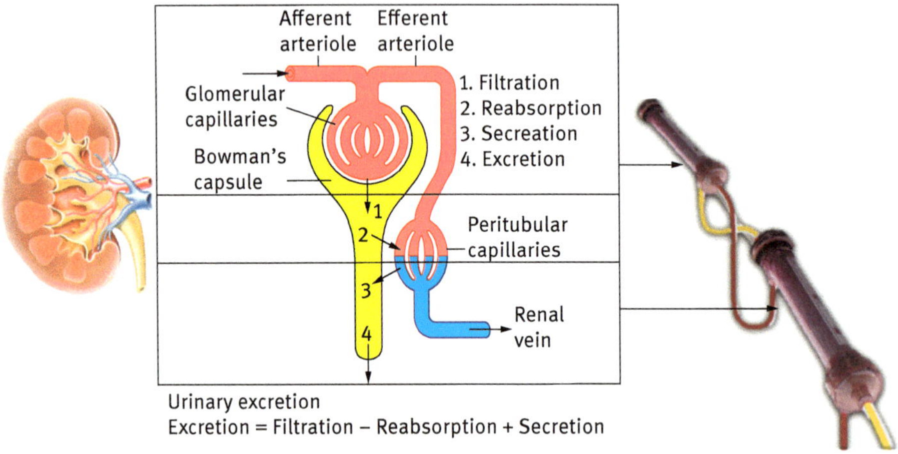

Fig. 7.6: Bioartificial kidney-like human kidney.

Initially several efforts focused on the development of an extracorporeal BAK consisting of a conventional synthetic hemofilter with a renal tubule cell assist device (RAD) in an acute extracorporeal blood circuit.

A bioartificial tubule has been constructed using renal tubule progenitor cells [101, 102] cultured on hollow-fiber membranes coated with extracellular matrix to enhance the attachment and growth of epithelial cells. The hollow fiber synthetic membranes provide immunoprotection and also an architectural scaffold for cells in the long-term implantation into a xenogeneic host [103].

Humes et al. scaled-up the single hollow-fiber device to a multifiber bioartificial RAD utilizing porcine renal proximal tubule cells grown along the lumen of polysulfone hollow fiber membranes (Figure 7.7 (a)) [104]. The device consists of hollow fibers with membrane surface areas as large as 0.7 m^2, containing up to 10^8 renal tubule cells into the lumen. *In vitro* studies demonstrated the retention of differentiated active transport of sodium, bicarbonate, glucose, and organic anions as well as important differentiated metabolic processes of the kidney such as ammoniagenesis, glutathione metabolism and synthesis of 1,25-dihydroxyvitamin D3 [104]. The RAD was used in series with a hemofilter in an extracorporeal hemoperfusion circuit into an acutely uremic dog. The blood was pumped out and entered the fibers of a hemofilter, being ultrafiltrated. Then the ultrafiltrate was delivered into the fibers of the tubule lumens within the RAD and discarded as "urine". The processed blood exiting from the RAD was delivered back to the animal. The tubule unit was able to maintain differentiated renal functional performance because metabolic substrates and low molecular weight growth factors were delivered to cells from the hemofilter and the blood in the extracapillary space. Membranes of suitable MWCO protect cells seeded into the lumen from immunoglobulins and immunocompetent cells present in the blood.

Clinical experience has been collected with RAD in patients with acute kidney injury. Human kidney cells were used and initial results in the first 10 treated patients in phase I/II trial demonstrated that this device is efficient when used in conjunction with hemoperfusion [105]. Cardiovascular stability of ten patients was maintained, and an increased native renal function was observed. Six of the patients survived past 28 days with renal function recovery.

In the phase II clinical study involving 58 patients with AKI continuous veno-venous hemofiltration, 40 patients received hemoperfusion with RAD, and 18 patients received only hemoperfusion. RAD treatment for up to 72 h promoted a statistically significant survival advantage over 180 days of follow-up.

7.4.2 Cell source

Research related to the development of BAK has focused on the use of renal proximal tubule-derived cells (Table 7.6). Human primary renal proximal tubule cells (HPTCs) have been used in clinical trials [105, 106]. However, most of the experimental work with animals have been performed with porcine primary renal proximal tubule cells [107–110]. In addition, the proximal tubule-like porcine cell line LLC-PK1 [111, 112], and other animal-derived cell lines such as Madin–Darby canine kidney (MDCK) cells have been used in the BAK system. The formation of a confluent differentiated epithelium sealed by tight junctions of renal cells on the porous membranes is critical, in order to maintain the cellular functions.

Tab. 7.6: Cell source to be used in a bioartificial kidney.

Cell source	Drawbacks
Human primary renal proximal tubule cells (HPTCs)	Limited availability, interdonor variability, dedifferentiation risk
Animal cell lines LLC-PK1, MDCK	Different requirements for growth and differentiation
Primary animal cells	Different requirements for growth and differentiation
Transgenic human proximal tubule-derived cell lines	Change over time
Stem cells	Need to develop differentiation protocols
Induced pluripotent stem (iPS) cells	Safety concerns for the use of oncogenes and integrating viral vectors for reprogramming

Animal cell lines and primary animal cells have different requirements for growth and differentiation with respect to the primary human cells. On the other hand, animal cell lines grow well on several types of hollow fiber membranes without any use of ECM. For example MDCK cells form a polarized epithelium on hollow fiber membranes consisting of polyethersulfone/polyvinylpyrrolidone (PES/PVP) without any precoating. These results are difficult to obtain when using HPTCs on membranes. These cells would not grow and survive on PES/PVP membranes. In contrast, MDCK and LLC-PK1 cells form confluent monolayers on these materials.

In many cases, ECM coatings of either laminin or collagen IV were used for seeding HPTCs [113]. The cell performance is mainly influenced by the physico-chemical characteristics of the native membrane, and not always a single ECM coating sufficiently improves cell-membrane interactions.

The cell source is a serious problem, as the primary cells have a limited lifespan. HPTCs are obtained from nontransplantable human kidneys. It is difficult to obtain a sufficient amount of cells, because it is not possible to use material from diseased kidneys due to possible alteration of cell functions. Thus, only nondiseased kidneys can be used. Primary proximal tubule cells show functional changes during passaging, and dedifferentiation [113–115]. Furthermore intrinsic interdonor variability and dedifferentiation risk make standardization difficult.

Moreover, it seems that HPTCs spontaneously form large and functional kidney tubules on 2D surfaces and within tubular substrates [116]; this is interesting for applications in nephrotoxicology, but for BAKs they compromise device functions and lead to clogging of the hollow fibers.

Alternatively, permanent human proximal tubule-derived cell lines like HK-2 cells could be used. HK-2 cells show some differentiated functions of proximal tubule cells, but are functionally and morphologically not equivalent to primary cells [113], and the use of oncogenes poses safety concerns. The human proximal tubule-derived cell lines did not express some important functional proteins, and further functional characterization would be required.

Another approach is to use hTert transgenic human proximal tubule-derived cell lines that show characteristic features and functions of differentiated HPTCs; however, these cells might change over time.

Stem cell-based approaches are the most attractive for achieving an unlimited and less variable cell source for BAKs. It has been shown that murine embryonic stem (ES) cells cultured *in vitro* can be induced to express specific markers specific for renal precursors [117], and *in vitro* conditioned medium from injured proximal tubule cells induced epithelial differentiation of human adipose-derived adult mesenchymal stem cells [118]. Although the results suggest a potential usefulness of stem cells in kidney bioengineering, further work is required to develop protocols for the differentiation of stem cells *in vitro* into mature human renal proximal tubule cells and other renal cell types.

Applications of induced pluripotent stem (iPS) cells might be the most attractive, but the use of oncogenes and integrating viral vectors for reprogramming brings up some safety concerns. The recently developed iPS cells free of vectors and transgene sequences may provide a solution to this problem [119].

In conclusion, among the cell types currently available, HPTCs appear as the most appropriate for clinical applications. However, the use of HPTCs poses many problems, and it would be important to explore stem cell-based and other alternative approaches.

7.4.3 Membranes

The membranes used for the development of a bioartificial kidney are mainly commercial hollow fiber membranes of PSF/PVP, which are designed for hemodialysis/hemofiltration. These membranes seem to be inappropriate for cell adhesion, probably due to the presence of PVP that generates problems in HPTC growth and survival. For this reason the membranes are coated with ECM in order to improve cell adhesion [120]. Membranes that are currently employed in hemodialysis/hemofiltration are optimized for contact with blood, and usually hydrophobic membranes such as those made of PSF are modified with hydrophilic additives in order to prevent protein adhesion. On the other hand the protein adsorption promotes cell adhesion. For this reason Ueda et al. suggested the use of asymmetric membranes with hemocompatible and cytocompatible surfaces [121]. These authors described a membrane consisting of PSF blended with a phospholipid polymer which was asymmetrically distributed between the skin and the sponge layer of the membrane. In general, it would be preferable to use relatively rough sponge layer for cell growth and to expose the smooth skin layer to the blood.

Alternatively asymmetric membranes can be realized by coating the surfaces with antifouling agents such as polyethylene glycol on the blood exposed side and adhesive coatings on the cell-exposed side. Dual-layered membranes with each layer composed of a different material could be interesting. Recent findings revealed problems with HPTC survival and differentiation on a variety of commercially available membrane materials and surface treatments, while ECM coatings lead improvements. HPTCs formed confluent epithelia on membranes consisting of PSF blended with Full-Cure (FC) under perfusion conditions. Single or double coatings did not induce further improvement of cell performance on PSF-FC membranes. Growth and differentiation of primary human cortical tubular epithelial cells were also observed on collagen IV-coated thin film and nanostructured materials [122].

7.4.4 Preclinical bioartificial kidney devices

Research on bioartificial kidneys has been performed by different groups [107, 108, 123, 124]. Clinical trials with BAKs have been performed by Humes et al. [105]. In their device, the patient's blood (red) first enters the hemofiltration unit (left), which contains ultrafiltration hollow fiber membranes (Figure 7.7 (a)) [105]. The blood and the ultrafiltrate (yellow) leaving the hemofiltration unit flow into the bioreactor unit (right), which contains hollow fiber membranes with an epithelium of renal cells (green) on the inner surfaces. The cells secrete molecules, which become enriched in the ultrafiltrate during processing and in the blood flowing on the outside of the hollow fiber membranes. The enriched blood flows back into the patient, and the processed ultrafiltrate (orange) is discarded. An enlarged cross-section of a hollow fiber membrane from the bioreactor unit is shown in the lower left corner. The ultrafiltrate flows in the lumen of the hollow fiber membrane, and the blood flows on the outside. The inner surface of the hollow fiber membrane is covered with secreting renal cells, enriched ultrafiltrate, and blood. BMP-7 in the ultrafiltrate would regulate HPTC performance, whereas BMP-7 in the bloodstream would be delivered to the patient.

One of the bioartificial kidneys developed in recent years is a device realized by the group of the University Medical Center Groningen called BioKid (Figure 7.7 (b)).[7] The bioreactor consists of hollow fiber membranes coated in the lumen side where nephron cells are cultured. The bioreactor mimics the function and operation of nephrons ensuring the removal of toxins that remain after hemodialysis. Thus the device improves the quality of hemodialysis treatment by reducing the risk of complications such as cardiovascular problems resulting from the accumulation of toxic waste products.

The bioartificial kidney developed by Vanderbilt (Figure 7.7 (c)) consists of two parts: a filter side and a cellular side [125]. On the filter side, silicone membranes with microscopic pores separate toxins from the blood, in a more efficient way than dialysis. Body pressure forces blood through silicone membrane, so a pump is not needed. On the cellular side, the filtered blood is pumped over a bed of cells taken from either the patient's own failing kidneys or from a donor. The cells sense the chemical makeup of the filtered blood and trigger the body to maintain appropriate levels of salt, glucose, and water. After entering the device the patient's blood is filtered by a silicone membrane that removes toxins. In the second stage a bed of cells regulates the chemical balance of the filtrated blood, reabsorbing water, sugar, and salts. The toxins and excess of water are passed into the waste outlet connected to the bladder. In Table 7.7 the main characteritics of membrane bioartificial kidney are provided.

7 www.bmm-program.nl/SITE/PUBLIC/GO/article.aspx?id=87&title=Bio-artificial+kidney

Tab. 7.7: Characteristics of a membrane bioartificial kidney.

Type of bioartificial system	Membrane configuration	Membrane material	Cell capacity	Cell type	Cell compartment
Renal tubule assist device RAD	Hollow fiber	Polysulfone-PVA	10^8	Proximal tubule renal derived cells	Intraluminal space
BAK	Hollow fiber	PSF/PVP ECM coated	10^8	HPTCs	Intraluminal space
Bioartificial renal epithelial cell system BRECS	Flat	Niobium-coated carbon disk	10^8	Renal epithelial cells	Disk surface
Wearable bioartificial kidney WEBAK	Flat unit & cartridge	Niobium-coated carbon disk, sorbent-based cartridge	10^8	Renal epithelial cells	Disk surface
Bioartificial renal tubule devices BTD	Hollow fiber	Ethylene vinyl alcohol (EVAL)	$3–7 \times 10^8$	Human renal proximal tubular cells	Intraluminal space

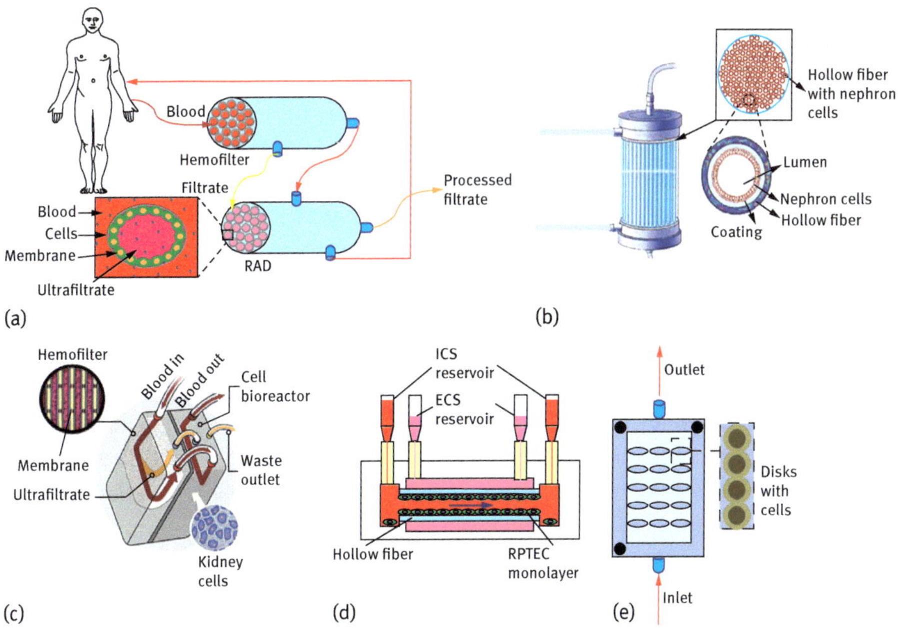

Fig. 7.7: Bioartificial kidney configurations: (a) BAK; (b) Biokid; (c) implantable bioartificial kidney; (d) lab-on-a-chip bioreactor system; (e) WEBAK.

A wearable bioartificial kidney (WEBAK) combining peritoneal dialysis with a bioartificial renal epithelial cell system (BRECS) has been developed (Figure 7.7 (e)) [126]. The BRECS is a perfusion bioreactor that utilizes primary renal epithelial cells derived from the kidney and expanded from progenitor cells during *in vitro* culture. Cells are seeded on porous disks, which are placed within a media flow path within the BRECS. A WEBAK utilizes peritoneal fluid to maintain cell viability and functionality and comprises the use of sorbent-based technologies to replace the excretory function of the kidney and of the compact BRECS to replace the metabolic function of the kidney.

The lab-on-a-chip bioreactor system has been developed by using human proximal tubule epithelial cells (RPTEC) cultured in the luminal space of a single hollow fiber of PES-PVP precoated with fibrin (Figure 7.7 (d)) [127]. To mimic a tubule, a tuneable hollow fiber membrane with an exterior skin layer was used with the aim to provide immunoprotection for cells from extracapillary blood flow, and a coarse inner surface that facilitates a hydrogel coating for cell attachment. The hollow fiber membrane is located inside a polydimethylsiloxane (PDMS) body and a glass substrate. The inner surface of the hollow fiber is featured with pores up to 0.3 μm in size, facilitating the attachment of extracellular matrix coating on the membrane. Together with a fibrin coating, a confluent monolayer of human kidney renal proximal tubule epithelial cells is successfully formed on the fiber inner surface under microfluidic flow conditions in a "lab-on-a-chip" bioreactor system.

7.5 Bioartificial lungs

Lung disease accounts for 9 million deaths per year worldwide: it ranks third in the list of causes of death in Europe[8], and is the fourth most common cause of death worldwide. It is expected to be third in rank due to high rate of tobacco smoking and demographic expansion leading to increased exposure to irritants arising from mining, agriculture, and urbanization. Beside malignancies, the predominant underlying disease is chronic obstructive pulmonary disease (COPD) with more than 15 million patients in Europe. COPD is a leading indication together with other pulmonary diseases, such as pulmonary fibrosis, pulmonary hypertension, and cystic fibrosis, for the development of a biohybrid lung.

In contrast to kidney or heart failure, at present no lung substitute is able to fulfil the main lung function, which is carbon dioxide removal and blood oxygenation. In the event of lung failure, mechanical ventilation has to be applied. This attempts to provide life-saving gas exchange, but at the same time causes ventilation-associated lung injury (VALI). In addition to VALI, other potentially life-threatening complications associated with mechanical ventilation develop frequently, such as ventilation-

8 World Health Organisation, www.who.int

associated pneumonia, and secondary injury to other organ systems such as the kidney, pancreas, and small intestine, which finally leads to remote organ failure. This scenario is of significance for critically ill patients in intensive care unit, and even more important for lung transplant patients with end-stage lung failure. Ventilation or even conventional extracorporeal gas exchange such as extracorporeal membrane oxygenation (ECMO) is only possible for bedridden patients and, most commonly, sedated patients. ECMO and mechanical ventilation are known to be significant risk factors for post-lung-transplant mortality. Several medical centers worldwide do not consider these patients as candidates for lung transplantation.

Membrane oxygenators currently in clinical use commonly employ packed hollow fiber membranes. Protein and cellular deposits on the synthetic surfaces lead to patient complications and the need for anticoagulation therapy. The native lung provides approximately 70 m^2 of gas transfer area in alveoli with actively anticoagulant endothelial surfaces lining the blood vessels. The limits of the membrane oxygenators currently used arise at the interface where oxygen and carbon dioxide pass between the blood and gas sides of synthetic hollow-fiber membranes. Blood proteins adsorb to the polymeric membrane surfaces, which can trigger the activation of immune cells and deposition of clots onto the fibers. This can result in immune responses and in a tendency toward bleeding, respectively, when the blood reenters the patient. Further, to achieve adequate gas transfer, membrane surface areas on the order of 1 m^2 are required. The size of current membrane oxygenators is incompatible with placement within the body [128]. Hitherto, lung transplantation is the only treatment option for patients with therapy-refractory, irreversible lung failure. However, the number of lung transplantations is limited by the shortage of donor organs. A pumpless interventional lung assist system developed by the European company Novalung® (Hechingen, Germany) became available on the market and which focuses on protective ventilation. This device is characterized by a microstructured gas exchange membrane that mimics the alveolar-capillary bed of the human lung; flow resistance is sufficiently low that a normally beating heart is sufficient to move blood through the device for oxygenation. No additional mechanical pump is required. Researchers in Medizinizsche Hochschule of Hannover proved that the Novalung® iLA™ could be used as a bridge to sustain patients with lung failure until transplant [129, 130]. This device was sufficient to fully replace lung function for a limited period of time failing in three to four weeks. Apparently, the surface of the gas exchange membrane triggers protein binding, blood cell aggregation, and thrombus formation. The improved hemo-compatibility of the Novalung® iLA™ by seeding endothelial cells onto the membrane gives it a more natural endothelial surface. Primary endothelial cells are seeded on the membrane, thereby preventing direct contact between the blood and the synthetic surface of the membrane.

7.5.1 Engineering the multiscaled architecture of the bioartificial lung

The functional unit of the lung is the air–blood interface between epithelium and endothelium. In order to facilitate efficient gas exchange between the gas on the alveolar side and blood on the capillary side, this interface must have a minimal diffusion length. A key role in the physiologic gas exchange process is played by hierarchical branching airways and vascular networks, which allow efficient perfusion and ventilation. Therefore, a crucial step in the design of a bioartificial lung is to mimic functional units and networks. In an adult lung there are about 300–500 million alveoli, which provide a total surface area around 100 m^2 [131, 132]. Considering that the alveolar wall is ≈ 0.5 μm thick, and the pulmonary vessels can be as small as 5 μm in diameter, it is very challenging to reproduce the complex macro- and microarchitecture of the human lung. Most of the studies focusing on the production of lung scaffolds led to the development of decellularization methods of the cadaveric organ. In this approach detergents and enzymes are perfused in order to remove all the cellular components of the tissue, producing a nonimmunogenic scaffold with the lung architecture. Petersen et al. and Ott et al. applied whole organ decellularization to the lung [133, 134]. Their approach achieved the removal of the cells from isolated adult rat lungs in a manner which preserved the structural characteristics including the alveoli. The remaining lung "scaffolds" were repopulated with epithelial and endothelial cells in a bioreactor. Lung tissue was successfully regenerated. Ott et al. attempted the perfusion decellularization of lungs to create whole lung scaffolds with a perfusable vascular bed and with preserved airway and alveolar geometry. A recent report of clinical implantation of a tissue-engineered conducting airway based on an acellular matrix scaffold underlines the clinical potential of this approach in the treatment of respiratory disorders. However this method requires preexisting native lung tissue, and therefore it cannot solve the problem of the human donor shortage. Alternatively xenogeneic (animal) scaffolds can be clinically used as biological scaffolds. For example porcine lung scaffolds are nonimmunogenic and pose no risk of zoonosis transmission, but they require extensive decellularization and rigorous testing of donor tissue prior to implantation. In this way an ECM whose structure closely matches that of native tissue is provided to the cells, retaining on the biological scaffolds a diverse set of proteins which can stimulate the proliferation of seeded cells and can serve as chemoattractants for host cells.

A different perspective involves the development of a bioartificial lung with an air-blood interface that is made from synthetic materials.

Recently the attention of scientists has focussed on the improvement of ECMO devices by coating the gas exchange surface with EPCs and ECs in order to realize a wearable bioartificial lung [135, 136]. This approach seeks to mimic the *in vivo* function of vascular endothelial cells (ECs) to yield a biocompatible surface, actively inhibiting platelet activation and deposition. A biohybrid lung prototype consisting of hollow fiber membranes modified at the surface with radio frequency glow discharge and protein adsorption to promote endothelial cell attachment and growth was developed by Polk et al. Both surface treatment and endothelialisation of HFs produced a more biocompatible surface. Considering that the diffusion of oxygen and carbon dioxide occurs across the alveolar-capillary barrier in the native lung, it is not clear what the impact of endothelialisation would be on gas transfer in the HF system. Increase of blood velocity to reduce the boundary layer thickness either by rotation or pulsation has been shown to improve gas diffusion. This biohybrid artificial lung prototype employs active mixing produced by rotation of a surface modified and endothelialized HF bundle. The novel aspects of the biohybrid artificial lung design are an endothelialized layer on the outer lumen of the HFs to present a more biocompatible blood-contacting surface, while the rotation of this fiber bundle should increase the effective gas transport per surface area.

Moreover, attempts are being made to produce miniaturized ECMO devices which are wearable, so that the patient can experience the benefits of mobility and exercise [137, 138]. These developments will produce a wearable bioartificial lung, which, by allowing longer term support prior to implantation, will represent a significant advance in patient care.

7.5.2 Cell types

Bioartificial lung constructs require appropriate cells in order to perform the same functions of the native lung. In this context cell types must be identified to achieve tissue formation. The most appropriate cells to cover the airway lumina are epithelial cells (EPCs) that *in vivo* provide mucociliary clearance of the airways. For the pulmonary vasculature, endothelial cells (ECs) are the main candidate since they downregulate the thrombogenic response. EPCs and ECs cooperate through signaling pathways to regulate the amount of interstitial fluid and to prevent the pulmonary edema in the native lung. Other important cells for the performance of lung functions include alveolar macrophages, which play an important role in lung clearance, and myofibroblasts, which contribute to the healing of the epithelium. A population of stem cells which provides self-renewal in response to injury is also resident in the native lung.

Therefore several cell types should be used to seed the scaffold. Alternatively, it can be used bone marrow from the patient that contains sufficient numbers of progenitor cells able to promote the repair mechanisms. Bone marrow is a source of mesenchymal stem cells (MSCs) which can be expanded and differentiated into EPCs and ECs (Table 7.8).

EPC and EC can be also differentiated from human embryonic stem cells (hESCs) which divide an unlimited number of times [139, 140]. There are still impediments to hESCs clinical use arising from ethical, legal, and religious concerns, although a limited number of clinical trials involving their use has been performed [141]. Some disadvantages about the use of these cells are related to the potentially tumorigenic-

Tab. 7.8: Cell source used in bioartificial lung.

Cell source	Drawbacks
Patient-derived ECs, EPCs	Invasive procedure for harvesting, difficult to propagate the required cell numbers
Patient-derived MSCs	Low yield of differentiated cells, restricted cell fate
Bone marrow cells	Restricted cell fate
hESCs	Immunogenicity and tumourigenicity, complex differentiation pathways, low yield of differentiated cells, ethical concerns.
iPSCs	Tumourigenicity

ity of hESCs and cells differentiated from them, and their nonautologous origin may make them immunogenic to the patient. Pluripotent stem cells generated from adult cells represent an appealing cell source for clinical translation [142]. The risk of tumorigenicity for iPSCs and differentiated iPSCs remains a concern for their clinical use.

7.5.3 Biomaterials for the bioartificial lung

Alternatively to biological materials, artificial scaffolds have the advantage of being nonimmunogenic; in addition, they can be produced rapidly in standardized conditions to meet the specific needs of a patient while overcoming the variability linked to individuals. The choice of the material for the scaffold fabrication is very important. It must be biocompatible, i.e. nontoxic, chemically stable, and should not cause adverse reactions when implanted. The material must have physico-chemical, transport, and mechanical properties as well as micro- and nanoscale structures mimicking those of the native ECM. Several biomaterials can be used to fabricate artificial scaffolds (Figure 7.8).

Different properties are required from the scaffold used for a bioartificial lung with respect to the artificial device. Hydrophobic polymers, which are typically used to make the hollow fibres in ECMO, do not allow cell adhesion and, therefore, cannot be used in a bioartificial lung. Surface modification through physiochemical and/or chemical treatment is needed to guarantee uniform cell attachment [143] and to create a cell monolayer resistant to high fluid shear stress due to the blood flow [135].

Tracheal scaffolds have been developed by using electrospun fibers of polyethylene terephthalate (PET) or nanocomposites of PET and polyurethane (PU) [144], which replicate closely the fibrous nanoscale structures formed by collagen and elastic fibers in the native tracheal wall (Figure 7.9).

Synthetic hydrogels [145] such as poly(ethylene glycol) (PEG) and poly(vinyl alcohol) (PVA), and synthetic elastomers [146] such as poly(glycerol sebacate) (PGS) [147] can be moulded to reproduce the complex architecture of the native lung. Synthetic

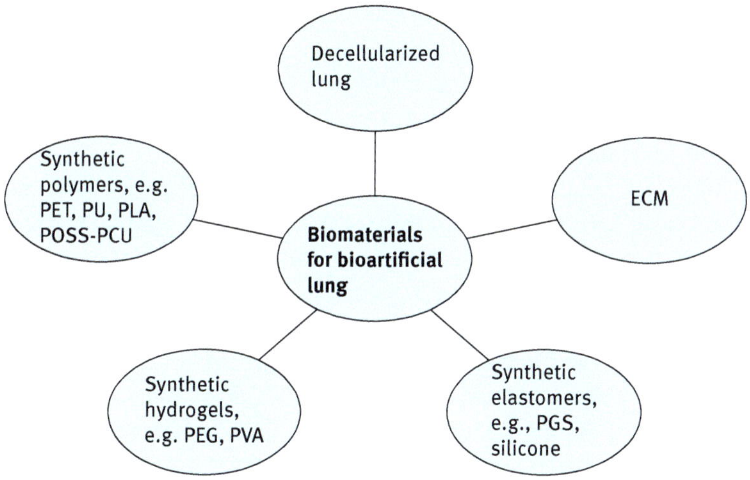

Fig. 7.8: Biomaterials for bioartificial lungs.

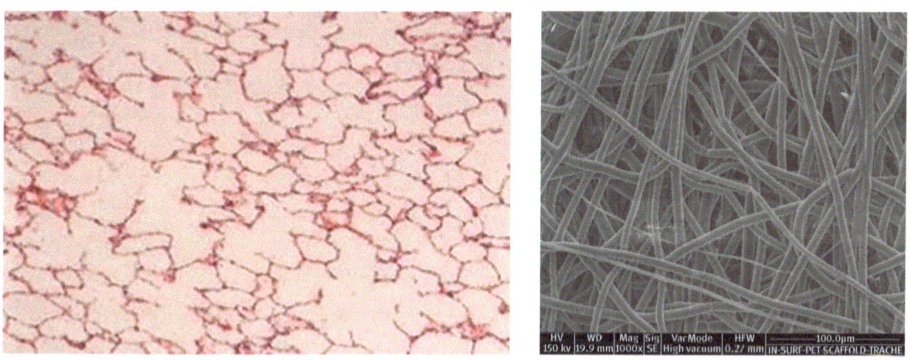

(a) (b)

Fig. 7.9: (a) Histological cross section of human lung tissue, and (b) a SEM image of electrospun PET trachea scaffold.

foam-moulded scaffolds address the issue of availability, since the organs which could be custom designed and manufactured have had a limited use as airway implants due to cell seeding issues. Once the dimensions of the patient are determined through a computer tomography scan, a dimensionally identical trachea is manufactured as needed. Several concerns had to be overcome. First, the foam proved to be stiffer than a natural trachea. This made it difficult for the surgeon to penetrate with sutures. Also, the stiffness was not conducive to optimal cell seeding and thus building a suitable cell mass require a longer time than that supposed. Synthetic scaffolds should reproduce the lung's functional unit represented by the air-blood interface between epithelium and endothelium, which must have a minimal diffusion length to facilitate efficient gas exchange between gas on the alveolar side and blood on the capillary side. Physiologic gas exchange depends also on hierarchical branching airways and vascular networks which allow both efficient perfusion and ventilation. A synthetic scaffold should be fabricated in order to reproduce networks which provide physiologic shear stress and functional gas transfer.

The scaffold must have sufficient pore size and interconnectivity to allow the transfer of dissolved gases across the interstitium [148]. The scaffolds should be free-permeable to liquids, because interstitial fluid is essential for providing aqueous diffusion of nutrients to the EPCs on the alveolar surface and for permitting chemical signalling between the EPCs and ECs. The porosity and structural properties of the scaffold should also permit the infiltration and migration of circulating host cells. The bioartificial lung must be engineered with appropriate weight and scaffold materials that exhibit elastic properties, which enable the lung to be deformed by the diaphragm. Furthermore, the material must be capable of undergoing many repeated cycles of large deformation without loss of integrity and mechanical properties.

With regard to materials for bioartificial lungs, the most appropriate are those matching closely composition and structure of native lung ECM. However, further developments in the field are required before artificial lung scaffolds can be approved for clinical use.

7.5.4 Bioengineered trachea

In 2008, the first tissue-engineered airway was successfully implanted by Macchiarini and colleagues in a 30-year-old woman with end-stage left-main bronchus malaria to restore lung function of a patient without immunosuppression therapy [149]. An airway from a deceased human donor was used to create a human decellularized matrix that was structurally and mechanically similar to a native trachea, with chemotactic and proangiogenic properties.[2] All loose connective tissue was removed from the donor trachea, reseeded with autologous epithelial cells and chondrocytes of mesenchymal-stem-cell origin. The *in vitro* expansion and differentiation of autologous cells occurred in a bioreactor designed ad hoc to address the requirements of: culturing different cell types on each side of a tubular matrix; supply nutrients and remove wastes; biomechanical cues in the form of hydrodynamic shear stress; autoclavability, ease of sterile handling, reliability, and precision compatible with good laboratory practice. 4 months after implant of recellularized scaffold, the patient was healthy, active, with normal lung function, and did not require immunosuppressive treatment. Although the airway was successfully implanted, some critical issues remained unaddressed such as (1) the feasibility of obtaining a viable, recellularized, and functional engineered airway, and its maintenance once implanted; (2) long-term stability of the decellularized natural matrix; (3) fate and tumorigenic risks associated with the implanted stem cells. Recently, Baiguera et al., using a nanocomposite polymeric material, obtained a tailor-made artificial tracheal scaffold with physical and mechanical properties similar to native tissue [150]. The artificial scaffold, which had been seeded ex vivo with autologous progenitor cells in the bioreactor, has been successfully implanted into a patient affected by a primary recurrent tracheobronchial tumor [151]. After 3 postoperative months, the nanocomposite was lined with a well-developed healthy mucosa, and no distal ischemic necrosis was observed [151].

Important questions, such as the evaluation of the long-term biomechanical properties of the bioengineered grafts, the optimization of tracheal synthetic graft and of bioreactor design, and the function and fate of the seeded cells, remain open [152, 153]. However, combining living cells with biocompatible and biodegradable scaffolds, represents an attractive approach to obtain an anatomical, physiological, and biomechanical airway replacement which could improve significantly functional outcomes for airway patients.

7.6 References

[1] Kizilel S, Garfinkel M, Opara E. The bioartificial pancreas: Progress and challenges. Diabetes Technol Ther. 2005; 7: 968–974.

[2] Martin Y, Vermette P. Bioreactors for tissue mass culture: design, characterization, and recent advances. Biomaterials. 2005; 26, 7481–7503.

[3] L. Giorno, L. De Bartolo, E. Drioli. In: Bhattacharyya D, Butterfield (eds.). New Insights into the Membrane Science and Technology: Polymeric and Biofunctional Membranes, vol. 8. Amsterdam: Elsevier; 2003. pp. 187–217.

[4] Freed LE, Vunjak-Novakovic G. Methods of Tissue Engineering. San Diego: Academic Press; 2002. pp. 97–110.

[5] De Bartolo L, Morelli S, Giorno L, Campana C, Rende M, Salerno S, Maida S, Drioli E. Polyethersulfone membrane biohybrid system using pig hepatocytes: effect of diclofenac on cell biotransformation and synthetic functions. J Memb Sci. 2006; 278: 133–143.

[6] De Bartolo L, Salerno S, Giorno L, Morelli S, Barbieri G, Curcio E, Rende M, Drioli E. Membrane bioreactor using pig hepatocytes for in vitro evaluation of antinflammatory drug. Catal Today. 2006; 118: 172–180.

[7] De Bartolo L, Morelli S, Rende M, Campana C, Salerno S, Quintiero N, Drioli E. Human hepatocyte morphology and functions in a multibore fiber bioreactor. Macromol Biosci. 2007; 7: 671–680.

[8] Hubbell JA and Langer R. Tissue Engineering, C&EN, vol. 13. Washington, DC: American Chemical Society. pp. 42–53, 1995.

[9] Aebischer P, Goddard M, Signore P, Timpson R. Functional recovery in hemiparkinsonian primates transplanted with polymer encapsulated PC12 cells. Exp Neurology. 1994; 126: 1.

[10] Arias IM, Boyer JL, Chisari FV, Fausto N, Schacter D, Shafritz DA. The Liver: Biology and Pathobiology. Philadelphia: Lippincott Williams & Wilkins, p. 591; 2001.

[11] Hansen LK, Mooney DJ, Vacanti JP, Ingber DE. Integrin binding and cell spreading on extracellular matrix act at different points in the cell cycle to promote hepatocyte growth. Molecular Biology of the Cell. 1994; 5(9): 967 –975.

[12] Crossin KL. Cell adhesion molecules activate signaling networks that influence proliferation, gene expression, and differentiation. Annals of New York Academy Science. 2003; 961(1): 159.
[13] te Velde AA, Ladiges NCJJ, Flendrig LM, Chamuleau RAFM. Functional activity of isolated pig hepatocytes attached to different extracellular matrix substrates. Implication for application of pig hepatocytes in a bioartificial liver. J Hepatology. 1995; 3: 184–192.
[14] Pasher A, Sauer IM, Neuhaus P. Analysis of allogeneic versus xenogeneic auxiliary organ perfusion in liver failure reveals superior efficacy of human livers. International J Artificial Organs. 2002; 25: 1006–1012.
[15] Allen JW, Bhatia SN. Engineering Liver Therapies for the Future. Tissue Engineering. 2002, 8: 725–737.
[16] Mitaka T, Mizuguchi T, Sato F, Mochizuki C, Mochizuki Y. Growth and maturation of small hepatocytes. J Gastroenterology and Hepatology. 1998; 13(Suppl): S70 –S77.
[17] Faris RA, Konkin T, Halpert G. Liver stem cells: a potential source of hepatocytes for the treatment of human liver disease. Artificial Organs. 2001; 25(7): 513 –521.
[18] Rippon HJ, Bishop AE. Embryonic stem cells. Cell proliferation. 2004; 37(1): 23–30.
[19] Dunn JCY, Yarmush ML, Koebe HG, Tompkins RG. Hepatocytes function and extracellular matrix geometry: long-term culture in a sandwich configuration. Faseb J. 1989; 3: 174–177.
[20] Koide N, Sakaguchi K, Koide Y, et al. Formation of mulncellular spheroids composed of adult rat hepatocytes in dishes with positively charged surfaces and under other non-adherent environments. Experimental Cell Researck. 1990; 186: 227–235.
[21] Yagi K, Tsuda K, Serada M, Yamada C, Kondoh A, Miura Y. Rapid formation of multicellular spheroids of adult rat hepatocytes by rotation culture and their immobilization within calcium alginate. Artificial Organs 1993; 17(11): 929–34.
[22] Kong LB, Chen S, Dememou AA, Rozga J. Matrix-induced liver cell aggregates (MILCA) for bioartificial liver use. International Journal of Artificial Organs. 1996; 19: 72–78.
[23] Liu ZC, Chang TMS. Coencapsulation of hepatocytes and bone marrow cells: In vitro and in vivo studies. Biothecnology Annual Review. 2006; 12: 137–151.
[24] Dixit V, Darvasi R, Arthur M et al. Restoration of liver functions in gunn rats without immunosuppression using transplated microencapsulated hepatocytes. Hepatology. 1990; 12: 1342–1349.
[25] Kino Y, Sawa M, Kasai S, Mito M. Multiporous cellulose microcarrier for the development of a hybrid artificial liver using isolated hepatocytes. Journal of Surgical Research. 1998; 79: 71–76.
[26] Wu C, Pan J, Bao Z, Yu Y. Fabrication and characterization of chitosan microcarrier for hepatocytes culture. J Material Science Material in Medicine. 2007; 18(11): 2211–2214.
[27] Jindal R, Nahmias Y, Tilles AW, Bethiaume F, Yarmush ML. Amino acid-mediated heterotypic interaction governs performance of hepatic tissue model. The Faseb Journal. 2009; 23(7): 2288–2298.
[28] Krause P, Sagatolislam F, Koenig S, Unthan-Fechner K, Probst I. Maintening hepatocytes differentiation in vitro through co-culture with hepatic stellate cells. In Vitro Cell Dev Biol Anim. 2009; 45(5/6): 205–212.
[29] Demetriou AA, Arnaout WS, Backfish G, Moscioni AD. In: Brunner G, Mito M. Artificial Liver Support, 2nd edition. Berlin: Springer; 1993. pp. 283–295.
[30] De Bartolo L, Morelli S, Bader A, Drioli E. Evaluation of cell behaviour related to physicochemical properties of polymeric membranes to be used in bioartificial organs. Biomaterials. 2002; 23(12): 2485–2497.

[31] De Bartolo L, Catapano G, Della Volpe C, Drioli E. The effect of surface roughness of microporous membranes on the kinetics of oxygen consumption and ammonia elimination by adherent hepatocytes. J Biomaterial Science, Polymer Edition 1999; 10: 641–655.
[32] Morelli S, Salerno S, Piscioneri A, Campana C, Drioli E, De Bartolo L. Membrane bioreactors for regenerative medicine: an example of the bioartificial liver. Asia-Pacific J Cheml Eng. 2010; 5: 146–159.
[33] Dionne KE, Cain BM, Li RH, Bell WJ, Doherty EJ, Rein DH, Lysaght MJ, Gentile FT. Biomaterials. 1996; 17: 257–266.
[34] McClelland RE, MacDonald JM, Coger RN. Biotechnol Bioeng. 2003; 82: 12–27.
[35] Curcio E, De Bartolo L, Barbieri G, Rende M, Giorno L, Morelli S, Drioli E. Diffusive and convective transport through hollow fiber membranes for liver cell culture. J. Biotechnol. 2005; 117: 309–321.
[36] Fung YC. Biomechanics. Mechanical properties of living tissues. New York: Springer-Verlag; 1993. pp. 66–108.
[37] Beck RE, Schultz JE. Biochem Biophys Acta, 1972; 255: 273–279.
[38] Mulder M. Basic Principles of Membrane Technology, Dordrecht, Boston, London; Kluwer Academic Publisher, pp. 110–144; 1991.
[39] Matsumara KN, Guevara GR, Huston H, Hamilton WL, Rikimare M, Yamasaki G, Matsumura MS. Hybrid bioartificial liver in hepatic failure: preliminary clinical report. Surgery. 1987; 101: 99–103.
[40] Margulis MS, Erukhimov EA, Andreiman LA, Viksna LM. Temporary organ substitution by hemoperfusion through suspension of active donor hepatocytes in a total complex of intensive therapy in patients with acute hepatic insufficiency. Resuscitascion. 1989; 18; 85–94.
[41] Sussman NL, Chong MG, Koussayer T, He DE, Shang TA, Whisennand HH, Kelly JJ. Reversal of fulminant hepatic failure using an extracorporeal liver assist device. Hepatology. 1992; 16: 60–65.
[42] Demetriou AA, Rozga J, Podesta L, Lepage E, Morsiani E, Moscioni AD, Hoffman A, McGrath M, Kong L, Rosen H, Villamil F, Woolf G, Vierling J, Makowka L. Early clinical experience with a hybrid bioartificial liver. Scandinavian Journal of Gastroenterology. 1995; 208: 111–117.
[43] Gerlach JC, Encke J, Hole O, Muller C, Ryan CJ, Neuhaus P. Bioreactor for larger scale hepatocyte in vitro perfusion. Transplantation. 1994; 58: 984–988.
[44] Patzer JF, Mazariegos GV, Lopez R. Preclinical evaluation of the Excorp Medical, Inc, bioartificial liver support system. J American College of Surgeons. 2002; 195(3): 299–310.
[45] Flendrig LM, la Soe JW, Jorning GG, et al. In vitro evaluation of a novel bioreactor based on an integral oxigenator and a spirally wound non woven polyester matrix for hepatocyte culture as a small aggregate. J Hepatology. 1997; 26: 1379–1392.
[46] Rozga J, Holzman MD, Ro MS, Griffin DW, Neuzil DF, Giorgio T, Moscioni AD, Demetriou AA. Development of a hybrid bioartificial liver. Ann Surg. 1993; 217, 502–511.
[47] Xue YL, Zhao SF, Luo Y, et al. TECA hybrid artificial liver support system in the treatment of acute liver failure. World J Gastroenterol. 2001; 7: 706–709.
[48] Ding YT, Qiu YD, Chen Z, Xu QX, Zhang HY, Tang Q, Yu. The development of a new bioartificial liver and its application in 12 acute liver failure patients. World J Gastroenterol. 2003; 9, 829–832.
[49] Hu WS, Friend JR, Wu FJ, Sielaff T, et al. Development of a bioartificial liver employing xenogeneic hepatocytes. Cytotechnology, 1997; 23: 29–38.
[50] Shiraha H, Koide N, Hada H, Ujike K, Nakamura M, Shinji T, Gotoh S, Tsuji T. Improvement of serum amino acid profile in hepatic failure with the bioartificial liver using multicellular hepatocyte spheroids. Biotechnology and Bioengineering. 1996; 50: 416–421.

[51] Naka S, Takeshita K, Yamamoto T, Tani T, Kodama M. Bioartificial liver support system using porcine hepatocytes in a three dimensional hollow fibre module with collagen gel: an evaluation in the swine acute liver failure model. Artificial Organs. 1999; 23: 822–828.
[52] De Bartolo L, Jarosch-Von Schweder G, Haverich A, Bader A. A novel full-scale flat bioreactor utilizing porcine hepatocytes: cell viability and tissue specific functions. Biotechnology Progress. 2000; 16: 102–108.
[53] Nagaki M, Miki K, Kim Y, Ishijama H, Hirahara I, Takahashi H, Sugiyama A, Moriwaki H. Development and characterization of a hybrid bioartificial liver using primary hepatocytes entrapped in a basement membrane matrix. Digestive Diseases and Sciences. 2001; 46: 1046–1056.
[54] Naka S, Takeshita K, Yamamoto T, Tani T, Kodama M. Bioartificial liver support system using porcine hepatocytes entrapped in a three-dimensional hollow fiber module with collagen gel: an evaluation in the swine acute liver failure model. Artif Organs. 1999; 23; 822–828.
[55] Jasmund L, Langsch A, Simmoteit R, Bader A. Cultivation of primary porcine hepatocytes in an OXY-HFB for use as a bioartificial liver device. Biotechnol Prog. 2002; 18: 839–846.
[56] Mizumoto H, Funatsu K. Liver regeneration using a hybrid artificial liver support system. Artif Organs. 2004; 28: 53–57.
[57] Nakazawa K, Matsushita T, Funatsu K. Prolonged lidocaine metabolizing activity of primary hepatocytes with spheroid culture using polyurethane foam as a culture substratum. Cytotechnology. 1997; 24: 235–242.
[58] Schmitmeier S, Langsch A, Jasmund I, Bader A. Development and characterization of a small-scale bioreactor based on a bioartificial hepatic culture model for predictive pharmacological in vitro screenings. Biotechnol Bioeng. 2006; 95: 1198–1206.
[59] De Bartolo L, Morelli S, Rende M, Campana C, Salerno S, Quintiero N, Drioli E. Human hepatocyte morphology and functions in a multibore fiber bioreactor. Macromol Biosci. 2007; 7: 671–680.
[60] Lu HF, Lim WS, Zhang PC, Chia SM, Yu H, Mao HQ, Leong KW. Galactosylated poly(vinylidene difluoride) hollow fiber bioreactor for hepatocyte culture. Tissue Eng. 2005; 11: 1667–1677.
[61] Memoli B, De Bartolo L, Favia P, Morelli S, Lopez LC, Procino A, Barbieri G, Curcio E, Giorno L, Esposito P, Cozzolino M, Brancaccio D, Andreucci VE, d'Agostino R, Drioli E. Fetuin-A gene expression, synthesis and release in primary human hepatocytes cultured in a galactosylated membrane bioreactor. Biomaterials. 2007; 28: 4836–4844.
[62] Curcio E, Salerno S, Barbieri G, De Bartolo L, Drioli E, Bader A. Mass transfer and metabolic reactions in hepatocyte spheroids cultured in rotating wall gas-permeable membrane system. Biomaterials. 2007; 28: 5487–5497.
[63] Ye SH, Watanable J, Takai M, Iwasaki Y, Ishihara K. High functional hollow fiber membrane modified with phospholipid polymers for a liver assist bioreactor. Biomaterials. 2006; 27: 1955–1962.
[64] Sauer IM, Scwartlander R, Schmid J, Efimova E, Vondran FWR, Kehr D, Pless G, Spinelli A, Brandeburg B, Hildt E, Neuhaus P. The Slide Reactor – a simple hollow fiber based bioreactor suitable for light microscopy. Artif Organs. 2005; 29: 264–267.
[65] Ostrovidov S, Jiang J, Sokai Y, Fujii T. Membrane-based PDMS microbioreactor for perfused 3D primary rat hepatocyte cultures. Biomed Microdevices. 2004; 6: 279–287.
[66] Roy P, Baskaran H, Tilles AW, Yarmush M, Toner M. Analysis of oxygen transport to hepatocytes in a flat-plate microchannel bioreactor. Ann Biomed Eng. 2001; 29: 947–955.
[67] De Bartolo L, Salerno S, Curcio E, Piscioneri A, Rende M, Morelli S, Tasselli F, Bader A, Drioli E. Human hepatocyte functions in a crossed hollow fiber membrane bioreactor. Biomaterials. 2009; 30: 2531–2543.

[68] Beger HG, Warshaw AL, Büchler MW, Kozarek RA, Lerch MM, Neoptolemos JP, Shiratori K, Whitcomb DC, Rau BM (eds.). The pancreas: an integrated textbook of basic science, medicine, and surgery. 2nd edition. Oxford, UK: Blackwell Publishing Ltd.; 2008. Front matter.
[69] Mathis D, Vence L, Benoist C. [beta]-Cell death during progression to diabetes. Nature. 2001; 414(6865): 792–798.
[70] Costa A, Conget I, Gomis R. Impaired glucose tolerance: is there a case for pharmacologic intervention? Treat Endocrinol. 2002; 1: 205–210.
[71] Opara EC, Mirmalek-Sani SH, Khanna O, Moya ML, Brey EM. Design of a bioartificial pancreas. J Investig Med. 2010; 58(7): 831–837.
[72] Furth ME, Atala A. Stem cell sources to treat diabetes. J Cell Biochem. 2009; 106: 507–511.
[73] Ramiya VK, Maraist M, Arfors KE, Schatz DA, Peck AB, Cornelius JG. Reversal of insulin-dependent diabetes using islets generated in vitro from pancreatic stem cells. Nat Med. 2000; 6: 278–282.
[74] Minami K, Okuno M, Miyawaki K, Okumachi A, Ishizaki K, Oyama K, et al. Lineage tracing and characterization of insulin-secreting cells generated from adult pancreatic acinar cells. Proc Natl Acad Sci USA. 2005; 102: 15116–15121.
[75] Zhou Q, Brown J, Kanarek A, Rajagopal J, Melton DA. In vivo reprogramming of adult pancreatic exocrine cells to beta-cells. Nature. 2008; 455: 627–632.
[76] Lumelsky N, Blondel O, Laeng P, Velasco I, Ravin R, McKay R. Differentiation of embryonic stem cells to insulin-secreting structures similar to pancreatic islets. Science. 2001; 292: 1389–1394.
[77] Devaskar SU, Giddings SJ, Rajakumar PA, Carnaghi LR, Menon RK, Zahm DS. Insulin gene expression and insulin synthesis in mammalian neuronal cells. J Biol Chem. 1994; 269: 8445–54.
[78] Zhang D, Jiang W, Liu M, Sui X, Yin X, Chen S, et al. Highly efficient differentiation of human ES cells and iPS cells into mature pancreatic insulin-producing cells. Cell Res. 2009; 19: 429–438.
[79] Chick WL, Like AA, Lauris V. Beta-cell culture on synthetic capillaries: An artificial endocrine pancreas. Science. 1975; 187: 847–856.
[80] Colton CK. Implantable biohybrid artificial organs. Cell Transplantation. 1995; 4(4): 415–436.
[81] Chang C-C. Analysis of transport phenomena and kinetics in an extravascular bioartificial pancreas. AICHE J. 1996; 42: 2668–2682.
[82] Calafiore R, Luca G, Calvitti M, et al. Cellular support systems for alginate microcapsules containing isltes, as composite bioartificial pancreas. Annals of the New York Academy of Sciences. 2001; 944: 240–252.
[83] De Vos P, Hillebrands JL, De Haan BJ, et al. Efficay of prevascularised expanded polytetrafluoroethylene solid support system as a transplantation site for pancreatic islets. Transplantation. 1997; 63(6): 824–830.
[84] Petersen P, Lembert N, Wesche J, et al. Improved diffusion properties of a new polysulfone membrane for the development of a bioartificial pancreas. Transplantation Proceedings. 2001; 33: 1952–1953.
[85] Sakai S, Ono T, Ijima H, Kawakami K Newly developed aminopropyl-silicate immunoisolation membrane for a microcapsule-shaped bioartificial pancreas. Annals of the New York Academy of Sciences. 2001; 944: 277–283.
[86] Silva AI and Mateus M. Development of a polysulfone hollow fiber vascular bioartificial pancreas device for in vitro studies. J Biotechnology. 2009; 139: 236–249.

[87] Kang J, Erdodi G, Kennedy JP, et al. Toward a bioartificial pancreas: Diffusion of insulin and IgG across immunoprotective membranes with controlled hydrophilic channel diameters. Macromolecular Bioscience. 2010. doi: 10.1002/mabi.200900386.
[88] Lee SH, Hao E, Savinov AY, et al. Human beta cell precursors mature into functional insulin-producing cells in an immunoisolation device: Implications for diabetes cell therapies. Transplantation, 2009; 87: 983–991.
[89] Lim F, Sun AM. Microencapsulated islets as bioartificial endocrine pancreas. Science. 1980; 210(4472): 908–910.
[90] de Vos P, de Haan BJ, Van Schilfgaarde R. Upscaling the production of microencapsulated pancreatic islets. Biomaterials. 1997; 18(16): 1085–1090.
[91] Tendulkar S, McQuilling JP, Childers C, Pareta RA, Opara EC, Ramasubramanian MK. A scalable microfluidic device for the mass production of microencapsulated islets. Transplantation Proc. 2011; 43(9): 3184–3187.
[92] Omer A, Duvivier-Kali V, Fernandes J, Tchipashvili V, Colton CK, Weir GC. Long-term normoglycemia in rats receiving transplants with encapsulated islets. Transplantation. 2005; 79(1): 52–58.
[93] Teramura Y, Iwata H. Bioartificial pancreas micorcapsulation and conformal cotaig of islets of Langerhans. Advanced drug delivery reviews. 2010; 62: 827–840.
[94] Maki T, Lodge JP, Carretta M, et al. Treatment of severe diabetes mellitus for more than one year using a vascularized hybrid artificial pancreas. Transplantation. 1993; 55(4): 713–717.
[95] Lacy PE, Hegre OD, Gerasimidi-Vazeou A, Gentile FT, Dionne KE. Maintenance of normoglycemia in diabetic mice by subcutaneous xenografts of encapsulated islets. Science. 1991; 254: 1782–1784.
[96] Storrs R, Dorian R, King SR, et al. Preclinical development of the islet sheet. Annals of the New York Academy of Sciences. 2001; 944: 252–266.
[97] Kaysen GA. The microinflammatory state in uremia: causes and potential consequences. J Am Soc Nephrol. 2001; 12: 1549–1557.
[98] Cozzolino M, Brancaccio D, Gallieni M, Slatopolsky E. Pathogenesis of vascular calcification in chronic kidney disease. Kidney Int. 2005; 68: 429–436.
[99] Pino CJ, Humes HD. Stem cell technology for the treatment of acute and chronic renal failure. Transl Res. 2010; 156(3): 161–168.
[100] Fissell WH, Fleischman AJ, Humes HD, Roy S. Development of continuous implantable renal replacement: past and future. Transl Res. 2007; 150(6): 327–336.
[101] Humes HD, Cieslinski DA. Interaction between growth factors and retinoic acid in the induction of kidney tubulogenesis in tissue culture. Exp Cell Res. 1992; 201: 8–15.
[102] Humes HD, Krauss JC, Cieslinski DA, Funke AJ. Tubulogenesis from isolated single cells of adult mammalian kidney: clonal analysis with a recombinant retrovirus. Am J Physiol. 1996; 271(40): F42–F49.
[103] O'Neil JJ, Stegemann JP, Nicholson DT, Mullon CJ-P, Maki T, Monaco AP, Solomon BA. Immunoprotection provided by the bioartificial pancreas in a xenogeneic host. Transplant Proc. 1997; 29: 2116–2117.
[104] Humes HD, MacKay SM, Funke AJ, Buffington DA. Tissue engineering of a bioartificial renal tubule assist device: in vitro transport and metabolic characteristics. Kidney Int. 1999; 55: 2502–2514.
[105] Humes HD, Weitzel WF, Bartlett RH, Swaniker FC, Paganini EP, Luderer JR, Sobota J. Initial clinical results of the bioartificial kidney containing human cells in ICU patients with acute renal failure. Kidney Int. 2004; 66: 1578–1588.

[106] Tumlin J, Wali R, Williams W, Murray P, Tolwani AJ, Vinnikova AK, Szerlip HM, Ye J, Paganini EP, Dworkin L, Finkel KW, Kraus MA, Humes HD. Efficacy and safety of renal tubule cell therapy for acute renal failure. J Am Soc Nephrol. 2008; 19: 1034–1040.
[107] Humes HD, Buffington DA, MacKay SM, Funke AJ, Weitzel WF. Replacement of renal function in uremic animals with a tissue-engineered kidney. Nat Biotechnol. 1999; 17: 451–455.
[108] Humes HD, MacKay SM, Funke AJ, Buffington DA. Tissue engineering of a bioartificial renal tubule assist device: in vitro transport and metabolic characteristics. Kidney Int. 1999; 55: 2502–2514.
[109] Fissell WH, Lou L, Abrishami S, Buffington DA, Humes HD. Bioartificial kidney ameliorates gram-negative bacteria-induced septic shock in uremic animals. J Am Soc Nephrol. 2003; 14: 454–461.
[110] Fissell WH, Dyke DB, Weitzel WF, Buffington DA, Westover AJ, MacKay SM, Gutierrez JM, Humes HD. Bioartificial kidney alters cytokine response and hemodynamics in endotoxin-challenged uremic animals. Blood Purif. 2002; 20: 55–60.
[111] Fujita Y, Terashima M, Kakuta T, Itoh J, Tokimasa T, Brown D, Saito A. Transcellular water transport and stability of expression in aquaporin 1- transfected LLC-PK1 cells in the development of a portable bioartificial renal tubule device. Tissue Eng. 2004; 10: 711–772.
[112] Inagaki M, Yokoyama TA, Sawada K, Duc VM, Kanai G, Lu J, Kakuta T, Saito A: Prevention of LLC-PK1 cell overgrowth in a bioartificial renal tubule device using a MEK inhibitor, U0126. J Biotechnol. 2007; 132: 57–64.
[113] Zhang H, Tasnim F, Ying JY, Zink D. The impact of extracellular matrix coatings on the performance of human renal cells applied in bioartificial kidneys. Biomaterials. 2009; 30: 2899–2911.
[114] Vesey DA, Qi W, Chen X, Pollock CA, Johnson DW. Isolation and primary culture of human proximal tubule cells. Methods Mol Biol. 2009; 466: 19–24.
[115] Verhulst A, Sayer R, De Broe ME, D'Haese PC, Brown CD: Human proximal tubular epithelium actively secretes but does not retain rosuvastatin. Mol Pharmacol. 2008; 74: 1084–1091.
[116] Zhang H, Lau SF, Heng BF, Teo PY, Alahakoon PK, Ni M, Tasnim F, Ying JY, Zink D. Generation of easily accessible human kidney tubules on twodimensional surfaces in vitro. J Cell Mol Med. 2010; 15(6): 1287–1298.
[117] Kim D, Dressler GR. Nephrogenic factors promote differentiation of mouse embryonic stem cells into renal epithelia. J Am Soc Nephrol. 2005; 16: 3527–3534.
[118] Baer PC, Bereiter-Hahn J, Missler C, Brzoska M, Schubert R, Gauer S, Geiger H. Conditioned medium from renal tubular epithelial cells initiates differentiation of human mesenchymal stem cells. Cell Prolif. 2009; 42: 29–37.
[119] Fusaki N, Ban H, Nishiyama A, Saeki K, Hasegawa M. Efficient induction of transgene-free human pluripotent stem cells using a vector based on Sendai virus, an RNA virus that does not integrate into the host genome. Proc Jpn Acad Ser B Phys Biol Sci. 2009; 85: 348–362.
[120] Tasnim F, Deng R, Hu M, Liour S, Li Y, Ni M, Ying J Y, Zink D. Achievements and challenges in bioartificial kidney development. Fibrogenesis & Tissue Repair. 2010; 3: 1–14.
[121] Ueda H, Watanabe J, Konno T, Takai M, Saito A, Ishihara K. Asymmetrically functional surface properties on biocompatible phospholipid polymer membrane for bioartificial kidney. J Biomed Mater Res A. 2006; 77: 19–27.
[122] Fissell WH, Manley S, Westover A, Humes HD, Fleischman AJ, Roy S. Differentiated growth of human renal tubule cells on thin-film and nanostructured materials. ASAIO J. 2006; 52: 221–227.
[123] Saito A. Research into the development of a wearable bioartificial kidney with a continuous hemofilter and a bioartificial tubule device using tubular epithelial cells. Artif Organs. 2004; 28: 58–63.

[124] Saito A, Aung T, Sekiguchi K, Sato Y: Present status and perspective of the development of a bioartificial kidney for chronic renal failure patients. Ther Apher Dial. 2006; 10: 342–347.

[125] Ronco C, Davenport A, Gura V. A wearable artificial kidney: dream or reality? Nat Clin Pract Nephrol. 2008; 4: 604–605.

[126] Buffington DA, Pino CJ, Hageman GM, Chen L, Humes HD. Bioartificial renal epithelial cell systems (BRECS): a compact, cryopreservable extracorporeal renal replacement device. Cell Transplant. 2012. doi: 10.3727/215517911X653328.

[127] Ng CP, Zhuang Y, Lin AWH, Teo JCM. A Fibrin-Based Tissue-Engineered Renal Proximal Tubule for Bioartificial Kidney Devices: Development, Characterization and In Vitro Transport Study. International J Tissue Engineering. 2013. doi: 10.1155/2013/319476.

[128] Wagner WR, Griffith BP. Reconstructing the lung. Science. 2010; 329: 520–522.

[129] Fischer S, et al. Bridge to lung transplantation with the extracorporeal membrane ventilator Novalung in the veno-venous mode: the initial Hannover experience. Asaio J. 2007; 53: 168–170.

[130] Fischer S, Simon AR, Welte T, et al. Bridge to lung transplantation with the novel pumpless interventional lung assist device NovaLung. J Thorac Cardiovasc Surg. 2006; 131: 719–723.

[131] Macchiarini P, Jungebluth P, Go T, et al. Clinical transplantation of a tissue-engineered airway. Lancet. 2008; 372: 2023–2030.

[132] Galletti PM, Colton CK. Artificial lungs and blood–gas exchange devices. In: Bronzino JD (ed.). The Biomedical Engineering Handbook. Boca Raton: CRC; 2000. pp. 1–19.

[133] Petersen TH, Calle EA, Zhao L, Lee EJ, Gui L, Raredon MB, Gavrilov K, Yi T, Zhuang ZW, Breuer C, Herzog E, Niklason LE. Tissue-engineered lungs for in vivo implantation. Science. 2010; 329: 538–541.

[134] Ott HC, Clippinger B, Conrad C, Schuetz C, Pomerantseva I, Ikonomou L, Kotton D, Vacanti JP. Regeneration and orthotopic transplantation of a bioartificial lung. Nat Med. 2010; 16: 927–933

[135] Fritsche CS, Simsch O, Weinberg EJ, et al. Pulmonary tissue engineering using dual-compartment polymer scaffolds with integrated vascular tree. Int J Artif Organs. 2009; 32: 701–710.

[136] Polk AA, Maul TM, McKeel DT, et al. A Biohybrid artificial lung prototype with active mixing of endothelialized microporous hollow fibers. Biotechnol Bioeng. 2010; 106: 490–500.

[137] Wu ZJ, Zhang T, Bianchi G, et al. Thirty-day in-vivo performance of a wearable artificial pump-lung for ambulatory respiratory support. Artif Organs. 2012; 93: 274–281.

[138] Garcia JP, Iacono A, Kon ZN, et al. Ambulatory extracorporeal membrane oxygenation: a new approach for bridge-to-lung transplantation. J Thorac Cardiovasc Surg. 2010; 139: 137–139.

[139] Vittet D, Prandini MH, Berthier R, et al. Embryonic stem cells differentiate in vitro to endothelial cells through successive maturation steps. Blood. 1996; 88: 3424–3431.

[140] Banerjee ER, Laflamme MA, Papayannopoulou T, et al. Human embryonic stem cells differentiated to lung lineage-specific cells ameliorate pulmonary fibrosis in a xenograft transplant mouse model. PLoS One. 2012; 7: e33165.

[141] Vats A, Tolley NS, Bishop AE, et al. Embryonic stem cells and tissue engineering: delivering stem cells to the clinic. J Roy Soc Med. 2005; 98: 346–350.

[142] Warren L, Manos PD, Ahfeldt T, et al. Highly efficient reprogramming to pluripotency and directed differentiation of human cells with synthetic modified mRNA. Cell Stem Cell. 2010; 7: 618–630.

[143] Yoshikawa HY, Cui J, Kratz K, et al. Quantitative evaluation of adhesion of osteosarcoma cells to hydrophobic polymer substrate with tunable elasticity. J Phys Chem B. 2012; 116: 8024–8030.

[144] Lemon G., Lim ML., Ajalloueian F., Macchiarini P. The development of the bioartificial lung. British Medical Bulletin. 2014; 110: 35–45.

[145] Lutolf MP, Hubbell JA. Synthetic biomaterials as instructive extracellular microenvironments for morphogenesis in tissue engineering. Nat Biotechnol. 2005; 23: 47–55.

[146] Chen QZ, Liang SL, Thouas GA. Elastomeric biomaterials for tissue engineering. Prog Poly Sci. 2013; 38: 584–671.

[147] Wang YD, Ameer GA, Sheppard BJ, et al. A tough biodegradable elastomer. Nat Biotechnol. 2002; 20: 602–606.

[148] Chung S, King MW. Design concepts and strategies for tissue engineering scaffolds. Biotechnol Appl Biochem. 2011; 58: 423–38.

[149] Macchiarini P, Jungebluth P, Go T, Asnaghi MA, Rees LE, Cogan TA, Dodson A, Martorell J, Bellini S, Parnigotto PP, Dickinson SC, Hollander AP, Mantero S, Conconi MT, Birchall MA. Clinical transplantation of a tissue-engineered airway. Lancet. 2008; 372: 2023–2030.

[150] Baiguera S, Jungebluth P, Burns A, Mavilia C, Haag J, De Coppi P, Macchiarini P. Tissue engineered human tracheas for in vivo implantation. Biomaterials. 2010; 31(34): 8931–8938.

[151] Jungebluth et al. Tracheobronchial transplantation with a stem-cell-seeded bioartificial nanocomposite: a proof-of-concept study. Lancet. 2011; 378: 1997–2004.

[152] Curcio E, Macchiarini P, De Bartolo L. Oxygen mass transfer in a human tissue engineered trachea. Biomaterials. 2010; 31: 5131–5136.

[153] Curcio E, Piscioneri A, Morelli S, Salerno S, Macchiarini P, De Bartolo L. Kinetics of oxygen uptake by cells potentially used in a tissue engineered trachea. Biomaterials. 2014; 35: 6829–6868.

8 Regulatory framework and ethical issues

8.1 Regulatory framework of cellular products

Cells are the core products in tissue engineering and regenerative medicine, and their huge therapeutic potential for human diseases mandatorily requires the precise definition of a specific regulatory framework. In the United States, cellular therapy products – i.e. cellular immunotherapies and other types of both autologous and allogeneic cells for certain therapeutic indications (including adult and embryonic stem cells) are regulated by the Center for Biologics Evaluation and Research (CBER), a subdivision of the American Food and Drug Administration (FDA), one of the world's most influential regulatory authorities. As a consequence of the rapid growth of cellular and gene therapy-related research and development, CBER receives several requests each year from medical researchers and manufacturers to evaluate and approve cellular and gene therapy products. To date, a short list of some cellular therapy products have been approved by CBER[1].

In 2007, Genzyme Biosurgery marketed CARICEL®, the first licensed cellular product based on *in vitro* autologous cultured chondrocytes indicated for the repair of symptomatic cartilage defects of the femoral condyle (medial, lateral, or trochlea), caused by acute or repetitive trauma.

PROVENGE®, manufactured by Dendreon Corporation (2010), was the first autologous cellular immunotherapy indicated for the treatment of asymptomatic or minimally symptomatic metastatic castrate resistant (hormone refractory) prostate cancer.

In 2011, marketing authorization was issued to LAVIV® from Fibrocell Technologies, Inc., an autologous cellular product indicated for improvement of the appearance of moderate to severe nasolabial fold wrinkles in adults.

Organogenesis Incorporated received approval in 2013 for marketing GINTUIT®, an allogeneic cellularized scaffold product (Allogeneic Cultured Keratinocytes and Fibroblasts in Bovine Collagen) indicated for topical (nonsubmerged) application to a surgically created vascular wound bed in the treatment of mucogingival conditions in adults.

In 2013, SSM Cardinal Glennon Children's Medical Center received approval for ALLOCORD®, an allogeneic cord blood hematopoietic progenitor cell therapy indicated for use in unrelated donor hematopoietic progenitor cell transplantation procedures in conjunction with an appropriate preparative regimen for hematopoietic and immunologic reconstitution in patients with disorders affecting the hematopoietic system. For the same purpose, approval was issued to Duke University School of

1 www.fda.gov/BiologicsBloodVaccines/CellularGeneTherapyProducts/ApprovedProducts/ucm354689.htm

DOI 10.1515/9783110268010-009

Medicine for DUCORD® (2012), to New York Blood Center Inc. for HEMACORD® (2013), and to Cleveland Cord Blood Center for CLEVECORD® (2016).

8.2 Development and approval process in the US and the EU

In the United States, cellular products are regulated under a variety of authorities: the Public Health Service Act (PHS Act), the Food Drug and Cosmetic Act (FD&C Act), regulations such as 21 Code of Federal Regulations (CFR), and Food & Drug Administration (FDA) guidance. Prior to administration of any new drug or biological product to humans, authorization must be issued by the FDA after submission of and investigational new drug application (IND) or investigational device exemption (IDE). The FDA also provides information and recommendations on drug and biological product development when human efficacy studies are not ethical or feasible; the regulations that set forth the pathway for approval of these products under 21 CFR 314 (drugs) or 21 CFR 601 (biological products) are commonly referred to as the *Animal Rule*[2].

Since 1970s, FDA allows the expanded access (also called "compassionate use") to experimental drugs and biologics in clinical trials (studies under development and not yet approved by FDA); under their personal responsibility, patients enroll in clinical trials to gain access to investigational therapies and contribute to finding out how well an investigational therapy works, and how safe it is for patients.

The introduction to the market of a new biologic or an antibiotic drug for human use requires a permission through submission of a biologics license application (BLA). The BLA, regulated under 21 CFR 600–680, includes information about the applicant, product, manufacturing procedures, preclinical and clinical studies, and labeling.

The premarket approval (PMA), to be submitted to the Center for Biologics Evaluation and Research (CBER), is made with the aim of demonstrating that a Class III device (a device that supports or sustains human life) to be marketed is at least as safe and effective – or substantially equivalent – to a legally marketed device (21 CFR 807.92 (a) [1]) which is not subject to a PMA. Human cells or tissues to be used for implantation, transplantation, infusion, or transfer into a human recipient (for example: bone, skin, corneas, ligaments, tendons, heart valves, hematopoietic stem/progenitor cells, etc.) are regulated as a human cell, tissue, and cellular and tissue-based product (HCT/P). 21 CFR Parts 1270 and 1271 regulate HCT/P in compliance with current good tissue practice under the PHS Act section 361. Parts 1270 and 1271 require tissue establishments to screen and test donors, to prepare and follow written procedures for the prevention of the spread of communicable disease, and to maintain records.

2 www.fda.gov/Drugs/GuidanceComplianceRegulatoryInformation/Guidances/default.htm

Tab. 8.1: List of guidelines provided by the European Medicines Agency (EMA) of relevance for advanced therapy medicinal products/cell-therapy and tissue engineering (www.ema.europa.eu).

Section	Guideline	Reference
Cell-therapy and tissue engineering	The overarching guideline for human cell-based medicinal products is the guideline on human cell-based medicinal products	EMEA/CHMP/410869/2006
	Reflection paper on stem cell-based medicinal products	EMA/CAT/571134/2009
	Reflection paper on in-vitro cultured chondrocyte containing products for cartilage repair of the knee	EMA/CAT/CPWP/568181/2009
	Guideline on xenogeneic cell-based medicinal products	EMEA/CHMP/CPWP/83508/2009
	Guideline on potency testing of cell based immunotherapy medicinal products for the treatment of cancer	CHMP/BWP/271475/06
	Reflection paper on clinical aspects related to tissue engineered products	EMA/CAT/573420/2009
Gene therapy	Quality, nonclinical and clinical aspects of medicinal products containing genetically modified cells	CHMP/GTWP/671639/2008
Biologicals: drug product	Guidance on the use of bovine serum in the manufacture of human biological medicinal products	CPMP/BWP/1793/02
	Minimising the risk of transmitting animal apongiform encephalopathy agents via human and veterinary medicinal products	EMA/410/01
	CHMP/CAT position statement on Creutzfeldt-Jakob disease and advanced therapy medicinal products	CHMP/CAT/BWP/353632/2010
	Position paper on the reestablishment of working seeds and working cell banks using TSE compliant materials	EMEA/22314/02
	Guideline on the use of porcine trypsin used in the manufacture of human biological medicinal products	EMA/CHMP/BWP/814397/2011
Biologicals: drug substance	Note for guidance on plasma derived medicinal products	CPMP/BWP/269/95

Tab. 8.1: (continued)

Section	Guideline	Reference
Quality: Excipients	Guideline for excipients in the dossier for application for marketing authorisation of a medicinal product	EMEA/CHMP/QWP/396951/2006
Quality: ICH	ICH Q2 (R1) Validation of analytical procedures: text and methodology	CPMP/ICH/381/95
	ICH Q5A (R1) Viral safety evaluation of biotechnology products derived from cell lines of human or animal origin	CPMP/ICH/295/95
	ICH Q5C Stability testing of biotechnological/biological products	CPMP/ICH/138/95
	ICH Q5D Derivation and characterisation of cell substrates used for production of biotechnological/biological products	CPMP/ICH/294/95
	ICH Q5E Comparability of biotechnological/biological products	CPMP/ICH/5721/03
	ICH Q7 Good manufacturing practice for active pharmaceutical ingredients	CPMP/ICH/4106/00
	ICH Q8 (R2) Pharmaceutical development	CHMP/ICH/167068/04
	ICH Q9 Quality risk management	EMA/CHMP/ICH/24235/2006
	ICH Q10 Pharmaceutical quality system	EMA/CHMP/ICH/214732/2007
Safety: ICH	ICH S6 (R1) Preclinical safety evaluation of biotechnology-derived pharmaceuticals	CHMP/ICH/731268/1998
Safety and efficacy: Biostatistics	Guideline on clinical trials in small populations	CHMP/EWP/83561/2005
	Points to consider on applications with 1. Meta-analyses; 2. One pivotal study	CPMP/EWP/2330/99
Efficacy: ICH	ICH E1 The extent of population exposure to assess clinical safety	CPMP/ICH/375/95
	ICH E3 Structure and content of clinical study reports	CPMP/ICH/137/95
	ICH E4 Dose response information to support drug registration	CPMP/ICH/378/95
	ICH E6 (R1) Good clinical practice	CPMP/ICH/135/95
	ICH E7 Geriatrics	CPMP/ICH/379/95
	ICH E8 General considerations for clinical trials	CPMP/ICH/291/95
	ICH E11 Clinical investigation of medicinal products in the paediatric population	CPMP/ICH/2711/99

In the European Union, relevant guidelines for advanced therapy medicinal products (ATMP) have been developed by the supranational European Medicine Agency (EMA), formerly known as European Agency for the Evaluation of Medicinal Products (EMEA – until 2004). Regulations for cell-based medicinal products (CBMP) have begun to be defined in Annex I to Directive 2001/83/EC of the European Parliament and Council, and then amended by the Regulation on Advanced Therapy Medicinal Products 1394/2007/EC [2]. Substantially, ATMP are manipulated cellular tissues intended to be used as biological medicinal products for the treatment and prevention of human diseases. The standards of quality and safety for the donation, procurement, testing, processing, preservation, storage and distribution of human cells are defined by the Directive 2004/23/EC.

Three different types of medicinal products are comprised within the ATMP: gene therapy medicinal products, tissue engineered products, and somatic cell therapy medicinal products. Table 8.1 lists relevant scientific guidelines provided by the EMA concerning cell therapy and tissue engineering.

8.3 Good manufacturing practice

The production process of manipulated ATMP and HCT/P are required to meet current standards of quality and safety defined by good tissue practice (GTP) and good manufacturing practice (GMP), concerning the design, the access, the use and maintenance, the quality assurance and validation of facilities, equipments, materials, and control strategies.

In Europe, GTP for human cells, tissues, and cellular and tissue-based products are established by the general Directive 2004/23/EC [3] and technical directives drawn from it, Directives 2006/17/EC [4] and 2006/86/EC [5]. In the United States, GTP is regulated by Title 21 of the CGFR, part 1271, which has the purpose of creating a unified registration and listing system for establishments that manufacture human cells, tissues, and HCT/P's, and to establish donor eligibility, current good tissue practice, and other procedures to prevent the introduction, transmission, and spread of communicable diseases by HCT/P's.

The European standards of GMP concerning the manufacture of cell-based medicinal products should be in compliance with the principles set out in Directive 2003/94/EC [6], while the US GMP are defined in 21 CFR 210 and 211.

Substantially, specific requirements of GMP and GTP cover three specific areas: the human starting material, the manufacturing process, and the characterization and control of cells (Figure 8.1).

The specific cell source is the basis of the classification of cell-based medicinal products (CBMP) as autologous or allogeneic. Autologous products are derived from the patient's own body. The main advance of this approach is the avoidance of immune rejection; however, risks related to the potentially diseased state of the patient have to be considered. Moreover, the individual manufacturing of autologous products requires a period of time that might be incompatible with an urgent treatment. Allogeneic CBMP are inherently off-the-shelf products and specifically designed to treat a large number of patients; on the other hand, they have the potential risk of inducing immune response and rejection, with controversial exceptions reported for mesenchymal stem cells [7–9].

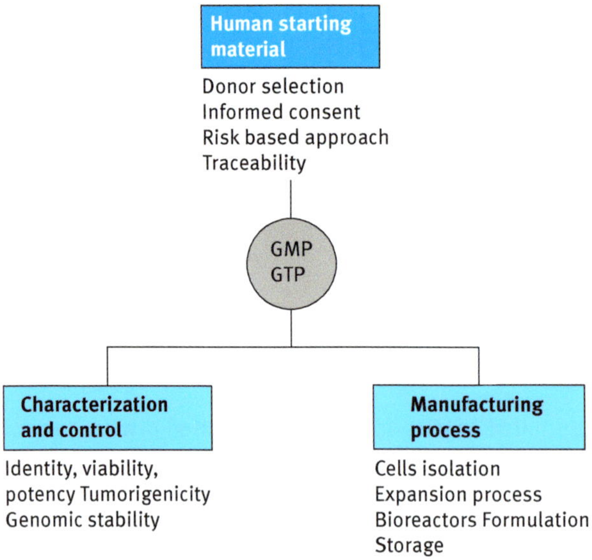

Fig. 8.1: Specific requirements of good manufacturing practice (GMP) and good tissue practice (GTP).

Directive 2006/17/EC regulates requirements and procedures for donation and procurement. The cellular material consists of primary cells to be used – directly or after few culture passages – in CBMP, or of cell lines based on a well-defined cell bank system (ICH guideline Q5D [10]).

The donor selection, based on body examination, the presence or likelihood of infection and medical history, should be carried out according to a risk-based approach in order to reduce the probability of adverse immunological responses in the recipient that might compromise the therapeutic activity of the CBMP. The potential use of donated human materials, risks associated with the donation, data protection and commercial background should be acknowledged by the donor through the informed consent form.

Collection, selection, culture, or modification procedures requiring the use of other materials, reagents, excipients (e.g. other cells, enzymes, antibodies, cytokines, growth factors, sera, antibiotics, etc.) should be clearly specified in detail and well documented with respect to identity, purity, sterility, and biological activity. Whenever appropriate, guidelines on viral safety must be taken into account [11, 12]. All procedures implemented to obtain cells from the organs and tissues should be described and validated. A manufacturing process that includes cryopreservation and long-term storage of cells should comply with GMP procedures with the aim to preserve the phenotype and the functionality of cells. Low-toxicity cryoprotectants, such as routinely used dimethyl sulfoxide (DMSO), polyethylene glycol, glycerol, methylcellulose, etc. reduce the amount of ice formed at low temperature simply by increasing the total concentration of solutes in the cell [13, 14].

In general, manufacturing process is critical and requires the proper implementation of validation procedures, the specification of process parameters, and the design of appropriate quality control protocols in order to ensure an acceptable level of standardization and reproducibility.

For CMBP incorporating structural components such as matrices, fibers, beads, devices, scaffolds, and any material used in addition or in combination with cells, requirements laid down in Directive 90/385/EEC on the approximation of the laws of the member states relating to active implantable medical devices [15] and subsequent modifications in Directive 93/42/EEC concerning medical devices [16] must be met. Materials used to manufacture this structural component should be validated with respect to their biocompatibility with cells, since tissue differentiation and functionality are highly dependent on the local environment. Therefore, studies should be carried out to verify the performance of biomaterials in a CBMP and, in particular, attention should be paid to

- release kinetics and/or rate of degradation of biodegradable or leachable materials, which might induce environmental alterations for the cells in time;
- chemico-physical and mechanical properties (e.g. topography, surface chemistry, strength) of structural components with respect to cellular viability, desired cell differentiation, functionality and genotype.

Standards for characterization and testing of medical devices to assess the potential for an adverse biological reaction to occur as a result of exposure to a given biomaterial are set out in international standard ISO 10993, Part I.

Manufacturing machines and procedures should also be suitable and qualified for aseptic production.

In vitro culture steps must be properly designed, specified, and validated to ensure an acceptable level of growth of the cellular population, of expansion to achieve the amount necessary for therapeutic use in CMBP, of differentiation. Relevant genotypic and phenotypic characteristics of the primary cell cultures, cell lines, and derived cell clones, and their stability, function, and integrity with respect to culture permanence and conditions should be assessed, while minimizing the risk of microbiological contamination and infectious decease transmission [17, 18]. Media and supplements are the major challenges for the cellular expansion process. Use of fetal bovine serum should follow the recommendations of the Note for Guidance on the "Use of Bovine

Serum in the Manufacture of Human Biological Medicinal Product" [19]; use of human alternatives such as pooled human sera or pooled platelet lysate appears feasible [20–22]. Although widely used against microbiological contamination, antibiotics should preferably be avoided due to the risk of allergic reactions in patients.

The characterization of a CBMP should include all the components present in the final product throughout each the development and/or manufacturing step. The characterization – to be made *in vitro* and *in vivo* – should be designed to allow setting up routine controls in the product and throughout the several steps of the process to guarantee efficacy, safety, reproducibility, and consistency. The identity of the cellular components should be characterized in terms of phenotypic and/or genotypic profiles by relevant markers based on gene expression, antigen presentation, response to exogenous stimuli, biochemical activity, etc. Source, manufacturing procedures, and culture conditions have an impact on the expression panel of cell surface markers [23].

Human starting material contains other cell types of different lineages and/or differentiation stages or that may be unrelated to the unique cell population of interest for the development of a CBMP. It is therefore necessary to set acceptance criteria for the amounts of contaminating cells, and implement necessary procedures to ensure purity against the unwanted cells that might have a negative impact on product efficacy and safety.

Potency, defined as the quantitative measure of biological activity based on the attribute of the product, is basically assessed by two typologies of assays: (1) *in vitro* assays using cell systems and (2) *in vivo* assays using animal models.

In vitro assays are usually based on the evaluation (1) of the expression of markers (e.g. cell surface markers, activation markers, expression pattern of specific genes) directly or indirectly related to the intended biological activity, and (2) of the physiological response under defined conditions (e.g. differentiation in specific cell types, secretion of specific proteins, etc.).

The establishment of adequate potency is complicated by the significant impact of the microenvironment on the phenotype and expression profile *in vitro*, and on the trafficking profile and functional behavior *in vivo* [24].

The clinical implementation of cell therapies suffers the potential risk of tumor formation induced by transplanted cells. The capability of human pluripotent embryonic stem (ES) cells and induced pluripotent stem (iPS) cells to form or promote the growth of existing tumors or to transform themselves into tumor cells has been proven, as well as the insurgence of genomic aberrations acquired by culture adaption [25]. The tumorigenic potential should be assessed by an analysis of the proliferative capacity of cells, dependence on exogenous stimuli, chromosomal integrity, etc.

A complete traceability of the patient and product is essential to monitor the safety and efficacy of CBMP; requirements are laid down in Directives 2004/23/EC, 2006/17/EC, and 2006/86/EC, and Art. 15 of the Regulation (EC) No. 1394/2007. The manufacturers should also establish the analytical tools necessary for the required comparability studies throughout development of their products. Appropriate guidance can be found in ICH Q5E "Comparability of Biotechnological/Biological Products" and related guidance documents [26].

8.4 From the laboratory to the market

The introduction into the market of new drugs or new clinical treatments requires the implementation of a specific development strategy and methodology. According to data provided by the European Federation of Pharmaceutical Industries and Associations (EFPIA) in 2012, the journey from laboratory bench to marketing approval is extremely difficult, and "chances of new substances becoming a marketable medicine remain relatively small: several studies have produced figures ranging from 1 in 5000 to 1 in 10 000"[3]. This is largely due to the high investment costs and strict screening procedures. Figure 8.2 provides a general overview of the development process from bench scale investigation to postmarketing surveillance.

In the first decade of the 21st century, the overall success rate for new drug applications getting FDA approval was near 1 to 10[4]. In order to address the financial challenges related to an inherently high risk market, the pharmaceutical industry is recently shifting towards therapeutic areas characterized by high therapeutic need and high return on investment (i.e. cancer treatment). Consequently, the FDA and EMA offer incentives to companies supporting the development of innovative diagnosis tools and medical treatments for rare diseases or disorders.

The development of a new drug or treatment begins at the identification of a specific disease to prevent or cure. The knowledge of the biological processes underlying the pathological status is critical in the definition of the target to be addressed by the new drug or treatment. In the case of new drug development, the traditional approach to identify and screen – at the laboratory scale – potentially active molecules is based on high throughput screening (HTS): hundreds of candidate compounds are put in contact with enzymes, cell cultures, or other biological substances in microliter-sized multi-wells plates. HTS is costly and time consuming, although automation of sample preparation and analytical procedures by computer-controlled robots partially mitigates these drawbacks. More recently, driven by the impressive development of computing tools and molecular-modelling software – which led to a deeper understanding of interactions at the atomic level, electronic structure, and charge density, bonding energy, activity of functional groups, etc. –, HTS has been complemented by virtual screening: here, the assessment of potential compounds is done in silico.

Preclinical studies cover all research activities, testing, and assessment required by health authorities before human clinical trials. *In vitro* laboratory studies and *in*

3 www.efpia.eu
4 http://fiercebiotech.com/story/study-rate-drug-approvals-dropping/2011-02-14.2012

vivo animal models should be designed to demonstrate the product's mechanism of action, its effectiveness and potency, safety, and toxic and pharmacologic effects in the case of new drugs.

The wide use of rodents as first-line *in vivo* models is mostly due to their broad availability and to the development of well-established methods for de novo generation of genetically modified mice and rats. These animal models are successfully applied for the verification of *in vivo* efficacy, for the discovery of new pharmacological targets, and for the study of pharmacological mechanisms of action [27]. However, growing ethical concerns, including the Directive No. 2010/63/EU promoting the "3R principle" (replace, reduce, and refine), are now limiting the overusing of animal experiments.

Regulatory guidelines of the FDA and EMA usually require safety testing in at least two mammalian species, including one nonrodent species, prior to human trial authorization [28].

Nonrodent mammalian species such as rabbits, dogs, swine, and monkeys are also widely used in pharmacological safety and pharmacokinetic studies [29].

Appropriate investment in preclinical development programs may lead to significant availability of study results, operational protocols, and statistically relevant data for a more rapid progress through clinical trials.

The FDA requires researchers to use good laboratory practices (GLP) in 21 CFR Part 58.1: "Good Laboratory Practice for Nonclinical Laboratory Studies", setting the minimum basic requirements for study conduct, personnel, facilities, equipment, protocols, operating procedures, study reports, and quality assurance. EMA provides ICH guideline S6 (R1) "preclinical safety evaluation of biotechnology-derived pharmaceuticals" with the aim of identifying an initial safe dose and subsequent dose escalation schemes in humans, assessment of the toxicity potential, and safety parameters for clinical monitoring.

In the United States, the investigational new drug (IND) application (21CFR312) and the investigational device exemption (IDE) application (21CFR812) must be submitted to the FDA agency for new drugs and new devices, respectively. In the European Union, the authorization of clinical trials occurs at member state level (clinical trial authorization, CTA), although the way clinical trials are conducted will undergo a major change when the clinical trial regulation takes effect in 2018, with the aim of harmonizing the assessment and supervision processes for clinical trials.

In general, applications include the description of the product and manufacturing processes, with sufficient details to assess the risks and potential benefits, dosing and toxicity levels, a proposal for a clinical protocol, including candidate selection/exclusion criteria, treatment/therapy procedures, follow-up methods and clinical trial stopping rules, and ethical committee clearance.

In a few exceptions, the normal clinical practice guidelines can be bypassed with the aim of speeding up the availability of drugs and devices for treating serious diseases. In this respect, the FDA has developed four distinct approaches: priority review

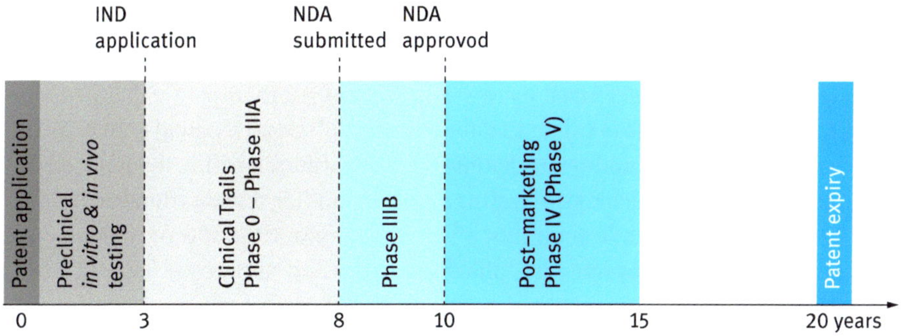

Fig. 8.2: General overview and timing (on average) of the development process for new drugs or new clinical treatments.

(fast FDA's review within six months), breakthrough therapy (a process designed to expedite the development and review of drugs which may demonstrate substantial improvement over available therapy), accelerated approval (regulations allowing drugs and treatments for serious conditions that filled an unmet medical need to be approved based on a surrogate endpoint), fast track (a process designed to facilitate the development and expedite the review of drugs and devices to treat serious conditions and fill an unmet medical need).[5]

Similarly, accelerated assessment is a procedure implemented by the EMA Committee for Medicinal Products for Human Use (CHMP) to reduce the timeframe (usually to 150 evaluation days instead of 210) for the review of a marketing-authorization application for a rapid assessment of products of major interest for public health, especially ones that are therapeutic innovations.[6]

Once the new drug or device proves an acceptable level of genotoxicity, carcinogenicity, reprotoxicity, and appropriate pharmacokinetic and pharmacodynamic parameters, having collected all necessary information required for the IND/IDE or CTA application, testing in humans can begin upon approval by the responsible authorities.

With the aim of increasing the chance of success in later stages of drug and treatments development, with particular attention to anticancer products (only 5 % of applications for new oncology drugs submitted to the FDA under an IND application are successful [30]), in 2006 the FDA released a document entitled "Guidance for Industry, Investigators, and Reviewers: Exploratory IND Studies," offering recommendations regarding safety testing, manufacturing, and clinical approaches to be used in very

5 www.fda.gov/forpatients/approvals/fast/ucm20041766.htm

6 www.ema.europa.eu/ema/index.jsp?curl=pages/regulation/general/general_content_000955.jsp&mid=WC0b01ac05809f843a

early studies, the so-called exploratory or *phase 0* trials, as guidelines to distinguish promising products from those that are not.

Phase 0 trials are characterized by the utilization of subtherapeutic doses (generally less than 1/100 of the dose calculated to produce a pharmacological effect, and no higher than 100 µg) of the product under investigation, (thereby allowing reduced initial pharmacologic and toxicological testing requirements) in a few patients in order to reduce the cost and time while acquiring a better understanding of the mechanism of action, pharmacokinetics or imaging, pharmacodynamics, and target localization of a new compound, or a series of related compounds, before undertaking phase I trials [31]. Phase 0 trials exploring microdosages have been also considered in the EMA document "Guideline on the evaluation of anticancer medicinal products in man" issued in December 2012 (EMA/CHMP/205/95/Rev. 4).

Phase I trials (also known as "entry in a human") are designed in order to assess the safety, tolerability, pharmacokinetics (PK), and pharmacodynamics (PD) of the new product. Specifically, the aim of phase I is to determine the maximum tolerated dose (MTD) of the new treatment, according to methodologies briefly schematized in Table 8.2.

Tests are usually carried out for 6–12 months on a group of approx. 20 healthy volunteers (to avoid confusing data in the presence of comorbidities); involvement of patients occurs in case of anticancer therapies, cytotoxic drugs, AIDS therapy or for those having exhausted all other therapeutic options to treat their disease. Informed consent is a basic prerequisite for human volunteers or patients.

Clinical trials for marketing authorization for human medicines in the European Economic Area (EEA) have to meet the requirements set out in Annex 1 of Directive 2001/83/EC, i.e. compliance with European Union clinical trial legislation as expressed in Directive 2001/20/EC or, for clinical trials conducted outside the EEA, compliance with international good clinical practice (GCP) and with the Declaration

Tab. 8.2: Phase I methodological approaches to identify the maximum tolerated dose (MTD).

Single ascending dose (SAD)	Multiple ascending dose (MAD)	Food effect study
A single dose of a drug is administered to a small number of subjects. After extended observation (24–48 h), if no adverse effects are detected, the dose is increased and given to a new group of subjects. The procedure is repeated until the occurrence and confirmation (same dose administered to a new group) of toxic effects.	Until the occurence toxic effects, a group of subjects receives multiple low doses of the tested drug. Samples of blood and urine are collected over time and analyzed in order to study PK and PD. The stopping rules are similar to SAD procedure.	Assess the impact of food on absorption, metabolism, and excretion of the drug. Subjects are administered either before, during or after a standard meal.

of Helsinki on Ethical Principles for Medical Research Involving Human Subjects. In the EEA, approximately 4000 clinical trials are authorized yearly; approximately 61 % of clinical trials are sponsored by the pharmaceutical industry and 39 % by noncommercial sponsors, mainly academic institutions.[7]

The main limitations of phase I consist in the preliminary nature of the tests conducted and in their restriction to a low number of homogeneous subjects. Therefore, a *phase II* is designed and implemented in order to confirm the effectiveness, safety, and tolerability of the product, and to monitor any side effects on patients with the disease being studied. The primary objective of phase II trials, conducted on a small number of patients (usually around 50–200 subjects), is to determine the efficacy of the product (optimal therapeutic dose regimen, duration, frequency of administration etc.); and secondly, to further investigate safety aspects [32].

Phase II is usually divided into two subphases, often referred two as *phase IIA* and *phase IIB*, each of which addressing different objectives associated with the new treatment. Phase IIA indicates exploratory, nonpivotal, multiple-stage single arm studies (a single arm study means that every patient enrolled in the clinical trial is treated in the same way, and no data are collected from people with a similar condition who do not get the drug) of the clinical efficacy, mechanism, dose range, and including pilot studies. Trials are single blind: i.e. information about the test is masked from the participant, to reduce or eliminate bias, a placebo effect, or conscious deception.

The phase IIB trials are pivotal since a "go/no go" decision must be made before proceed into phase III testing. In phase IIB, randomized multiple arms studies are employed to examine the impact of subgroups (inducing heterogeneity) on the treatment efficacy, safety, and dosage amount and/or frequency. Trials are double-blind to eliminate subjective, unrecognized biases carried by an participants and conductors of an experiment.

Phase III trials are the final hurdle to be overcome before entry into the market. Multicenter (and, often, multinational) tests are designed in order to investigate the efficacy and safety in a patient population varying from a few hundred to thousands of subjects, and can take up several years. Phase III includes randomized and double-blind trials as a standard, but also uncontrolled trials, historical controls, nonrandomized concurrent trials, factorial designs, and group sequential designs [33]. Traditional endpoints for trials have included overall survival as a priority, as well as the disease progression rate, and the time to treatment failure. For an accelerated time-to-market in case of potentially life-saving treatments, surrogate markers are possible.

Phase III is subdivided in two subphases: (1) *phase IIIA* is designed to get sufficient and appropriate data on efficacy and safety before NDA submission; (2) *phase IIIB* allows patients to continue treatment after NDA but prior to product approval and market launch, investigates possible label expansions (checking if the new product

7 www.ema.europa.eu

works for additional types of patients/diseases beyond the original target) and provides additional safety data. Phase III closely studies the clinical trial activities and determines the "go/no go" decision for marketing.

Even after release and marketing of a new product, the FDA, EMA, and relevant regulatory agencies require continued evaluation of efficacy and safety, and benefit-risk ratio. *Phase IV* trials, conducted after the new product has been marketed (post-marketing surveillance, PSM), aim at evaluating the real effectiveness and safety profile in various and large populations, and the eventual occurrence of side/harmful effects, rare or associated to long-term use, that may necessitate regulatory actions or that may results in a product being withdrawn. These studies have an inherently observational, nonexperimental, and noninterventional nature. Phase IV also assesses new indications, formulations, dosages, duration of treatment, medication interactions, etc. [33]. Table 8.3 compares the specific objectives and peculiarities of phases I–IV.

Tab. 8.3: Comparison of clinical trial phases I–IV.

	Phase I	**Phase II**	**Phase III**	**Phase IV**
Objectives	Study of metabolic and pharmacological actions, identification of maximum tolerated dose	Evaluate effectiveness, short-term side effects, risks for a specific population and disease	Evaluate effectiveness on clinical outcomes, risk-benefit ratio in a demographically diverse sample	Postmarketing surveillance, identify eventual long-term or rare side effects
Factors studied	Bioavailability; dosage; metabolic mechanisms; pharmacodynamics; pharmacokinetics	Bioavailability; efficacy and safety; drug-disease/drug-drug interactions; pharmacodynamics; pharmacokinetics	Drug-disease/drug-drug interactions; dosage; risks-benefit ratio; efficacy and safety in subgroups	Epidemiological data; efficacy and safety
Sample size and population	20–30, healthy volunteers or individuals with the target disease (cancer, HIV)	200–300, individuals with the target disease	Hundreds to thousands, individuals with target disease	Large population, individuals with target disease

The term *phase V* does not officially appear in either FDA or EMA lexica; this term indicates transnational research designed to "move from bench to bedside" and refers to comparative effectiveness research and community-based research.

8.5 Ethics

Since tissue engineering deals with *in vitro* regeneration or reconstruction of human tissues for clinical applications – including implantation of human tissue engineered products – it is increasingly a subject for ethical inquiry. Different philosophical and religious positions influence the ethical debate, focusing on the impact that new technological developments may have on the principles and norms of human communities, and analyzing the conditions for their acceptance or rejection. In Europe, the legal framework is defined by EU directives that ratify the respect for fundamental human rights, the informed consent for tissue and organ donors, the respect for the privacy, anonymity, and confidentiality of the donor, the voluntary and gratis donation, and the protection of the safety of recipients through quality testing, risk management, good manufacturing, and clinical practices.

Biomedical research involving human beings is also driven by fundamental ethical principles, as enunciated in the Charter of Fundamental Rights of the European Union (incorporated as part II of the draft treaty establishing a Constitution of Europe), and the Convention on Human Rights and Biomedicine and additional protocols issued by the Council of Europe; both these documents provide the background for national legislation on medical applications. Additional contributions are given by the European Group on Ethics in Science and New Technologies (EGE), an independent advisory body of the President of the European Commission. Since its inception in 1991, the EGE has provided the Commission with high quality and independent advice on the bioethical aspects of new technologies in connection with EU legislation or policies.

The EGE addressed in depth ethical aspects related to human tissue banks, emphasizing the principles of respect for human beings, their dignity and autonomy, safety and protection, the protection of privacy, the confidentiality and safety of the donor and the recipient, the prevention of possible discrimination in terms of equal access to the availability of tissues. Concerning the specific issue of umbilical cord blood banking, the EGE raised some ethical concerns regarding cord blood banking for autologous uses, recommending complete and transparent information to consumers, cord blood storage or families at risk of specific diseases or with rare HLA types in public cord blood banks in order to ensure fair access to healthcare services, fair access to transplantation for any citizen whatever his/her ethnic origin by promoting specific measures to have enough donations from different ethnic groups with different HLA patterns [34].

Some critical issues are the ethical aspects on human stem cell research and use, risk-benefit assessment, safety and security in the transplantation of genetically modified cells, and when stem cells are derived from somatic cells. It is also recommended that procurement and storage of stem cells in cell banks should be regulated at the European level in order to facilitate traceability, and that measures to prevent commercialization of cadaver fetal tissue should be implemented [35].

Opinions expressed by the EGE also cover ethics concerning the patenting of inventions involving elements of human origin, with an ethical dilemma arising due to the fact that patents can encourage scientific progress and, at the same time, can also impair access to health care. The EGE belief is that unmodified stem cell lines are non-patentable, while stem cell lines modified by *in vitro* treatments or genetically modified so that they have acquired characteristics for specific industrial application, and fulfil the legal requirements for patentability. Uses of human embryos for industrial or commercial purposes are contrary to public policy and morality and not patentable according to the principle of the noncommercialization of the human body [35].

Among the recent emerging technologies reviewed by EGE, we mention here: synthetic biology (distinction between *natural* and *artificial* life; distinction between life in a biological sense and its use in a social and cultural context, deserving particular care and respect which are at the core of the concept of human dignity; "anthropocentric"/"biocentric"/"ecocentric" visions; biosafety risks related to synthetic microorganisms released into the environment; biosecurity arising from possible use of synthetic lethal and virulent pathogens for terrorist attacks, biowar, or other maleficent uses; IPR for processes, such as the production of chimeras from germ cells, totipotent cells from plants and animals, process for cloning human beings and modified germ-line cells, etc.); nanomedicine (safety of nanomedical products, nanotechnology-based diagnostics and therapies; risk assessment, also with respect to the toxicity of nanoparticles in humans and in the environment; need for prospective technology assessment, including consideration of social effects; legal regulatory issues; IPR issues; nanomedicine tests on the market; information and consent; transparent communication and public trust; clinical research and adequate ethical review process for projects involving studies of nanomedical devices on human beings; medical and non-medical uses), germline genome editing technology research (suspension or moratorium for clinical application) [36, 37].

In general, the complexity of the ethical issues raised by the use of human cells and biological material in regenerative medicine and tissue engineering poses different questions and dilemmas form anthropological, social and economic point of view (Figure 8.3).

Concerns, usually reinforced whenever embryonic stem cells or animal guinea pigs are employed, are based on the variety of religious and philosophical visions and are often strengthened by moral prescriptions with regard to the meaning, the function, and the limits of the human body [38–40].

The most popular method for harvesting stem cells is still extraction from a fertilized embryo, which, consequently, will die without the possibility of generating a complete human being. Do the benefits of developing tissues prevail over the life of a potential individual?

The ethical dimensions of material exchange – especially in the case of the exchange of biospecimens – are likely to be subject to ethical guidelines. The Charter of Fundamental Rights and the Convention on Biomedicine both emphasize the gratuity

Ethics in TE and HTEPs

Antropological issues	**Social issues**	**Economic issues**
– Meaning of human being and life – Cell sources – Animal research	– Access to therapy – Exchange of cellular material – Risk-benefit balance, informed consent	– Material value – Clinical trials and follow-up – Sustainability of heath care system

Fig. 8.3: Categorized ethical issues in tissue engineering (TE) and human tissue engineered products (HTEPs).

and altruistic donation of body material according to the principle that trading human biological material is at odds with the human dignity of a person. Due to the increasing value of cellular material, the eventual transition from free donation to commercial ownership will require a careful investigation on the typology and characteristics of body material in order to alleviate concerns about the commodification of the body and/or its parts [41].

Despite the evidence that animal models cannot fully mimic the complexity of the human body, thus resulting in a limited and partial source of information about the efficacy and safety of ATMP, animal trials are considered to be a necessary step in the research activity leading up to a clinical trial [42]. Use of animals in experiments poses ethical concerns and is the subject of heated debate, exacerbated by increased public awareness of animal welfare issues and the belief that the intrinsic value of a single animal is higher than the value of knowledge obtained from animal sacrifice. Reduction or replacement of animals in preclinical laboratory tests, although considered to be a slow and difficult process, is today a crucial objective: suffering caused by an investigation must be weighed against the real benefits [43–45].

Due to the long-term nature of biomedical implants, clinical trials on humans require careful monitoring of events in posttrial follow-ups over a long period of time in order to minimize the risks for the trial participants. The assessment of safe interaction between the receiver's body and the implanted living material (to detect and prevent undesired cell propagation, migration, differentiation, or inflammatory responses, and even rejection), the evaluation of the efficacy of the regenerative action, the confirmation of the therapeutic value of human tissue engineered products are essential for a correct calculation of the risk/benefit ratio [46, 47]. A risk/benefit balance is the premise for the ethically acceptability of innovative practices and for the truly informed consent of the patient.

Ultimately, beyond the different visions and perceptions, the bottom line for ethical guidelines must remain the commitment to primary and inalienable values: the fundamental principle of respect for the dignity of human beings, the equitable access of patients to safe and effective products, and the respect for the autonomy and the rights of cell donors and recipients.

8.6 References

[1] Directive 2001/83/EC of the European Parliament and of the Council of 6 November 2001 on the Community code relating to medicinal products for human use.
[2] Regulation (EC) No. 1394/2007 of the European Parliament and of the Council on advanced therapy medicinal products and amending Directive 2001/83/EC and Regulation (EC) No. 726/2004.
[3] Directive 2004/23/EC of the European Parliament and of the Council of 31 March 2004 on setting standards of quality and safety for the donation, procurement, testing, processing, preservation, storage and distribution of human tissues and cells.
[4] Commission Directive 2006/17/EC of 8 February 2006 implementing Directive 2004/23/EC of the European Parliament and of the Council as regards certain technical requirements for the donation, procurement and testing of human tissues and cells.
[5] Commission Directive 2006/86/EC of 24 October 2006 implementing Directive 2004/23/EC of the European Parliament and of the Council as regards traceability requirements, notification of serious adverse reactions and events and certain technical requirements for the coding, processing preservation, storage and distribution of human tissues and cells.
[6] Directive 2003/94/EC laying down the principles and guidelines of good manufacturing practice in respect of medicinal products for human use and investigational medicinal products for human use.
[7] Benichou G, Yamada Y, Yun SH, Lin C, Fray M, Tocco G. Immune recognition and rejection of allogeneic skin grafts. Immunotherapy. 2011; 3(6): 757–770.
[8] Griffin MD, Ryan AE, Alagesan S, Lohan P, Treacy O, Ritter T. Anti-donor immune responses elicited by allogeneic mesenchymal stem cells: what have we learned so far? Immunology and Cell Biology. 2013; 91: 40–51.
[9] Rong Z, Wang M, Hu Z, Stradner M, Zhu S, Kong H, Yi H, Goldrath A, Yang Y-G, Xu Y, Fu X. An Effective Approach to Prevent Immune Rejection of Human ESC-Derived Allografts. Cell Stem Cell. 2014; 14(1): 121–130.
[10] ICH Q5D, Derivation and Characterisation of Cell Substrates Used for Production of Biotechnological/Biological Products (CPMP/ICH/294/95).
[11] ICH Q5A Guideline on Quality of Biotechnological Products: Viral Safety Evaluation of Biotechnology Product Derived From Cell Lines in of human or animal origin (CPMP/ICH/295/95).
[12] EMEA/CPMP Note for Guidance on Virus Validation Studies: The Design, Contribution and Interpretation of Studies validating the Inactivation and Removal of Viruses (CPMP/BWP/268/95).
[13] Pegg DE. Principles of cryopreservation. Methods Mol Biol. 2007; 368: 39–57.
[14] Thirumala S, Gimble JM, Devireddy RV. Evaluation of methylcellulose and dimethyl sulfoxide as the cryoprotectants in a serum-free freezing media for cryopreservation of adipose-derived adult stem cells. Stem Cells Dev. 2010; 19(4): 513–522.
[15] Council Directive 90/385/EEC of 20 June 1990 on the approximation of the laws of Member States relating to Active Implantable Medical Devices.
[16] Council Directive 93/42/EEC of 14 June 1993 concerning Medical Devices.
[17] Unger C, Skottman H, Blomberg P, Dilber MS, Hovatta O. Good manufacturing practice and clinical-grade human embryonic stem cell lines. Human Molecular Genetics. 2008; 17: R48–R53.

[18] Giancola R, Bonfini T, Iacone A. Cell therapy: cGMP facilities and manufacturing. Muscles, Ligaments and Tendons Journal. 2012; 2(3): 243–247.

[19] EMEA/CHMP. Note for Guidance on Use of Bovine Serum in the Manufacture of Human Biological Medicinal Product (CPMP/BWP/1793/02).

[20] Bieback K, Hecker A, Kocaömer A, Lannert H, Schallmoser K, Strunk D, Klüter H. Human Alternatives to Fetal Bovine Serum for the Expansion of Mesenchymal Stromal Cells from Bone Marrow. Stem Cells. 2009; 27(9): 2331–2341.

[21] Rauch C, Feifel E, Amann EM, Spötl HP, Schennach H, Pfaller W, Gstraunthaler G. Alternatives to the use of fetal bovine serum: human platelet lysates as a serum substitute in cell culture media. ALTEX. 2011; 28(4): 305–316.

[22] Witzeneder K, Lindenmair A, Gabriel C, Höller K, Theiß D, Redl H, Hennerbichler S. Human-Derived Alternatives to Fetal Bovine Serum in Cell Culture. Transfus Med Hemother. 2013; 40(6): 417–423.

[23] Camilleri ET, et al. Identification and validation of multiple cell surface markers of clinical-grade adipose-derived mesenchymal stromal cells as novel release criteria for good manufacturing practice-compliant production. Stem Cell Res Ther. 2016; 7: 107.

[24] Schneider CK. Challenges with advanced therapy medicinal products and how to meet them. Nature Reviews Drug Discovery. 2010; 9: 195–201.

[25] Ben-David U, Benvenisty N. The tumorigenicity of human embryonic and induced pluripotent stem cells. Nature Reviews Cancer. 2011; 11: 268–277.

[26] ICH Q5E, Comparability of Biotechnological/Biological Products (CPMP/ICH/5721/03).

[27] Harrison C. Patenting natural products just got harder. Nat Biotechnol. 2014; 32: 403–404.

[28] Atanasov AG, et al. Discovery and resupply of pharmacologically active plant-derived natural products: A review. Biotechnol Adv. 2015; 33(8): 1582–1614.

[29] Pellegatti M. . Dogs and monkeys in preclinical drug development: the challenge of reducing and replacing. Expert Opin Drug Metab Toxicol. 2013; 9: 1171–1180.

[30] Dickson M, Gagnon JP. The cost of newdrug discovery and development. Discovery Medicine. 2004; 4: 172–179.

[31] Eliopoulos H, Giranda V, Carr R, Tiehen R, Leahy T, Gordon G. Phase 0 Trials: An Industry Perspective. Clin Cancer Res. 2008; 14(12): 3683–3688.

[32] Rai SN, Ray HE, Pan J, Barnes C, Cambon AC, Wu X, Bonassi S, Srivastava DK. Phase II Clinical Trials: Issues and Practices. Biometrics & Biostatistics International J. 2014; 1(2): 1–3.

[33] Mahan VL. Clinical Trial Phases. Int J Clinical Medicine. 2014; 5: 1374–1383.

[34] European Group on Ethics in Science and New Technologies (EGE), Opinion No. 19 – 16/03/2004 – Ethical aspects of umbilical cord blood banking.

[35] European Group on Ethics in Science and New Technologies (EGE), Opinion No. 16 – 07/05/2002 – Ethical aspects of patenting inventions involving human stem cells

[36] European Group on Ethics in Science and New Technologies (EGE), Opinion No. 25 – 17/11/2009 – Ethics of synthetic biology;

[37] European Group on Ethics in Science and New Technologies. Opinion No. 21 – 17/01/2007 – Ethical aspects of nanomedicine.

[38] Carvalho AS, Ramalho-Santos J. How can ethics relate to science? The case of stem cell research. European J Human Genetics. 2013; 21(6): 591–595.

[39] Karpowicz P, Cohen CB, van der Kooy D. It is ethical to transplant human stem cells into non-human embryos. Nat Med. 2004; 10: 331–335.

[40] de Vries R. Ethical concepts regarding the genetic engineering of laboratory animals: a confrontation with moral beliefs from the practice of biomedical research. Med Health Care Philos. 2006; 9: 211–225.

[41] Dickenson D. Commodification of human tissue; implications for feminist and development ethics. Developing World Bioethics. 2002; 2(1): 55–63.

[42] Olson H, et al. Concordance of the Toxicity of Pharmaceuticals in Humans and in Animals. Regulatory Toxicology and Pharmacology. 2000; 32(1): 56–67.

[43] Rollin BE. The Regulation of Animal Research and the Emergence of Animal Ethics: A Conceptual History. Theoretical Medicine and Bioethics. 2006; 27: 285–304.

[44] Festing S, Wilkinson R. The ethics of animal research. EMBO reports. 2007; 8(6): 526–530.

[45] Rollin BE. The Moral Status of Invasive Animal Research. Animal Research Ethics: Evolving Views and Practices, Hastings Center Report Special Report. 2012; 42(6): S4–S6.

[46] Williams D. Benefit and risk in tissue engineering. Materials Today. 2004; 7(5): 24–29.

[47] Trommelmans L, Selling J, Dierickx K. Ethical reflections on clinical trials with human tissue engineered products. J Med Ethics. 2008; 34: 9 e1.

Index

activated platelet 122
acute kidney injury 217
acute respiratory failure 103
adhesion of platelets 121
advanced therapy medicinal product 245
albumin 132, 172, 189
alternative pathway 126
Alwall dialyzer 86
AN69® 128
anchorage-dependent cell 196
animal model 257
anticoagulant 133
antithrombin III 133
artificial scaffold 228

BAL system 203, 204
binodal and spinodal curves 20
bioartifical pancreas 211
bioartificial kidney 217
bioartificial liver 192
bioartificial lung 224
bioartificial organ 187
bioartificial renal epithelial cell system 224
biocompatibility of dialysis membranes 127
biocompatibility parameters 120
BioKid 222
biological polymers 1
biomedical polymers 1
bioreactor configuration 201
blood cell 131
blood clotting 134
blood flow rate 109
blood oxygenator 69, 112
blood-membrane interaction 119
blood/plasma compartment 201
bubble-point 36

cake filtration model 65
cake layer model 63
carbon dioxide 69, 102, 205, 224
cell adhesion 169
cell proliferation 166
cell survival 166
cellulose acetate 4, 98
cellulose diacetate 4, 130
cellulose triacetate 4, 130

chemical modification 177
chemical potential 49
chitosan 12
classical pathway 126
clearance 75, 90
clearance of the middle molecular weight molecules 93
clinical trial 257
coagulation activation 124
coagulation cascade 124
coagulation factor 124
coating 177
collagen 12
complement protein 125
computational fluid dynamics 139
concentration polarization 61
contact angle 39, 176
continuous stirred-tank reactor 140
convective flux 61
creatinine 95
critical O_2 threshold 158
crossed hollow fiber membrane bioreactor 207
culture model for hepatocytes 194
cumulative $F(t)$ response 145
Cuprophan® 128
cytoskeletal component 169

Darcy's law 67
DEAE-modified cellulose membrane 128
Debye–Huckel theory 72
decellularized matrix 230
degradability 38
degree of crystallinity 3
diabetes 209
dialysate flow 88, 90
dialysis patient 84
dialysis treatment 83
dialyzer 90
dialyzer performance 89
diffusion coefficient 57, 58
diffusive flux 61
diffusivity 68
Donnan exclusion 74
Donnan potential 72
dry heat sterilization 42
dry-wet spinning 29

$E(t)$ distribution 145
electrochemical potential 49, 71
electrospinning 33
embryonic stem cell 193, 211, 220
end-stage organ failure 187
end-stage renal disease 83
endothelization of biomaterials 134
enhancement factor 71
enthalpy of mixing 15
entropy of configuration 17
entropy of mixing 15
epithelial cell 180
ethical aspects 255
ethylenevinylalcohol 5
EVAL® 128
extracellular matrix 165
extracorporeal BAK 218
extracorporeal carbon dioxide removal 105
extracorporeal life support 106
extracorporeal liver-assist device 203
extracorporeal membrane oxygenation 103
extravascular device 212
extrusion 31

Ferry equation 36
fibrinogen 132, 172
fibrinogen receptor 122
fibroblast 180
fibronectin 168
finite element method 157
flat membrane bioreactor 205
flat plate microchannel bioreactor 207
Flory–Huggins theory 16
fluid-dynamic model 149
Food & Drug Administration 242
fouling 63
functionalization 177, 178
fundamental ethical principle 255

galactosylated membrane bioreactor 206
Gibbs free energy of mixing 15
glass transition temperature 2
glomerular capillary 82
good manufacturing practice 245
good tissue practice 245

Hagen–Poiseuille's law 53
heart-lung machine 104
hemodiafiltration 96
hemodialyzer 75, 86
hemofiltration 93
hemoglobin 70
Hemophan® 128
hepatic progenitor cell 193
hepatocyte 189, 192, 193
high efficiency hemodialysis 92
hollow fiber membrane bioreactor 154, 207
hyaluronic acid 12
hybrid bioartificial liver 204
hybrid liver-support device 195
hydraulic permeance 59
hydrodynamic solvent permeability 35
hydrophilicity 177

immobilization of galactose derivatives 179
immortalized hepatoblast carcinoma line 192
immuno-reaction 119
immunoglobulin G 132
immunoisolation barrier 196
induced pluripotent stem cell 212, 221
insulin-producing cell 211
integrin 121, 169
intermediate and complete blocking models 64
intravascular device 212
intravenous membrane oxygenator 108
intrinsic and extrinsic pathways of
coagulation 124
ion-exchange membrane 74
islet sheet 216
islets of Langerhans 209

Kedem–Katchalsky–Spiegler equations 52
kidney 81, 217
kidney function 83
Kiil dialyzer 86
kinetics of platelet adhesion 123
Knudsen diffusion 65
Knudsen number 65
Kolff kidney 85
Krogh cylinder 154

lab-on-a-chip bioreactor 224
laminin 166, 220
Langmuir isotherm 67
Laplace equation 37
lectin pathway 126
leukocyte 131
liquid chemical sterilization 43

liquid membrane 28
liver 189
liver failure 190
liver function 190
liver transplantation 191
low-temperature plasma modification 178
low-temperature sterilization 42
lung 102
lung disease 107, 224
lung function 102
lung scaffold 226
lung transplantation 225

mass transfer 58
mechanical properties 179
melt spinning 29
membrane bioartificial pancreas 212
membrane biocompatibility 98, 131
membrane characterization 35
membrane lung 111
membrane oxygenator 103, 110
membrane plasmapheresis 99
membrane receptor 166
Michaelis–Menten equation 150
microchannel bioreactor 207
microencapsulated islets of Langerhans 212
microfiltration 52
microlithography 33
micropatterned membrane 173
modified polyetheretherketone 5
modular extracorporeal liver system 204
morphological surface properties 173
multibore fiber bioreactor 206
multifiber bioartificial RAD 218

nanofiltration 73
Navier–Stokes equation 54
nephron 82
NH_3 glow discharge plasma 178
nonequilibrium thermodynamics 51
nonsolvent diffusion induced phase
separation 27
Novalung® 225

oxygen 70
oxygenating hollow fiber bioreactor 205
oxygenator 69, 102
ozone sterilization 46

PAN 128
pancreas 208
pancreas-derived cell 211
PancreAssist™ System 216
Peclet number 157
perm-porometry 37
phase I trial 252
phase II trial 253
phase III trial 253
phase inversion 15, 20
phase IV trial 254
physical structure of the chain 3
physico-chemical properties of membranes 175
plasma protein adsorption 132
plasmafilter 101
plug-flow reactor 140
poly(lactic-co-glycolic acid) 10
poly(p-dioxanone) 12
polyacrylonitrile 4
polyamide 11
polycaprolactone 11
polycarbonate 11
polyetheretherketone 4
polyethersulfone 4
polyethylene 4
polyglycolide or poly(glycolic acid) 10
polyimide 9
polylacide 10
polymer/solvent system 19
polymethylmethacrylate 5
polyphosphazene 11
polypropylene 5
polysulfone 4
polytetrafluoroethylene 5
polyurethane 11
polyvinylchloride 5
polyvinylidene fluoride 5
precipitation by solvent evaporation process 27
preclinical study 249
premarket approval 242
protein adsorption 170, 173
protein affinity 170
proximal tubule-derived cell line 220
PUF-HALSS 206
pulse injection 143

radial flow bioreactor 204
radiation sterilization 44
reactive oxygen species 131

regenerated cellulose 98, 127
regulatory framework 241
renal proximal tubule cell 219
Renkin equation 56
residence time distribution 142
Reynolds number 63
RGD moiety 179
RGD sequence 169

Schmidt number 63
selectivity 68
Sherwood number 63
sieving coefficient 89
sintering 32
solubility 68
solute sieving coefficient 94
space time 140
space velocity 141
standard blocking model 64
Starling flow 59
Stavermann reflection 57
stem cells 175, 193, 211
sterilization method 41
stimulus-response technique 143
Stokes–Einstein equation 56
stretching 32
surface free energy 39, 176
surface topography 172
synthetic polymers 1

TECA-hybrid artificial liver support system 204
temperature induced phase separation 27
template leaching 33
ternary system polymer/solvent/nonsolvent 23
therapeutic apheresis 101
thermal sterilization 42
thrombin 125
thrombin-antithrombin III 128
thrombogenicity 121
thrombogenicity of membranes 119
thrombosis 119
tissue-engineered airway 230
toxin 101
tracheal scaffold 228
track-etching 33
transport of gases 67
transport of nutrients and metabolites 196
type I collagen 168

ultrafiltration 55, 73
ultrafiltration coefficient 89
urea 83, 190
urea clearance 92
urine 82

van Oss–Chaudhury–Good theory 176
vascular hollow fiber BAP 215
vitronectin 168
Vroman effect 170

water sorption 38
wearable bioartificial kidney 224
wettability 39

Young's equation 39
Young's modulus 38, 180